Natural Resources and Environmental Justice

Australian Perspectives

Editors: Anna Lukasiewicz, Stephen Dovers, Libby Robin, Jennifer McKay, Steven Schilizzi and Sonia Graham

CSIRO
PUBLISHING

National Library of Australia Cataloguing-in-Publication entry

Natural resources and environmental justice : Australian perspectives / Anna Lukasiewicz, Stephen Dovers, Libby Robin, Jennifer McKay, Steven Schilizzi and Sonia Graham (editors).

9781486306374 (paperback)
9781486306381 (epdf)
9781486306398 (epub)

Includes bibliographical references and index.

Environmental justice – Australia.
Environmental protection – Social aspects – Australia.
Environmental policy – Social aspects – Australia.

Lukasiewicz, Anna, editor.
Dovers, Stephen, editor.
Robin, Libby, 1956– editor.
McKay, J. M. (Jennifer M.), editor.
Schilizzi, Steven, editor.
Graham, Sonia, editor.

363.70994

Published by
CSIRO Publishing
36 Gardiner Road, Clayton VIC 3168
Private Bag 10, Clayton South VIC 3169
Australia

Telephone: [+613] 9545 8555
Email: csiropublishing@csiro.au
Website: www.publishing.csiro.au

Cover: image by kstudija/Shutterstock

Set in 10.5/12 Minion & Stone Sans
Edited by Peter Storer Editorial Services
Cover design by Andrew Weatherill
Typeset by Thomson Digital
Index by Bruce Gillespie
Printed by Ingram Lightning Source

Mar26_RP_ILS

Contents

Acknowledgements v
Preface vi
List of contributors vii

1 **Environment and justice: defining the field** 1
Anna Lukasiewicz

PART 1: CONTEXT AND CONCEPTS 11

2 **A history of global ideas about environmental justice** 13
Libby Robin

3 **Justice research in practice: how can it influence environmental policy and planning?** 27
Geoffrey J. Syme

4 **Justice approaches: methods and methodology in environmental justice research** 39
Sonia Graham, Claudia Baldwin, Jennifer McKay and Sue Jackson

PART 2: FROM THE GROUND UP 59

5 **Concept of public interest in the Adelaide Desalination Plant delivery: lessons for public policy makers on improving social justice** 61
Elnaz Ettehad, Jennifer McKay and Ganesh Keremane

6 **Company–community relations in the mining context: a relational justice perspective** 79
Mirella Cobeleanschi Gavidia and Deanna Kemp

7 **Accounting for justice in local government responses to sea-level rise: evidence from two local councils in Victoria, Australia** 91
Sonia Graham and Jon Barnett

8 **Exploring the justice in forestry negotiations: trading justice for politics** 105
Lain Dare and Jacki Schirmer

PART 3: LENSES INTO JUSTICE 119

9 Enduring and persistent injustices in water access in Australia 121
Sue Jackson

10 Animal property rights: justice or conservation? 133
John Hadley

11 Justice in water resource management 143
Claudia Baldwin

12 Legal determinations, geography and justice in Australia's coal seam gas debate 155
David J. Turton

13 The justice implications of focusing on the economically efficient use of natural resources and their environmental impacts 169
Steven G.M. Schilizzi and Md. Sayed Iftekhar

PART 4: THE WAY FORWARD 183

14 Politics of innovation and the spirit of justice 185
Katherine A. Daniell, Ehsan Nabavi and Claudia F. Benham

15 How can environmental justice be assessed when different stakeholders disagree on what is just? A practical solution 205
Steven G.M. Schilizzi

16 Australian jurisprudence of justice in water management: present limitations, future issues and law reform suggestions 215
Jennifer McKay

17 The Social Justice Framework: untangling the maze of justice complexities 233
Anna Lukasiewicz

18 Exploring environmental justice and public policy 251
Stephen Dovers

19 Current status and future prospects for justice research in environmental management 263
Anna Lukasiewicz, Stephen Dovers, Libby Robin, Jennifer McKay, Steven G.M. Schilizzi and Sonia Graham

Index 267

Acknowledgements

This edited volume is an outcome of the 2015 'Justice, fairness and equity in natural resource management' workshop, funded by the Academy of the Social Sciences in Australia and hosted by the Fenner School for Environment and Society at the Australian National University. Held on the 12th and 13th of October 2015, this national workshop (the first of its kind) brought together 21 participants from 11 different institutions to discuss justice research in environmental management. It was convened by Dr Anna Lukasiewicz and Prof Stephen Dovers from ANU and Assoc Prof Claudia Baldwin from the University of the Sunshine Coast. Bringing together such a strong and diverse team of contributors has been a great learning journey and the calibre of researchers involved is reflected in the final product.

The publication of this volume would not have been possible without the support of the Academy of the Social Sciences in Australia and the Fenner School, especially the assistance of Clive Hilliker who produced most of the figures used. Finally, the editorial team would like to thank all the contributors for sharing their passions and research with us.

Preface

Natural Resources and Environmental Justice: Australian Perspectives brings together research from the intersection of justice, which has a long and rich tradition of inquiry in philosophy, law and the social sciences, and human interactions with their natural world, an equally impressive field of enquiry. Such research is fragmented, multi- and interdisciplinary and scattered throughout the social sciences, humanities and the ecological literature. In addition, researchers are isolated within their institutional and interdisciplinary settings, and have few interdisciplinary platforms to communicate with their peers in different research institutions. An aim of this book is to ensure that justice researchers, whether their motivation is based on rights for the environment or for humans, and independent of discipline, develop a way of communicating that fosters a sense of belonging rather than being at the periphery of more 'mainstream' concerns. The field is growing in importance, both in Australia and internationally, and researchers are actively creating a 'space' in academia that explores the many intersections between environment, natural resource management and justice. This book aims to draw together the disparate threads into a more coherent synthesis of a field with diverse applications, to critically evaluate direction, research priorities, and to advance research.

A main catalyst for the development of this book was a 2015 workshop 'Justice, fairness and equity in natural resource management', supported by the Academy of Social Sciences in Australia, in conjunction with the Fenner School of Environment and Society at the Australian National University (ANU), Canberra. This workshop was the first of its kind – an opportunity to provide consolidation, synthesis and a coordinated future research agenda for the field in Australia. The book reflects the emerging ideas from the workshop. The contributions here are not exhaustive but they illustrate the kind of research that is being done across Australia. Contributions range from PhD scholars describing their case studies to practitioners and scholarly experts reflecting on decades of experience in the field.

List of contributors

Editors

Anna Lukasiewicz is associated with the Fenner School of Environment and Society at the Australian National University. With an interdisciplinary background focusing on sustainability, Anna has been developing the Social Justice Framework – an empirically grounded guide incorporating justice and fairness into environmental and NRM.

Stephen Dovers is a Professor and the Director of the Fenner School of Environment and Society at the Australian National University. He focuses on generic policy and institutional dimensions in environment and sustainability, combining contemporary and historical perspectives across a diverse range of sectors, most of which represent environmental dilemmas with strong justice conflicts.

Libby Robin is a Professor in the Fenner School of Environment and Society at the Australian National University, and affiliated professor of KTH Stockholm and the National Museum of Australia. She is a historian of environmental ideas who has worked in interdisciplinary environmental studies since 1999. Libby's current project works with museums to give voice to communities facing the effects of global warming.

Jennifer McKay is a Professor of Business Law, University of South Australia, and an Adjunct Professor of environmental and water resources law at the University of Lincoln, UK. She has researched and taught on water law and justice issues in Australia relating to drought and floods and in India and in the US (Fulbright senior scholar UC Berkeley) and holds several Ministerial appointments.

Steven Schilizzi is an Associate Professor at the School of Agricultural and Resource Economics at the University of Western Australia. His expertise is mainly in environmental and resource economics, but his background also includes social psychology, anthropology and (ancient and modern) philosophy. One of his research topics is equity (distributive justice) in resource allocation.

Sonia Graham is a Lecturer in social research and policy in the School of Social Sciences at the University of New South Wales. Her research seeks to understand the ways in which environmental policies affect people, focusing on concepts such as collective action, trust, power, fairness, legitimacy and values.

Authors

Claudia Baldwin is an Associate Professor at the University of the Sunshine Coast. She has worked on policy and planning in government, consulting, and research about water management, as well as justice issues, using participatory and visual methods in Australia and overseas.

Jon Barnett is a Professor and Australian Research Council Future Fellow in the School of Geography at Melbourne University. He is a political geographer who researches the impacts of, and responses to, environmental change on social systems in Australia, East Asia and the South Pacific. Jon is also a co-editor of *Global Environmental Change.*

Claudia Benham is a Postdoctoral Research Fellow at the Global Change Institute, University of Queensland. Her PhD thesis, conducted through the Fenner School of Environment and Society and CSIRO Energy, examined equity and sustainability issues related to industrial development in coastal and marine social-ecological systems, exploring aspects of distributive, procedural, interactional, spatial and temporal justice.

Katherine Daniell is a Senior Lecturer in the Fenner School of Environment and Society at the Australian National University. Her current research work focuses on the challenges of implementing participatory approaches to policy and action for sustainable development. Much of this work concerns public engagement processes and thus has a strong underlying emphasis on distributive, procedural and interactional justice.

Lain Dare is a Senior Research Fellow at the Institute for Governance and Policy Analysis in the University of Canberra. She has a strong background in commercial resource management, focusing on issues of rural and regional governance, using her empirical experiences to enhance the understanding, impact and justice outcomes of political spaces in the rural and regional sectors.

Elnaz Ettehad is a PhD candidate within the Centre for Comparative Water Policies and Laws at the University of South Australia. Her PhD project seeks to understand how public interest has been considered in decision-making processes, which strongly affiliates with procedural justice.

Mirella Gavidia is a researcher at the Centre for Social Responsibility in Mining, at the University of Queensland. Her research explores ways to foster social justice by empowering communities affected by large development projects to negotiate their interests and to improve the ways they deal with related social and environmental impacts.

John Hadley is a Senior Lecturer in philosophy in the School of Humanities and Communication Arts at Western Sydney University. He has published on a wide range of topics in animal and environmental ethics, including recent papers on assisting wild animals in need, animal rights extremism, moral responsibility for harming animals and the ethics of animal confinement.

Sayeed Iftekhar is a Senior Research Fellow in the Centre for Environmental Economics and Policy (CEEP) of the University of Western Australia. He is an environmental economist with background in forestry and natural resource management. He is interested in equity-efficiency trade-offs and self-serving bias.

Sue Jackson is an Associate Professor and Australian Research Council Future Fellow at the Australian Rivers Institute, Griffith University. She is a geographer with over 20 years' experience in the social dimensions of NRM, particularly community-based conservation initiatives and institutions. She has written extensively on Indigenous water rights and requirements and the inequities in allocating and managing water in Australia.

Ganesh Keremane is an Adjunct Research Fellow at the School of Law, University of South Australia. He has a PhD in Business and Management from the University of South Australia. His research interests include institutional and policy analysis of surface and groundwater management and assessing community attitudes and perceptions towards 'new' water sources.

Deanna Kemp is an Associate Professor at the Sustainable Minerals Institute (SMI) at the University of Queensland (UQ). Her research focuses on company–community conflict, displacement and resettlement, and human rights and development challenges. She is one of a limited number of people who combine practice and scholarship in this area of mining and social justice.

Ehsan Nabavi is a PhD candidate in the School of Politics and International Relations at the Australian National University. With an interdisciplinary background (trained as a civil engineer), focusing on water and sustainability, Ehsan's research primarily concerns justice and 'conflict transformation' over water resources.

Jacki Schirmer is an Associate Professor at the University of Canberra. Her research explores the relationships between environmental health and human values, behaviour and wellbeing. Her research focuses on understanding the socio-economic and health impacts of changing access to natural resources, addressing NRM conflicts and supporting positive community engagement.

Geoff Syme is currently an Adjunct Professor of Planning at Edith Cowan University. He spent 34 years as a researcher at CSIRO. He pioneered social justice research in water management in Australia with several national studies of social justice issues pertaining to water planning, conservation and allocation.

David Turton is a PhD candidate in the Fenner School of Environment and Society at the Australian National University. David's PhD research is focused on the legal geography of unconventional gas in Australia where, as a legal geographer, David explores the extent of spatial justice.

1

Environment and justice: defining the field

Anna Lukasiewicz

Justice and the environment – what are we actually talking about?

When we talk about the intersection of environment and justice research, we refer to diverse but inter-related bodies of study that deal with: environmental management, including wildlife management as is prevalent in the US (Andrews 1999); community-based natural resource management, popular in Africa and Australia (Jones and Murphree 2004; Curtis and Lefroy 2010; Curtis *et al.* 2014; Lockwood *et al.* 2010), and climate change, which is a global concern (Ikeme 2003; Thomas and Twyman 2005). All of these bodies of literature refer to interaction of humans with natural systems or resources such as soil, water, biodiversity and the industries that are dependent on them, such as agriculture, forestry, mining, fisheries and tourism. In other words, we consider research that specifically examines these socio-ecological interactions through the prism of justice (Schlosberg 2007) at all scales. Such research examines the interactions between human groups as they endeavour to arrive at 'just' decisions in relation to the environment and allocation and use of natural resources.

At this point the reader may simply shrug their shoulder and exclaim 'but that is simply environmental justice – nothing new'. The problem with simply calling this field *environmental justice* is that the term is both all-encompassing and limiting, depending on which body of literature one draws on. Environmental justice research traditionally looked at distribution of environmental costs or negatives, such as pollution, inequitable access to environmental resources and relative lack of access to environmental decision making among vulnerable social groups (Agyeman *et al.* 2003) but in some disciplines it has since expanded to include a multitude of other concerns (Schlosberg 2013), although this expansion is not recognised in all disciplines (Reed and George 2011).

Our 'intersection point' between justice and environment thus definitely includes the distributional and procedural aspects of environmental justice, but it also includes research in *social justice*, which looks at the allocation of goods and benefits within a society, focusing on distributive, procedural and relational fairness, often from the viewpoint of marginalised or disadvantaged stakeholders (Syme and Nancarrow 2001; Whiteman 2009). Social justice sometimes considers human–environmental interactions (Foster *et al.* 2010), but the majority of its research remains firmly anthropocentric. Both environmental and social justice often view the environment as a passive background against which justice is played out. Recently, a trend has emerged to pick an environmental theme and to concentrate justice research around it: for example, water justice (Perreault 2014; Zwarteveen and Boelens 2014); climate justice (Posner and Weisbach 2010); or food justice (Gottlieb and Joshi 2013; Wittman 2009). In contrast, *ecological justice* identifies the environment as a subject of justice – a stakeholder to whom justice is owed (Driscoll and Starik 2004; Opotow and Clayton 1994; Starik 1995; White 2008). Other fields that contribute to our 'intersection point' include: environmental law (Le Bouthillier *et al.* 2012); ecological economics (Costanza 1989); environmental philosophy (Mathews 2014); and aspects of political theory and research into human rights (Gearty 2010; Hancock 2003).

Justice discussed in this book is inclusive of the *environmental*, *social* and *ecological* justice concerns mentioned above and involves, but is not limited to: research around justice concerns of governance (access to, participation and influence in decision making – procedural fairness); access to the resource (distributional fairness); public benefit and the costs and benefits of resource use on different stakeholders (alleviating unequal burdens); economic and socio-political power asymmetries between different stakeholder groups; the impacts of resource use on non-human life and ecosystems (inclusion of the environment as a stakeholder); and consideration of long-term effects (benefits and burdens across generations – inter-generational equity) and across and between species (inter-species equity). It is also worth noting here that most justice research starts by identifying and researching injustice.

A further point needs to be made about words commonly substituted for 'justice'. There is some confusion about whether words, such as fairness and equity, are indicative of nuanced differences, or whether they essentially mean the same thing. Again, the answer depends on the disciplinary context. Some disciplines, such as philosophy, differentiate between justice and fairness, while others, in the social sciences, tend to use the two terms interchangeably (see Finkel *et al.* 2001 for a discussion of the two terms). The term equity is commonly used in economics and is sometimes used interchangeably with fairness across the social sciences. We suggest that recent advancements in this field warrant a more theoretically sound basis for research that deconstructs or unravels components of justice.

Given the disciplinary diversity described above, the reader will find individual chapters referring to different (yet somewhat complementary) conceptualisations of justice. Figure 1.1 is a map of the different justice typologies presented in this book. It shows four groupings of justice types: based on scale (temporal and spatial justice); aspects of the decision-making process (distributive, procedural and interactional justice); and justice philosophes and human rights. The environmental and social justice grouping sits in the middle of the diagram because it encompass aspects of the other three. Here we must comment on the normative and descriptive nature of the different justice types. The justice types grounded in philosophy and law (normative philosophies and human rights approaches) are largely normative (i.e. they deal with how things ought to be). However, the types of justice arising out of psychology, geography and history (spatial and temporal justice, distributive, procedural and interactional justice) tend to be more descriptive, concerned with analysing what is, rather than what ought to be. These divisions are largely nominal, however, because all these types are inter-connected. For instance, a researcher may take spatial or temporal justice as a focus and through that lens deal with distributive and procedural justice issues in order to compare what they see with Sen's Idea of Justice. Social and environmental justice can thus be either normative or descriptive, depending

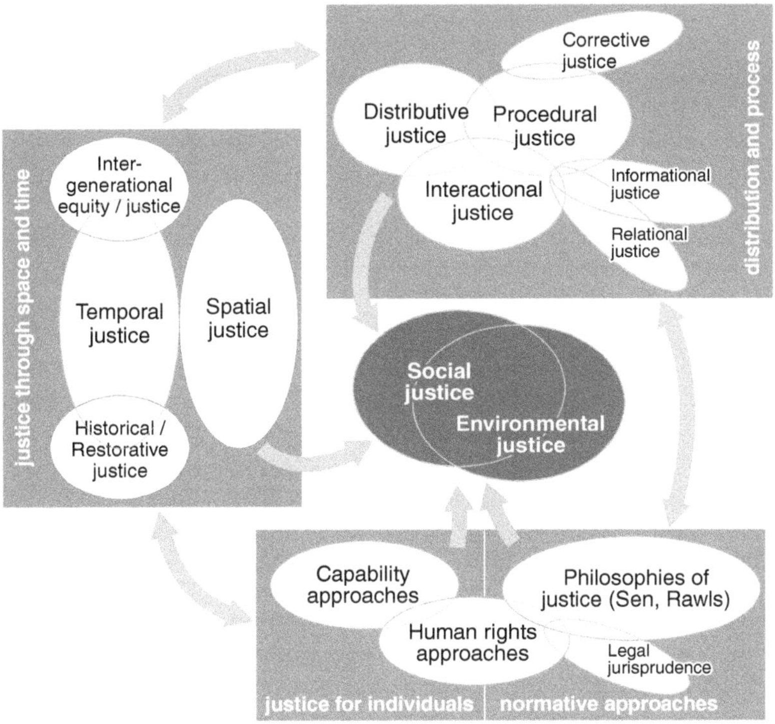

Figure 1.1. Justice typologies.

on the individual researcher's disciplinary background and how they choose to mix and match different justice types.

Chapters 2, 15 and 18 refer broadly to environmental justice (in its more inclusive definition), while Chapter 3 talks about social justice, within the environmental context. Chapters 8, 11, 14 and 17 all use the common three-dimensional conceptualisation of justice (distributive, procedural and interactional), which is based on the social psychology understanding of justice. While this concept is commonly used, it is not the only way of understanding justice. Chapter 7 also uses these three dimensions of justice but adds temporal and spatial, as well as informational, justice (which others might categorise as a mixture of procedural and interactional justice). Chapter 9 uses a somewhat different conceptualisation: historical justice (also sometimes called restorative justice); while Chapter 6 focuses on relational justice (a mixture of procedural and interactional justice). Chapter 16 examines different conceptualisations of justice, including the philosophies of Rawls (2009) and Sen (2009), as well as corrective and political justices. Other chapters use disciplinary understandings of justice: philosophy (Chapter 10), law (Chapter 12) and economics (Chapter 13 and 15). We leave it up to the reader to decide on the most appropriate name for this research field, be it environmental justice, social justice or a combination of any of the above. The last thing we want is to create additional terminology to add to the existing array of terms.

Resources and the environment

Now that we have considered what justice means, we need to describe what justice is applied to, which ranges from the prosaic *natural resources* through to *the natural world*, *the environment* and more holistic and spiritual *nature*. How we conceptualise the environment is partly determined by the type of justice we wish to apply to it. That the environment can be both a subject and object of justice has already been mentioned above. However, the environment can be purely an ecological construct of complex coupled ecosystems, or a more spiritual conceptualisation of nature as a divine entity. It can be a background to questions of human justice or a key life support system that is itself owed justice.

The term *resource* is also a relative construct. A resource is valuable to its user group but may have negative or neutral value to non-users. Furthermore, the status of a resource can vary over time or from place to place. The specific aspects of the environment discussed in this book range from general resources such as air, water and soil or focus on individual species or concepts. The latter include notions such as access to green spaces and sustainable living. Although this book aimed to cover a range of environmental and natural resource concerns, some are more prominent. The focus on natural resource management, particularly sharing among different stakeholder groups, is representative of the type of research that is commonly done in Australia. The chosen themes represent the most topical and conflict-ridden natural resources: water, mining, energy and forestry. We acknowledge that some areas of research, such as the religious dimension, are missing.

The very term *natural resource management* puts humans in an active role of making decisions about the environment, which is relegated to a set of resources (Crawhall 2015). One could simply use the term environmental management but that has similar conceptual limitations. Regardless of the term, we aim to be inclusive of the different conceptualisations of the environment and include research that places the environment as an object and a subject of justice. As for justice above, we have chosen not to construct new terminology and leave it up to individual authors (and the reader) to choose language appropriate to their applications.

Who does this kind of research in Australia?

Given the wide variety of contexts covered in this book, all authors feel as though their chapter contributes to the intersection point of justice and environment set out above. However, given the wide boundaries, what then are the commonalities that would enable ongoing identification with the group? Surprisingly, despite the disparate research approaches and disciplinary diversity, we have found a lot of commonalities in researcher attitudes and motivations. Before the 2015 workshop, all participants were asked to provide a biography that included how they describe themselves, where they see their research as fitting and what their motivations are. A common thread among the responses (further cemented in group discussions during the workshop) is that most do the research they do in order to give a voice to the disadvantaged (such as marginalised communities and the environment), to effect real change, or to 'make a difference', be that in policy processes or outcomes. The research is thus a moral issue in which researchers are personally invested, and for some it is regarded as a form of political action or advocacy. The research is not meant to be abstract but grounded in real life, with the goal of influencing actual processes and outcomes, which explains the prevalence of diverse empirical approaches.

Exploration of researcher motivation and their self-categorisation led to a 'spectrum of involvement' categorisation, based on how people defined themselves (Table 1.1).

Although this typology is based on a relatively small sample (of 28 invited respondents, 21 of whom attended the workshop), it could prove to be a useful tool in connecting

Table 1.1. A typology of justice researchers, based on how central justice is to their research

Level of involvement	Description
Interested	Those who find this intersection of research relevant but do not directly describe their work in these terms, such as public policy researchers.
Incidental	Those who do not set out to research justice but find it crops up in their results as an important topic or find that their results are applicable to justice.
Affiliated	Those who see themselves as working in the environmental space, although not on 'justice' but a similar field, such as public engagement or conflict resolution.
Core	Those who set out to directly engage with justice issues within the environmental space.

disparate researchers and forging new collaborations and networks (because the categorisation works regardless of whether a researcher comes to issues of justice from a background in the environment or vice versa). It also demonstrates why this field of research is so fragmented and disconnected, abounding in individual case studies; much of it is done incidentally or indirectly. Conversely, these results show that the relevance of justice research extends well beyond the relative few who make it the core of their work. This book represents a mix of these typologies. Positively, the diversity of disciplines engaged, to a greater or lesser degree, extends and in various ways, adds a theoretical, methodological and applied richness that is potentially a great strength and an invigorating research environment.

Factors to consider when applying justice to the environment

Regardless of the type of justice and the conceptualisation of the environment, achieving justice (or fairness or equity) in environmental dilemmas is inherently difficult and there are several common factors that create these difficulties. They are directly and indirectly referred to by all authors in this book and therefore only briefly described here.

Temporal scale

In any given conflict, tension or decision about a natural resource (or an aspect of the environment, or the rights or otherwise of a particular species) neither the decision maker nor the stakeholders start with a clean slate. There is already a history of interactions around the object in question: already perceptions exist about who is powerful and powerless. Therefore, any decision does not stand on its own but is inserted into an existing social process that is dealing with past consequences of decisions and will have to take account of possible future outcomes. Decisions of the past continue to affect current stakeholders and a decision deemed to be fair at the current time may be deemed unfair in the future. Also, any decision will produce its own (intended and unforeseen) consequences. The temporal scale is often overlooked in social and political processes that focus strongly on the present.

Spatial scale

The more we understand of ecology, the more environmental management and understanding is steered away from a reductionist simplification (managing small bits or individual aspects of the environment) to a holistic ecosystem-based approach and a focus on adaptive learning. However, different stakeholders at different spatial scales have different, and often incompatible, goals. Thus a decision perceived to be fair at one level may be seen as unfair at another. Spatial complexities of as yet not-fully-understood ecological systems continue to perplex scientists and create dilemmas. The issue of scales underlines Patrick's (2014) finding of justice as a continuous process, rather than a single static decision.

Nature of the resource

Natural capital (air, water or soil) is unlike human or built capital (investment stocks or buildings). Natural resources can be controlled to an extent, for example by damming rivers or fertilising soils but not totally; minerals cannot be mined from places where they do not naturally occur; ecosystem services are affected by natural phenomena such as droughts. Natural resources are essential for life, including human life. They are the basis of historical wealth and underpin human civilisation. Therefore their governance (and the benefits that are derived from their exploitation) is an arena where powerful stakeholders are created and dominant groups seek to cement their dominance (see Neal *et al.* 2014 who make this case for water). Any decision over natural resources must bear in mind the properties of the resource, meaning that some simple solutions (such as equal distributions) may be inappropriate or unworkable.

The unique properties of natural resources also mean that any decision making is made in the face of: constant uncertainty; perpetual lack of knowledge (as consequences of actions are often unknown); unpredictability (when decisions may need to be made quickly in the face of sudden changes); dependence of some stakeholders on the resource (as livelihoods built around an established usage of a resource or its use may make it unavailable to species that depend on it); and the domination of scientific understanding of nature that conflict with deeply held values and beliefs. Inter-species conflicts fall into this space as humans compete with non-human species over the same resource or humans have to manage different species' competing dependence on the same resource(s).

Recognition

Before we begin asking 'what is fair (just or equitable)?' we must establish to whom justice is owed. The concept of a moral community is described by Deutsch (1975). Justice applies to those inside the moral community but not to those excluded. If a moral community is expanded to include future generations, ecosystems and other species (Clayton 2000), justice judgments will differ to a moral community that only includes present-day humans (see Stone 2010 for a discussion on what would happen if trees were included in western society's current moral community). The extent to which justice is owed to future generations goes to the importance of temporal scales discussed above. Sometimes part of the battle is to accept that a previously unrecognised stakeholder is owed justice. Only then can society determine what is just for them.

The importance of values

A fairness judgment is often something felt in the gut rather than rationally explained (Gross 2014). *Un*fairness is particularly intuitive. Fairness/justice judgments depend on values and beliefs and 'moral mandates', which are deeply held beliefs that, if violated, will lead to a perception of unfairness, regardless of how fair the process was (Skitka 2002). Values and beliefs are not static – they can change over time and between cultures, once again reinforcing the idea of justice as a continual process of dealing with consequences.

Context

Ultimately, whether a decision has been fair or just, given the complexities described above, depends on whether it leads towards a desired, shared and understood societal goal. The problem is that the ultimate goals of any society are rarely clearly articulated by its leaders and endorsed by its members. Without having a shared, agreed-upon vision for our society, we cannot ultimately know whether the ongoing justice journey is heading in a fair or unfair direction. Failing such a shared vision, fairness judgments are left to perceptions of individual stakeholders who may share culturally similar universal justice principles but not apply these consistently in a situational justice context.

The short explanation of these complexities illustrates why justice is often deemed too hard in real life. It boils down to this: the way we interact with our environment results in complex dilemmas where justice is a process rather than a single decision that changes according to points of view, temporal and spatial scales and is based on diverse and contested (and changing) values and beliefs.

Overview of the book

The contributions of this book have been divided into four parts. Contributions in Part 1 'Context and concepts' are broad perspectives aimed to set the scene within the book. Chapter 2 provides a historical overview of environmental justice ideas, with a particular focus on how international conversations influenced the concept in Australia. Chapter 3 has a more practical focus on how justice research has influenced natural resource management to date in Australia. Chapter 4 completes Part 1 by examining the methodological approaches used by researchers in this book.

The following two parts contain either single or multiple case study contributions. Part 2 'From the ground up' showcases single, one-off case studies, mainly from the 'affiliated' and 'incidental' type of researchers exploring participative and relational aspects of justice in different environmental dilemmas – urban water supply, mining, climate change adaptation and forestry. Chapter 5 discusses the justice implications of urban water desalinisation in Adelaide, South Australia with a focus on public interest, while Chapter 6 examines the relational aspects of justice between a mining company and local communities in Brazil. Chapter 7 focuses on local government plans for climate change adaptation in Victoria, and Chapter 8 discusses the negotiation process for the end of the Tasmanian forestry conflict. Although the contexts differ greatly, all of these examples are concerned mainly with participative (and relational) justice as it plays out between affected stakeholders and an authoritative decision maker. In the Australian cases, this decision maker is either the local, state or national government, while in the Brazilian case it is an international mining company.

Part 3 'Lenses into justice' continues the focus on case studies, but they represent a particular thematic or disciplinary focus by ('core' and 'affiliated') researchers who consistently publish on their justice topic. This part starts with Chapter 9 on the enduring injustices that Indigenous people face in water access. Chapter 10 deals with animal property rights within the discipline of philosophy. Chapter 11 then discusses several case studies in water management through the specific application of the Social Justice Framework while Chapter 12 applies a legal geography perspective to the ongoing conflicts over coal seam gas. Part 3 ends with Chapter 13, which provides an economics approach to natural resource management.

Part 4 'The way forward' presents different disciplinary and thematic perspectives that seek to answer some of the persistent challenges of the field. Chapter 14 provides guidance on dealing with the political aspects of environmental dilemmas through an analysis of multiple international and Australian case studies, while Chapter 15 proposes a 'practical' economic solution to assessing actions in the face of conflicting values. Chapter 16 offers legal guidance on Australia's jurisprudence of justice in environmental management and Chapter 17 outlines the Social Justice Framework: a way for environmental resource managers to incorporate justice dimensions and principles into their everyday decision making. Finally, Chapter 18 brings public policy expertise to environmental dilemmas and offers guidance for the pursuit of environmental justice.

We have chosen not to change the different disciplinary styles nor to impose any overarching themes on the chapter contributions. Thus a reader familiar with social science contributions might find the economic, legal and philosophical chapters quite different in style. We hope that this book enables the readers to recognise the disciplinary and methodological diversity of the field and adds to the reader's understanding of ongoing justice research, and introduces a range of approaches to practical, policy and scholarly problems in Australia and beyond.

References

Agyeman J, Bullard RD, Evans B (Eds) (2003) *Just Sustainabilities: Development in an Unequal World*. MIT Press, Cambridge MA, USA.

Andrews RNL (1999) *Managing the Environment, Managing Ourselves: A History of American Environmental Policy*. Yale University, New Haven CT, USA.

Clayton S (2000) Models of justice in the environmental debate. *The Journal of Social Issues* **56**, 459–474. doi:10.1111/0022-4537.00178

Costanza R (1989) What is ecological economics? *Ecological Economics* **1**, 1–7. doi:10.1016/0921-8009(89)90020-7

Crawhall N (2015) Social and economic influences shaping protected areas. In *Protected Area Governance and Management*. (Eds GL Worboys, M Lockwood, A Kothari, S Feary and I Pulsford) pp. 117–144. ANU Press, Canberra.

Curtis A, Lefroy EC (2010) Beyond threat- and asset-based approaches to natural resource management in Australia. *Australasian Journal of Environmental Management* **17**, 134–141. doi:10.1080/14486563.2010.9725260

Curtis A, Ross H, Marshall G, Baldwin C, Cavaye JM, Freeman C, *et al.* (2014) The great experiment with devolved NRM governance: lessons from community engagement in Australia and New Zealand since the 1980s. *Australasian Journal of Environmental Management* **21**, 175–199. doi:10.1080/14486563.2014.935747

Deutsch M (1975) Equity, equality and need: what determines which values will be used as a basis of distributive justice? *The Journal of Social Issues* **31**, 137–149. doi:10.1111/j.1540-4560.1975.tb01000.x

Driscoll C, Starik M (2004) The primordial stakeholder: advancing the conceptual consideration of stakeholder status for the natural environment. *Journal of Business Ethics* **49**, 55–73. doi:10.1023/B:BUSI.0000013852.62017.0e

Finkel N, Harré R, Rodriguez Lopez J-L (2001) Commonsense morality across cultures: notions of fairness, justice, honor and equity. *Discourse Studies* **3**, 5–27. doi:10.1177/1461445601003001001

Foster JB, Clark B, York R (2010) *The Ecological Rift: Capitalism's War on the Earth.* Monthly Review Press, New York, USA.
Gearty C (2010) Do human rights help or hinder environmental protection? *Journal of Human Rights and the Environment* **1**, 7–22. doi:10.4337/jhre.2010.01.01
Gottlieb R, Joshi A (2013) *Food Justice.* MIT Press, Cambridge MA, USA.
Gross C (2014) *Fairness and Justice in Environmental Decision Making: Water Under the Bridge.* Routledge Explorations in Environmental Studies, London, UK and New York, USA.
Hancock J (2003) *Environmental Human Rights: Power, Ethics, and Law.* Ashgate, Burlington VT, USA.
Ikeme J (2003) Equity, environmental justice and sustainability: incomplete approaches tin climate change politics. *Global Environmental Change* **13**, 195–206. doi:10.1016/S0959-3780(03)00047-5
Jones BTB, Murphree M (2004) Community-based natural resource management as a conservation mechanism. In *Parks in Transition: Biodiversity, Rural Development, and the Bottom Line.* (Ed. B Child) pp. 63–105. Earthscan, London, UK.
Le Bouthillier Y, Cohen MA, Gonzalez Marquez JJ, Mumma A, Smith S (Eds) (2012) *Poverty Alleviation and Environmental Law.* Edward Elgar, Northampton MA, USA.
Lockwood M, Davidson J, Curtis A, Stratford E, Griffith R (2010) Governance principles for natural resource management. *Society & Natural Resources* **23**, 986–1001. doi:10.1080/08941920802178214
Mathews F (2014) The heartache of wildlife ethics. *Wildlife Australia* **51**, 34–39.
Neal MJ, Lukasiewicz A, Syme G (2014) Why justice matters in water governance: some ideas for a 'water justice framework'. *Water Policy* **16**, 1–18. doi:10.2166/wp.2014.109
Opotow S, Clayton S (1994) Green justice: conceptions of fairness and the natural world. *The Journal of Social Issues* **50**, 1–11. doi:10.1111/j.1540-4560.1994.tb02416.x
Patrick MJ (2014) The cycles and spirals of justice in water-allocation decision making. *Water International* **39**, 63–80. doi:10.1080/02508060.2013.863646
Perreault T (2014) What kind of governance for what kind of equity? Towards a theorization of justice in water governance. *Water International* **39**, 233–245. doi:10.1080/02508060.2014.886843
Posner EA, Weisbach D (2010) *Climate Change Justice.* Princeton University Press, Princeton, NJ, USA.
Rawls J (2009) *A Theory of Justice: Revised Edition.* Harvard University Press, Cambridge MA, USA.
Reed MG, George C (2011) Where in the world is environmental justice? *Progress in Human Geography* **35**, 835–842. doi:10.1177/0309132510388384
Sen A (2009) *The Idea of Justice.* Harvard University Press, Cambridge MA, USA.
Schlosberg D (2007) *Defining Environmental Justice: Theories, Movements, and Nature.* Oxford University Press, Oxford, UK.
Schlosberg D (2013) Theorising environmental justice: the expanding sphere of a discourse. *Environmental Politics* **22**, 37–55. doi:10.1080/09644016.2013.755387
Skitka LJ (2002) Do the means always justify the ends or do the ends sometimes justify the means? A value protection model of justice reasoning. *Personality and Social Psychology Bulletin* **28**, 588–597. doi:10.1177/0146167202288003
Starik M (1995) Should trees have managerial standing? Toward stakeholder status for non-human nature. *Journal of Business Ethics* **14**, 207–217. doi:10.1007/BF00881435
Stone C (2010) *Should Trees Have Standing? Law, Morality, and the Environment.* 3rd edn. Oxford University Press, New York, USA.

Syme GJ, Nancarrow BE (2001) Social justice and environmental management: an introduction. *Social Justice Research* **14**, 343–347. doi:10.1023/A:1014628827223

Thomas DSG, Twyman C (2005) Equity and justice in climate change adaptation amongst natural-resource-dependent societies. *Global Environmental Change* **15**, 115–124. doi:10.1016/j.gloenvcha.2004.10.001

White R (2008) *Crimes Against Nature: Environmental Criminology and Ecological Justice.* Willan Publishing, Portland OR, USA.

Whiteman G (2009) All my relations: understanding perceptions of justice and conflict between companies and indigenous peoples. *Organization Studies* **30**, 101–120. doi:10.1177/0170840608100518

Wittman H (2009) Reworking the metabolic rift: La Vı'a Campesina, agrarian citizenship, and food sovereignty. *The Journal of Peasant Studies* **36**, 805–826. doi:10.1080/03066150903353991

Zwarteveen MZ, Boelens R (2014) Defining, researching and struggling for water justice: some conceptual building blocks for research and action. *Water International* **39**, 143–158. doi:10.1080/02508060.2014.891168

PART 1

CONTEXT AND CONCEPTS

2

A history of global ideas about environmental justice

Libby Robin

Summary

This chapter puts the idea of justice in environmental and natural resource management in a historical context, particularly since the rise of 'the environment' and the global era in the 1940s. It argues that the future is understood by different experts at different times, and that justice cuts across many sectors: prominent in law, history and governance, but also invoked by natural scientists concerned about the environment and resources. Key global ideas in environmental justice include human rights, sustainability and intergenerational equity, and the precautionary principle, all of which became important in global settings, but have specific local and national applications for natural resources such as water, forests and fisheries.

Introduction

The interwoven history of ideas of globalisation, the environmental and justice provides a background against which the different authors in this book explore the practical work of just environmental management in local situations in Australia. International benchmarking brokered under the auspices of organisations such as the United Nations, the Commission on Human Rights, the International Union for the Conservation of Nature (IUCN), the International Council of Scientific Unions (ICSU) and new organisations such as Future Earth (ICSU 2016) provides the language for how justice and environmental management are negotiated at regional levels, including cities, catchment management areas, states and nationally. Global agreements often form a basis for 'best practice', which frames negotiations between levels – between a city and a state, or between the states and the Australian Government, over environmental questions and natural resources and their implications for social justice.

'Best practice' is a difficult thing to capture in scholarly terms, but it is an important guiding principle for managers. Justice is similarly abstract and implicit. Although environmental managers would agree that they would not deliberately choose a course of action that fostered *injustice*, justice is not easy to articulate in every situation or to enforce legally or legislatively. Although justice may be perceived through the work of the law, *injustice* has historical origins and practical implications. Justice is often not evident, even in key documents that might elect to be explicit about it: for example, neither the International Monetary Fund (Boughton 2004) nor the United Nations Environment Program (UNEP 2015) mention the word 'justice' anywhere in their governance planning programs.

People who 'struggle' against environmental injustice and inequities in natural resource distributions concentrate on enabling the voices of those overlooked in decision making to be heard (Armiero and Sedrez 2014). To create a place at the decision-making table for more stakeholders, the first step is including local parties whose livelihoods and sense of identity may be adversely affected by distant environmental decision making.

What is fair, equitable or just has changed remarkably over time and at different rates in different places. Here we consider justice as a journey, not a destination; its practice is embedded in local conditions, no matter how global and lofty its principles. Environmental justice includes both social justice for human stakeholders and the ethics of care with respect to non-human stakeholders: fellow 'citizens' of the planet. Although environmental managers begin with a famous medical principle 'first do no harm', there may come points where not causing injustice to humans may disadvantage non-human others. There is barely a policy language for managing this sort of dissonance, although the rise of the practical environmental humanities internationally (Ekström and Sörlin 2012; Griffiths 2015; Head *et al.* 2017) is beginning to engage with this.

Global history for practical management

This chapter's approach, using the history of ideas, reflects on present justice concepts and identifies those that have long histories, which in turn suggests they may have long futures. History provides context and perspective: it explains how and why some ideas have become guiding principles for policy making, and what sorts of applications they might have. History is also a useful tool for interdisciplinary engagement. It can be a way to facilitate discussions between people whose understandings of an environmental issue are very different. Acknowledging a shared or different history of relations with natural resources may be important to clearing the air, and to including stakeholders who may be excluded by highly expert situations. Telling stories is human, and history making is something that is inclusive of a wide range of people. Fairness begins with an inclusive discussion between stakeholders in natural resource management, and the common touch of history is important in just decision making.

Without history, the present is all we have, and the future is foreshortened by simply projecting forward from what is now. Getting 'beyond the present' is itself a precondition for justice, as social commentator and historian, Tony Judt argues:

> *Something is profoundly wrong with the way we live today. Much of what appears 'natural' today dates from the 1980s: the obsession with wealth creation, the cult of privatization and the private sector, the growing disparities of rich and poor. And above all, the rhetoric which accompanies these: uncritical admiration for unfettered markets, disdain for the public sector, the delusion of endless growth. We cannot go on living like this. The little crash of 2008 was a reminder that unregulated capitalism is its own worst enemy. (Judt 2010: 7)*

Judt's primary concern here is social change for human justice, rather than the future of non-human nature, but the default western (and global) ways of living, many of which date back just 30 years, seriously affect natural resources and their management. The over-consumption of planetary resources by the wealthy has accelerated in this period, despite the fact that most of the key global ideas in natural resource and environmental management – sustainability, resilience and the precautionary principle – have also arisen in that time (Fig. 2.1). The accelerating orthodoxy of growth and inequality between rich and poor drove the Brundtland report's definition of 'sustainable development' (WCED 1987), which became the touchstone phrase of the 1992 Earth Summit in Rio.

Also important to understanding environmental justice is the emergence of 'the environment' itself in the context of international discussions of post-war reconciliation in the 1940s. Scientists and technical experts were prominent in this era, where, as physicist and novelist C.P. Snow put it, 'the scientists have the future in their bones' (Snow 1959: 7). The United Nations encouraged experts from different disciplines and different countries to work together on practical and technical projects for peace and new knowledge. They approached such questions as desertification and soil erosion with the military logistics of their time (Robin 2007), and natural resource management became international diplomacy.

One important international initiative in this era was the 1955 Princeton Conference, 'Man's role in shaping the face of the Earth', sponsored by the Wenner-Gren anthropological foundation, which brought together 73 academics, policy makers and figures from industry. Under the guidance of three prominent United States professors – geographer Carl Sauer, entomologist Marston Bates and polymath historian, philosopher and city planning theorist Lewis Mumford – this was a conference envisioning futures for planet Earth nearly a decade before there were any 'environmental sciences'. Before the war, the common language of science was German. By 1955, it was (American) English. The term 'environmental sciences' emerged in the early 1960s to embrace both theory and practice and interdisciplinary understandings of the world, at least in part because of this conference.

The Universal Declaration of Human Rights

International ideas about injustice have emerged slowly over western history since the 'enlightenment'. When slavery was abolished in the 18th century, new justice movements turned to property rights and votes for women: something that was achieved in the late 19th and early 20th century in most of the west. While the recognition of the rights of people to be treated as human beings, and not to be discriminated against because of their race or religion, was a principle that 'civilised countries' accepted, the atrocities of the Nazi regime in the Second World War created a new need for an explicit acknowledgement of human rights.

The United Nations was established in 1945 as an organisation that upheld human rights and worked for more equal human opportunities in the world. In 1948, its foundational Charter was complemented by *The Universal Declaration of Human Rights*, which creates the basis for global justice that can transcend the national. Article 2 is explicit on this point:

> *Everyone is entitled to all the rights and freedoms set forth in this Declaration, without distinction of any kind, such as race, colour, sex, language, religion, political or other opinion, national or social origin, property, birth or other status. Furthermore, no distinction shall be made on the basis of the political, jurisdictional or international*

status of the country or territory to which a person belongs, whether it be independent, trust, non-self-governing or under any other limitation of sovereignty. (United Nations 1948c)

One early initiative within the United Nations was an educational and cultural organisation which, in November 1945, became UNESCO, the 'S' for Scientific being added in recognition of the importance of science to global peace making (Archibald 2006). Julian Huxley, an ecologist and the first Director-General of UNESCO, and Joseph Needham, a biochemist and sinologist and the first head of UNESCO's Natural Sciences section, were leading lights in an earlier inter-war movement for 'social responsibility of science'. Together they saw unique possibilities for international and 'universal' outreach in a scientific secretariat within the United Nations. Although science was a late addition, UNESCO programs established in partnership with ICSU nurtured the development of ideas about 'the environment' and placed them on a global footing.

UNESCO was established to create more equal opportunities between the developed and the developing world, through building institutions for education, science and culture. Even within countries where slavery had been abolished, there was continuing social injustice, (e.g. for African-Americans in the United States). Aboriginal people in Australia were not counted as citizens in the census until the Constitution was amended in 1967 following a national referendum. Although human rights are still evolving, even now, the question of rights for 'nature' or 'more than human others' arose during the environmental revolution of the late 1960s, when the metaphor of 'Spaceship Earth' and the actual pictures of the Earth from space reinforced a sense of limits to exploiting nature for human purposes without counting the cost (Meadows *et al.* 1972; Carson 1962). For example, Christopher Stone (1972) developed a legal case in an important paper 'Should trees have standing?' probing how nature might become a litigant in the US legal system.

Conservation: a first step towards justice in natural resource distribution

The International Union for the Protection of Nature (IUPN) was established with support from UNESCO in Fontainebleau, France, in October 1948. Its founding document marks a shift from 'nature in need of protection' (as in the title of the group) to 'the environment':

WHEREAS the term 'Protection of Nature' may be defined as the preservation of the entire world biotic community, or man's natural environment, which includes the earth's renewable natural resources of which it is composed, and the foundation of human civilization. (UNESCO 1948a: 1)

These 1948 documents laid the foundations for natural resource governance for much of the post-war period. This document also provides an early use of the term *environment*, here qualified with the words, 'man's natural', but soon simply 'the'. By 1956, the IUPN was renamed IUCN, the International Union for the *Conservation* of Nature. Conservation is more about *wise use* than protection. If natural resources are 'the foundation of human civilization', they are to be used. Use is defined according to the principles of a civil society, including justice to the management of natural resources.

The issue of the conservation of resources discussed through the 'United Nations scientific conference on the conservation and utilisation of resources' (UNSCCUR) conference at Lake Success in 1949 (UNESCO 1948a; IUPN 1950). This was an initiative of United States President Harry S. Truman via the United Nations Economic and Social Council. Truman made the point that natural resources had demonstrated their crucial importance during the war, and could be considered a basis for conflict:

> *The real or exaggerated fear of resource shortages and declining standards of living has in the past involved nations in warfare. Every member of the United Nations is deeply interested in preventing a recurrence of that fear and those consequences. (UNESCO 1948b: 1)*

The rebuilding of natural resources for a fairer world was a part of the post-war reconstruction agenda of the UN. Conservation was expressed as a 'tool of peace'. Alongside peace was the question of fairness in post-war development, explicitly a problem of access to natural resources, including atomic energy for peaceful purposes. Natural resources had been important in the war effort, and they had been diminished and destroyed by the war. Truman envisaged a meeting of minds among experts who would 'not necessarily represent the views of the government of their nations', but who understood natural resources through science and experience. This moment created a recognition of expertise for natural resource management. Truman wrote:

> *Warfare has taken a heavy toll of many natural resources; the rebuilding of the nations and the industrialization of under-developed areas will require an additional large depletion of them ... Conservation can become a major basis for peace. Modern science has itself become a major international resource which facilitates the use of other resources. Their adequate utilization can become a major basis of world prosperity. (IUPN 1950: vi)*

The Lake Success conference program began with atomic energy, but also included nature conservation. Sessions were scheduled so as to enable delegates to UNSCCUR and to the concurrent IUPN conference on nature conservation to attend both meetings.

Natural resources were linked to economics ('prosperity'), security and foreign policy, and crucially portrayed as limited and vulnerable. They could be 'wasted' and they were in need of 'management' by experts, who were practitioners of natural and technical sciences. The IUPN conference itself was explicitly 'technical' – 'The international technical conference for the protection of nature'. The management of justice in the distribution of limited resources was implicit, and assumed to be part of a technical brief.

The environment in the Space Age

If the era from 1945 to 1962 could be characterised as 'science for peacemaking', finishing with the new emergence of 'environmental scientists', the next step was the development of the new United Nations Environment Programme (UNEP), which was launched in Stockholm in 1972. Activist social movements for peace, justice and the environment merged in the late 1960s, informing and colouring the UNESCO sponsored initiative. UNEP now aims to contribute to the building of peace, the eradication of poverty, sustainable development and intercultural dialogue through education, the sciences, culture, communication and information, and all of those aspirations were included in its foundation documents for the Stockholm conference. One difficult issue was whether UNEP or UNESCO should be the body primarily responsible for environmental science, including such international programs as the 'Man and the biosphere' (MAB) (Bridgewater 2016).

The 1960s was also the 'Space Age'. The NASA and Soviet missions to reach the moon created new metaphors and images, reinforcing the idea that Earth's natural resources were limited and needed careful management. Economist Kenneth Boulding famously wrote of 'Spaceship Earth', while the photographs of the lonely blue planet beamed back from the Apollo 8 mission created a strong sense that there was 'only one Earth' (Höhler 2015). Rachel Carson's *Silent Spring* (Carson 1962), a parable about a spring without birdsong because of pesticides in the American landscape, was also widely read and very influential in building environmental activism, including in Europe where the 'flower power' protests of San Francisco were closely followed by similar actions in West Germany.

In Australia, the concern about humanity overwhelming nature was expressed by zoologist and writer, AJ (Jock) Marshall, whose 1966 polemic, *The Great Extermination: Anglo-Australian's Cupidity, Wickedness and Waste* was cited by many conservationists (Robin 1998). Marshall was concerned about the extinction of mammals and birds in Australia and he was also highly critical of suburban ugliness and urban design. His book presented an integrated sense of humanity's attack on the whole environment. The Australian Conservation Foundation (ACF), formally established in 1965, quickly became the national peak body for conservation. With a strong, scientifically informed leadership, it closely monitored international developments in environmental management, particularly for national parks and reserves (Mosley 1972; Robin 1994). In the urban setting, the unlikely Builders Labourers Federation withdrew their labour and placed 'green bans' on demolishing bush preventing unsympathetic development in Kellys Bush in Sydney (Bonyhady 1993). The ACF, through crucial leaders such as ecologist Francis Ratcliffe (foundation Secretary of ACF and former head of CSIRO Wildlife Section) and geographer Geoff Mosley (ACF's research officer in the 1960s, and Director from 1973 to 1986), kept abreast of international developments in all environmental sectors, including cities and heritage management, which together became 'the national estate' in the early '70s. UNESCO's Committee for the Protection of World Cultural and National Heritage, spoke of an 'International Estate' in 1975 legislation for the Australian Heritage Commission (later the Australian Heritage Council) and managed Australia's nominations to the World Heritage Commission.

The CSIRO also watched the UNEP and other global developments closely. In 1970, they sent a senior member of the executive, Dr Max Day on a major 'fact-finding mission' as they considered establishing an 'environmental division' (Day 1970). Although CSIRO Executive ultimately decided against a special division, the environment became important in the strategic planning of Australia's major public science organisation, and in its advice to governments of the day, and the international environment programs framed the research questions for public science in Australia. Max Day visited the ecology laboratory of C.S. (Buzz) Holling in Vancouver, who was already theorising 'resilience' (Holling 1973), and undertook leadership roles in European policy think tanks from the 1970s, and met with MIT mathematical modeller Denis Meadows, third author of the *Limits to Growth* report to the Club of Rome (Meadows *et al.* 1972). Day's international mission finished up in Stockholm, observing the global environment community as it came together to prepare to launch the UNEP in 1972.

The United Nations Conference on the human environment met at Stockholm from 5 to 16 June 1972, and its important Declaration presented a common outlook and principles to inspire and guide the people of the world in the preservation and enhancement of the human environment. It declared that both the natural and the man-made environments are essential to human 'well being and to the enjoyment of basic human rights, the right to life itself'. Crucially, it linked development of people and human rights – and recognised that the most industrialised countries had more work to do to protect the *environment*, while poorer countries had also to prioritise the wellbeing of their citizens, noting that many places could not yet offer 'minimum levels required for a decent human existence', because people were 'deprived of adequate food and clothing, shelter and education, health and sanitation' (Declaration of the United Nations Conference on the Human Environment 1972, Article 4). The gap between rich and poor was explicitly cited, even for wealthy nations. The question of justice and human rights was fundamental to the Declaration and to its definition of the 'human environment', and managing the environment embraced a broadening expertise.

From Stockholm to Rio: principles of intergenerational equity, sustainability and precautionary justice

In October 1987, after 3 years of work and consultation, the World Commission on Environment and Development set up by the United Nations published its report *Our Common Future* (WCED 1987). More commonly known as the Brundtland Report after the Commission's chair, Norwegian Prime Minister Gro Harlem Brundtland, this document made *sustainable development* a household word. This landmark document facilitated international cooperation in environmental policy in new ways and made sustainability one of the leading aspirations in late 20th-century politics. Its principle of *intergenerational equity* 'to ensure that [the world] meets the needs of the present without compromising the ability of future generations to meet their own needs'. 'The concept of sustainable development does imply limits', it continued, 'not absolute limits but limitations imposed by the present state of technology and social organisation on environmental resources and by the ability of the biosphere to absorb the effects of human activities' (WCED 1987: 16).

The idea of 'sustainable yield' had its origins 300 years earlier in forestry, where Hans Carl von Carlowitz (1645–1714) introduced the principle of planting forests to ensure wood resources for future generations. Carlowitz recognised that the trees of a forest required at least 40 years – essentially at the time a human lifetime – for renewal. Europe at this time endured 'timber famines' resulting in shortages of materials for ship-building, mines (a crucial interest to the Prussian king) and for heating, which led them to having to source and transport timber from faraway places around the Baltic Sea. Writing in 1713, as the 'Little Ice Age' had Europe in its grip, von Carlowitz wrote to the King, who owned all the forests, advocating deliberate artificial (*Kunstliche*) techniques to regrow and manage the forests of Prussia, which had hitherto been regarded as nature's providence (Robin *et al.* 2013). Planting forests and ensuring that they grew in ways that could be easily harvested required imagining the lives of future generations, and providing for their needs – a basic principle of intergenerational equity. Sustainability was the first principle of conservation, or 'wise use' of resources, and von Carlowitz persuaded his King that such management was an economically sound investment.

The *precautionary principle* also emerged in the 1980s, when the World Charter for Nature (1982) was adopted by the UN General Assembly. Its first application was through

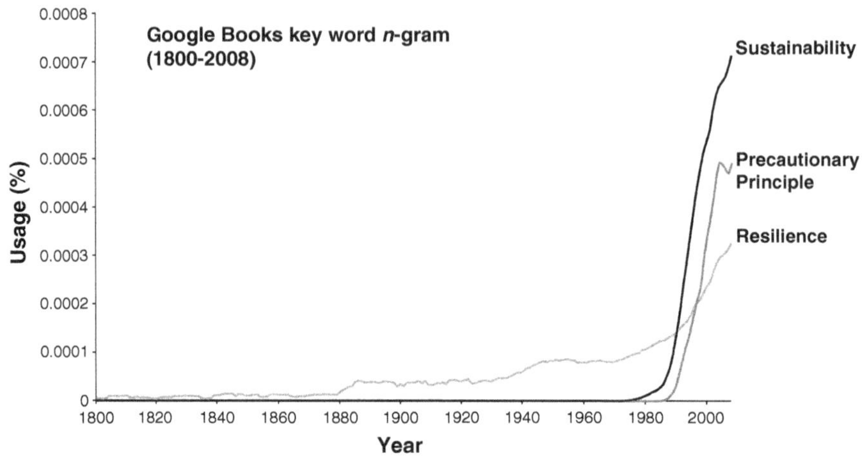

Figure 2.1. A short survey of Google Books (via n-gram) reveals the increase in key terms *sustainability*, *precautionary principle* and *resilience* in the period following the United Nations conference in Rio (1992).

the Montreal Protocol on Substances that Deplete the Ozone Layer (1987, ratified 1989). Linguistically, the use of the term accelerated after 1992 (Fig. 2.1), when it was an explicit principle of the Rio Declaration on Environment and Development:

> *Where there are threats of serious or irreversible damage, lack of full scientific certainty shall not be used as a reason for postponing cost-effective measures to prevent environmental degradation. (UNCED, Annex I, Article 15)*

Policy makers use this principle to underpin discretionary decisions in situations where there is the possibility of harm or injustice following an action, but a decision is needed before scientific knowledge on the matter can be amassed. The principle implies that there is a social responsibility to protect the public from exposure to harm, when scientific investigation has found a plausible risk. Later scientific findings providing sound evidence that no harm will result can allow a reversal of the decision. In some international legal systems (e.g. the law of the European Union), the application of the precautionary principle has been instilled as a statutory requirement.

The new global era

Environmental issues are important in conceptualising a global community and targets for global governance. A social imaginary enables 'common practices and a widely shared sense of legitimacy' (Taylor 2004: 23), which enable a society to be governed, regulated and managed. In the new global era, the idea of the nation has been stretched to include the whole planet. Global governance took form in 1995, when the Commission on Global Governance (CGG), another United Nations initiative, published *Our Global Neighbourhood* (CGG 1995), following the fall of the Berlin wall and the end of the Cold War. The CGG described the global era as one with a 'heightened sense of an endangered future' (Sinclair 2012: 20). The World Trade Organization (WTO) was established on 1 January of the same year out of the Uruguay Round of Trade Talks (1986–1994). WTO deals with the rules of trade between nations. Based in Geneva, Switzerland, it had 161 member countries as at 26 April 2015. (WTO 2015).

Although WTO comprises national members, global governance increasingly is conducted by intergovernmental organisations and non-governmental organisations (NGOs). Many intergovernmental organisations originated in the United Nations and its agencies, but other major ones include the Organization of American States and the Universal Postal Union. By 2001, there were 238 intergovernmental organisations engaging in global governance for the new millennium, serving very different purposes, including sorting disputes (International Court of Justice), providing services (UN High Commissioner for Refugees) and gathering data on planetary systems (UNEP). Their power is informal, resting on 'the mutual benefits from conformity to the system'. (Sinclair 2012: 21) Global NGOs are private, voluntary and usually focus on specific issues. Environment and development aid are prominent concerns in this sector. Adding a 'non-government' voice to debates about major problems that would otherwise be dominated by the richest and most powerful states – organisations such as the World Wildlife Fund, Conservation International, Oxfam, Transparency International and Amnesty International offer different models for global decision making and governance.

In terms of world science, the latest global group is Future Earth, 'a major international research platform providing the knowledge and support to accelerate our transformations to a sustainable world' according to its website. Director, Paul Shrivastava, was appointed in February 2015 (Future Earth 2015). Scientific initiatives have increasingly become concerned with human justice and equity as well as planetary accounting. For example, Rockström *et al.* (2009) consider all the major different biophysical pressures on the planetary system, but frame their research question in justice terms, as they seek a 'safe operating space for humanity' in a planet that is so changed that it is no longer part of the epoch of the Holocene – the period when agriculture and most world civilisations emerged.

The notion of boundaries has also shaped the justice agenda. In the run-up to Rio+20, Oxfam researcher Kate Raworth adapted the visual framework of Rockström and colleagues – linking the biophysical conditions put forward by the Stockholm Resilience Center group to 11 social conditions making for 'inclusive and sustainable development' (Raworth 2012). Raworth's 'doughnut' diagram, which brings the biophysical limits of the planet on the outside together with social conditions for sustainability on the inside is a powerful integrative tool for speaking about all the requirements for a safe and a just 'operating space' in which human societies can thrive. 'Moving into this space demands far greater equity – within and between countries – in the use of natural resources, and far greater efficiency in transforming those resources to meet human needs', Raworth argued.

Making the future cultural as well as planetary

Social theorist Arjun Appadurai (2013) has been critical of the technocratic construction of the future by experts from social and natural sciences, particularly neoclassical economics, whose concern is not the 'sentiment' of society, but rather the contractual and risk management strategies that now define society's possibilities (or limits). Modelling and prediction, the tools of the digital revolution, have been adopted by economics, environmental science, and planning and disaster management alike as the ways to visualise, rather than *imagine*, the future. Imagination is a collective practice that plays a vital role in finding a place in the world, where 'technical or neutral space' does not give sufficient weight to how people actually live in the future, in lives 'shot through with affect and with sensations'. Appadurai defines justice in terms of *hope*, arguing that 'the capacity to aspire is unequally distributed' in a world where the future is 'handed over to economics' (Appadurai 2013: 286–287).

The 'risk society' (Beck 1992) is a dominant western way to consider communities of all sizes. It is the brainchild of 'modern regimes of diagnosis, counting and accounting' (Burchell *et al.* 1991). A risk society is a dangerous place in new ways. An 'avalanche of numbers' leads to the quantification of future uncertainty and an 'ethics of probability' (Appadurai 2013: 295). So-called 'casino capitalism' (Lewis 2007) has profited from different catastrophes, from Hurricane Katrina (2005) onwards. Financial institutions and powerful individuals bet on disaster because they are concerned with profits rather than people. They work in worlds where issues of injustice (and people who suffer) are invisible. Emergency management itself can become part of a 'new economy of catastrophe' (Klein 2007: 58). Appadurai cautions against globalism's 'deep trend' in the 'broadening of risk-taking and risk-bearing as properties of human life' that increasingly link 'distant societies, cross-national and market boundaries' and connect 'institutions of power and the agencies of ordinary human beings worldwide' (Appadurai 2013: 3). Managing risk this way can force a divergence between economic and civil responsibilities that fosters both injustice and environmental disaster. The new global era demands a greater care about the environment and people in distant places, where local decisions can affect global outcomes. A concern for environmental justice recognises the 'slow violence' (Nixon 2011) dealt out by environmental disasters brought on by rapid development and growth without checks, particularly in poor places under pressure from rich partners.

The 'risk society' framework has been most recently revised in an emergency management context via the *Sendai Framework for Disaster Risk Reduction 2015–2030*, adopted at the Third UN World Conference in Sendai, Japan, on 18 March 2015 (United Nations 2015). In this document, there is a new emphasis on disaster *risk* management (as opposed to disaster management) and a justice-based approach (and risk-sensitive investment) to preventing new risks, reducing existing risk and strengthening resilience. Not just natural, but also anthropogenic and environmental, hazards are considered, and 'health resilience' is strongly promoted through support for health infrastructure, cultural heritage and workplaces (United Nations 2015: 5).

Conclusions

The 1940s were significant in conceptualising the environment. Before this, human affairs and nature were seen as separate realms, with nature being just a passive background against which the affairs of state played out. Natural resources had long been a feature of the trade between nations, but forests, fisheries and other natural resources became more explicitly part of international economics and governance when gross domestic product (GDP) was adopted as a way to measure and compare national economies in 1944 at the Bretton Woods conference.

Technologies and scientific discoveries opened up new possibilities. Undoubtedly, the nuclear threat, which had global ramifications for generations into the future, was significant in shaping the discourse of post-war governance and long-term planning. The environment was conceptualised as fragile, deserving of stewardship. It was no longer magical or frightening, as it had been in pre-'enlightenment' times, but now, rather, something that people talked about managing – and even 'saving' from wasteful practices. Its properties were measureable – and therefore limited, rather than being 'boundless' and providential. Natural resources such as forests, soil and water were reconceived in ways that mandated responsible and thrifty management strategies, and in ways that could foster development for poorer countries. The notions of *the environment* and *the global era* have co-evolved through the notion of governance (Christoff and Eckersley 2013) and, since the foundation of UNCED in 1972, the coupling of environmental and developmental justice has gathered pace. Governance, or the act of governing, is usually about society and its political mechanisms, and is typically on the level of a nation state or smaller. It becomes more difficult at a global level to make collective decisions that define expectations, grant power or verify performance. In a sense, the frameworks of governance define how the people governed by them see themselves, whether the community is local, national or global. Since the Brundtland report, the *needs of the world's poor* have been fundamental to the definition of sustainable development – they are the ones 'to which overriding priority must be given' in decisions that affect both environment and society (WCED 1987). This is a strong statement of justice right at the crux of sustainability. It is this that is visually expressed in the Oxfam 'doughnut' approach (Raworth 2012).

This chapter has explored how an understanding of history – what is possible, what has been tried – might support a more conscious justice agenda in natural resource management. In particular, it supports the emphasis of this book on *practice* – on negotiation and listening to situations in place, not just borrowing ideas from elsewhere without care for their context. Nonetheless, knowledge of international trends and ideas can provide fresh arguments that can assist local stakeholders to make better cases for just treatment in managing limited natural resources. After all, as ecologist and practical manager Richard Kenchington put it: 'We don't manage the environment. Only the human behaviours that affect its structure and process' (Kenchington 1994: 250).

Acknowledgements

The Ely Institute discussions with Paul Warde and Sverker Sörlin in the course of the Expertise for the Future Project have richly informed this chapter. I am also grateful to Stephen Dovers for some astute comments about how history is playing out for justice in present debates.

References

Appadurai A (2013) *The Future as Cultural Fact: Essays on the Global Condition*. Verso, London, UK.

Archibald G (2006) How the 'S' came to be in UNESCO. In *Sixty Years of Science at UNESCO – 1945 to 2005*. (Eds P Petitjean, V Zharov, G Glaser, J Richardson, B de Padirac and G Archibald) pp. 36–40. UNESCO, Paris, France, <http://unesdoc.unesco.org/images/0014/001481/148187e.pdf>.

Armiero M, Sedrez L (Eds) (2014) *A History of Environmentalism: Local Struggles, Global Histories*. Bloomsbury, New York, USA.

Beck U (1992) *Risk Society: Towards a New Modernity* (trans. Mark Ritter). Sage, London, UK.

Bonyhady T (1993) *Places Worth Keeping: Conservationists, Politics and Law.* Allen & Unwin, Sydney.

Boughton JM (2004) 'The IMF and the force of history: ten events and ten ideas that have shaped the institution'. IMF Working Paper WP/04/75, Washington DC, USA, <https://www.imf.org/external/pubs/cat/longres.aspx?sk=17199.0>.

Bridgewater P (2016) The Man and Biosphere Programme of UNESCO: rambunctious child of the sixties, but was the promise fulfilled? *Current Opinion in Environmental Sustainability* **19**, 1–6. doi:10.1016/j.cosust.2015.08.009

Burchell G, Gordon C, Miller P (Eds) (1991) *The Foucault Effect: Studies in Governmentality.* University of Chicago Press, Chicago IL, USA.

Carson R (1962) *Silent Spring.* Ballantyne, New York, USA.

CGG (1995) *Our Global Neighbourhood: The Report of the Commission on Global Governance.* Oxford University Press, New York, USA.

Christoff P, Eckersley R (2013) *Globalization and the Environment.* Rowman and Littlefield, Washington DC, USA.

Day M (1970) 'CSIRO and the environmental sciences'. Report to CSIRO Executive following international fact-finding mission in 1970. Archives of Maxwell Cooper Day, FAA (Fellowship file), Australian Academy of Science, Canberra.

Ekström A, Sörlin S (Eds) (2012) *Alltings Mått: Humanistisk Kunskap I Framtidens Samhälle.* Norstedts, Stockholm, Sweden (in Swedish).

Future Earth (2015) Future Earth website, <http://www.futureearth.org/>.

Griffiths T (2015) Environmental history, Australian style. *Australian Historical Studies* **46**(2), 157–173. doi:10.1080/1031461X.2015.1035289

Head L, Saltzman S, Setten G, Stenseke M (Eds) (2017) *Nature, Temporality and Environmental Management: Scandinavian and Australian Perspectives on Peoples and Landscapes.* Routledge, London, UK.

Höhler S (2015) *Spaceship Earth in the Environmental Age, 1960–1990.* Pickering and Chatto, London, UK.

Holling CS (1973) Resilience and Stability of Ecological Systems. *Annual Review of Ecology and Systematics* **4**, 1–23.

ICSU (2016) International Council of Scientific Unions website, Paris, France, <http://www.icsu.org/future-earth>.

IUPN (1950) *Proceedings of the United Nations Scientific Conference on the Conservation and Utilization of Resources.* IUPN, Vol. 1, United Nations, New York, USA.

Judt T (2010) *Ill Fares the Land: A Treatise on our Present Discontents.* WW Norton, New York, USA.

Kenchington RA (1994) Conservation and Coastal Zone Management. In *Conservation Biology in Australia.* (Eds C Moritz and J Kikkawa) pp. 245–250. Surrey Beatty, Sydney.

Klein N (2007) Disaster capitalism: the new economy of catastrophe. *Harper's Magazine*, October, pp. 47–58.

Lewis M (2007) In nature's casino. *New York Times*, 26 August.

Meadows DH, Randers J, Meadows DL (1972) *The Limits to Growth.* Universe Books, New York, USA.

Mosley JG (1972) *Wild Life Reserves.* Oxford University Press, Melbourne.

Nixon R (2011) *Slow Violence and the Environmentalism of the Poor.* Harvard University Press, Cambridge MA, USA.

Raworth K (2012) 'A safe and just space for humanity: can we live within the doughnut?' Oxfam Discussion Paper, Oxford, UK, <https://www.oxfam.org/sites/www.oxfam.org/files/file_attachments/dp-a-safe-and-just-space-for-humanity-130212-en_5.pdf>.

Robin L (1994) Nature conservation as a national concern: the role of the Australian Academy of Science. *Historical Records of Australian Science* **10**(1), 1–24. doi:10.1071/HR9941010001

Robin L (1998) *Defending the Little Desert: The Rise of Ecological Consciousness in Australia.* Melbourne University Press, Melbourne.

Robin L (2007) *How a Continent Created a Nation.* University of New South Wales Press, Sydney.

Robin L, Sörlin S, Warde P (2013) *The Future of Nature: Documents of Global Change*. Yale University Press, New Haven CT, USA.

Rockström J, Steffen W, Noone K, Persson Å, Chapin III FS, Lambin EF, *et al.* (2009) A safe operating space for humanity. *Nature* **461**, 472–475.

Sinclair TJ (2012) *Global Governance*. Polity, Cambridge, UK.

Snow CP (1959) *The Two Cultures*. Cambridge University Press, London, UK.

Stone C (1972) Should trees have standing? *Southern California Law Review* **45**, 450–501.

Taylor C (2004) *Modern Social Imaginaries*. Duke University Press, Durham NC, USA.

UNEP (2015) *Sourcebook of Opportunities for Enhancing Cooperation Among the Biodiversity Related Conventions at National and Regional Levels*. UNEP, Geneva, Switzerland, <http://www.unep.org/delc/Portals/119/publications/cooperation-sourcebook-biodiversity-conventions.pdf>.

UNESCO (1948a) *Conference for the Establishment of the International Union for the Protection of Nature*. UNESCO Archives NS/UIPN/12 Fontainebleau 5 Oct 1948, Paris, France, <http://unesdoc.unesco.org/images/0015/001547/154724eb.pdf>.

UNESCO (1948b) *The Scientific Conference on the Conservation and Utilization of Resources, UNSCCUR*. (A Memorandum by the Department of Exact and Natural Sciences, UNESCO), 10 November 1948.) UNESCO Archives, Paris, France, <http://unesdoc.unesco.org/images/0015/001547/154751eb.pdf>.

United Nations (1948c) *Universal Declaration of Human Rights*. United Nations, New York, USA, <http://www.un.org/en/universal-declaration-human-rights/>.

United Nations (2015) *Sendai Framework for Disaster Risk Reduction 2015–2030*. United Nations, New York, USA, <http://www.preventionweb.net/files/43291_sendaiframeworkfordrren.pdf>.

WCED (1987) 'Our common future'. Report of the World Commission on Environment and Development (Gro Harlem Brundtland, Chair) Oxford University Press, Oxford UK.

WTO (2016) World Trade Organization website. WTO, Geneva, Switzerland, <https://www.wto.org/>.

3

Justice research in practice: how can it influence environmental policy and planning?

Geoffrey J. Syme

Summary

This chapter describes the evolution of justice considerations in environmental planning. The rise of the public involvement movement in the 1970s led to a series of guidelines as to how government and industry sponsored public programs should be ethically conducted. In parallel, the environmental justice movement created a literature that clearly demonstrated that the negative effects of environmental degradation fell unequally on particular groups in society, such as the poor or those in an ethnic minority. The chapter explores the concepts of justice that have been used as the two interests merged. Through examples of the application of justice principles in environmental decision making, the dynamics of justice arguments are explored. These include: the issues of relative deprivation, deservingness and entitlement; evidence-based decision making; the issues relating to group interest; and the scales and levels of justice rhetoric. It concludes that the influence of justice research on decision making has been limited to this stage. Nevertheless, if opportunities relating to the creation of a common language, avoidance of the use of public discourse to reach a pre-determined outcome, and creating defensible evaluation programs within the political process are improved substantially, more influence will be achieved.

Introduction

The fundamental justice questions for natural resource management relate to how the benefits of access should be shared. In some cases, arguments relate to the amounts of the natural resource allocated to individuals, groups or nations. In other situations, arguments centre on defining the components of the concept of justice itself.

There have been centuries of philosophical discussion about the appropriate approaches to justice issues. These philosophies have varied from virtue theory whereby it is judged that those who have resources deserve them because of their social status through to ethically based theories such as those of Kant's moral imperative, ideas of procedurally based justice and a variety of utility-based theories (see Wenz 1988 for an excellent review). In many ways, these reflect alternative visions of how the world *should* be.

Such formulations provided an important background to the rise of the environment as an issue in the issue-attention cycle in the 1960s. In a decade of increased social activism, the public vented its displeasure on a variety of environmental issues such as pollution and the allocation of natural resources such as water. Formal environmental movements became entrenched and local community protests abounded.

This was disconcerting for the planning community who had, since the Second World War, conducted what was considered to be rational planning in which experts, generally on the behalf of government, planned using science and engineering logic for the public good. Protesters were often derisively labelled as emotional NIMBYs ('not in my backyard') and a variety of other suitably catchy acronyms. The assumption was that NIMBY's were selfish and against justice for the wider community. The question of the motivations of NIMBY's and whether there are good or bad outcomes from them for the community as a whole continues to be debated (e.g. Feldman and Turner 2014).

Despite antipathy from some planners, it became evident that there was some need for planners to include systematic public comment in the decision-making process, especially where allocation or environmental quality decisions could be controversial. Even worse, studies of the ethics of planners began to show that in practice they were not the detached rationalists they purported to be. In fact Hudson (1979) recognised five approaches to the planning role and many planners adopted any or all of those roles depending on the circumstances. These roles were: Synoptic (or rational); Incremental (muddling through according to the circumstances); Transactive (face-to-face discussions with people); Advocacy (enhancing the influence of an interest group) and Radical (where planning was used to empower spontaneous protest). In short, planners responded according to their role and their values. Detached rationality could not necessarily be assured. The planners' role in terms of ethics and justice has become even more complex with the outsourcing of planning practice by government to consultants and others in the current age of neoliberal thought (Steele 2009).

The 'public involvement', 'consultation', 'engagement' response

The response from planners to the increasing lack of trust in them getting it right if left to their own devices was to develop processes of public involvement whereby public input was to be heard. For many, however, the motivation for such programs has been instrumental in that it was hoped that, once information was given to the public, politically unwanted protest would die down. Sometimes the aim was to isolate the politicians who were not heeding professional advice by recruiting the community. As a consultant for government and the mining industry in the 1970s, often the tacit goal of the planner was to manipulate the community to see 'sense' and difficult negotiations had to be facilitated between the community and the planner.

The academic planning literature was quick to respond to this situation and Arnstein's ladder of participation (Arnstein 1969) became the touchstone against which professional tried to gauge progress in terms of perceived justice and reasonable influence of the community. Gradually formal organisations began to appear to create benchmarks for adequate professional performance (e.g. IPP 1981). Nonetheless, while issues of respect and information flow, resourcing and so on were discussed and manuals of good practice developed, progress towards achieving the level of influence required from both the planner and the public's perspective has been patchy. This is shown in Fig. 3.1 and indicates

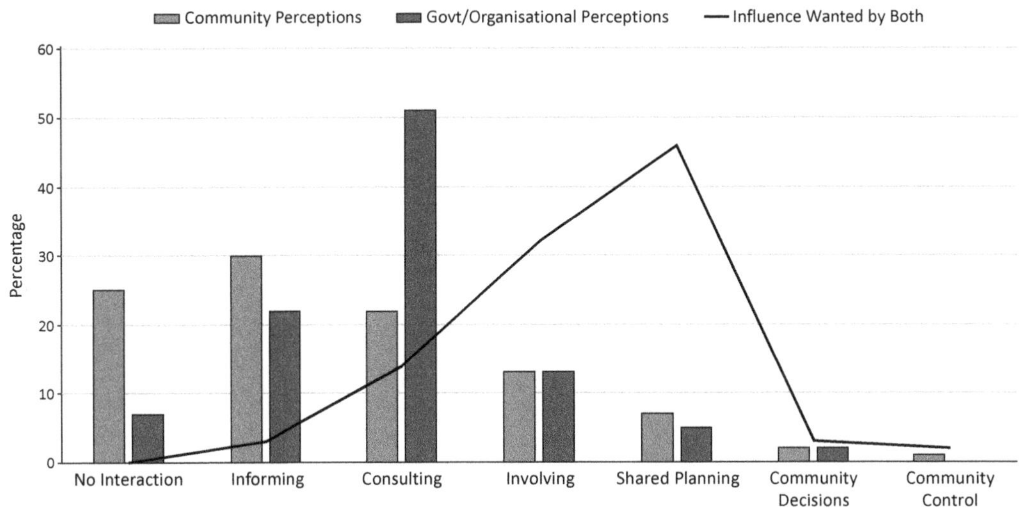

Figure 3.1. Perceived current and desired influence by the public and planners. The level of desired influence was the same for the public and the planners and is represented by a single line.

where people thought they had got to in Western Australia in the late 1980s (Syme and Nancarrow 2005) in regard to land-use planning. It seems at that point that the partnership required by both the public and governance organisations was still to be attained.

Applying principles of justice

In parallel, there was the emergence of the environmental justice movement, initially led by sociologists who clearly demonstrated that different sections of communities were consistently being exposed to greater costs of pollution or environmental amenity. These communities were often of lower socio-economic status or ethnic groups in the USA (e.g. Schnaiberg 1975). Predominantly white, middle-class male planners were seen to have failed to justly deliver on their behalf. Thus there were many reasons both moral and practical that required a better response from the planning community. The recent adoption of the concept of sustainability has made this an even greater challenge that includes the need to consider the environment and future generations in justice considerations.

This sociological literature provided the motivation for examining the justice principles that were inherent in the decision-making process itself. Social psychologically based research, which was not initially applied to the environmental field or public involvement programs, began to develop. Theoretical studies as to what were the dimensions of equity and how they were applied to justice considerations in organisations appeared (Rasinski 1987). These gradually became applied to environmental allocation decisions (Syme and Fenton 1993; Lawrence *et al.* 1997) in an attempt to create a normatively defensible set of criteria against which perceived justice could be assessed for environmental decision making. The rights of the environment also became included in such studies. In later years, the empirical measurement of lay justice philosophies and how they could be incorporated in decision making appeared (e.g. Syme *et al.* 1999). The breadth of concepts measured in the public arenas is shown in Box 3.1 (from Syme *et al.* 2015).

Box 3.1: Principles of justice

[1] Distributive justice (who gets what)
1.1 Equity (you are rewarded in proportion to effort)
1.2 Equality (people get equal distributions)
1.3 Need (people who need it for basic reasons get first preference)
1.4 Self interest (each individual prefers allocations that are best for themselves)
1.5 Efficiency (people who use the resource most efficiently are entitled to more)

[2] Procedural justice (how decisions are made)
2.1 Voice (having the opportunity to be listened to and have some influence)
2.2 Participation (being able to participate in decision-making processes)
2.3 Consistency (the rules are not abruptly changed)
2.4 Ethicality (lack of corruption, follows due process)
2.5 Impartiality (everyone has the same opportunities to participate)
2.6 Accuracy (the information provided for discussion is correct)
2.7 Error correctability (there is a chance to appeal in terms of facts)
2.8 Representativeness (there was a good range of stakeholders included)

[3] Combining distributive and procedural justice
3.1 Fairness (an overall judgement of both the process and the outcome)

[4] Interactive judgment (do people feel they have been treated appropriately?)
4.1 Trust (do people trust the government agency and other participants?)
4.2 Dignity (do people feel they have been treated with dignity?)
4.3 Respect (do people feel they have been appropriately respected?)

[5] Justice philosophies
5.1 Virtue theory (people who already have a resource should retain it because they are inherently good)
5.2 Prior rights (people who have used the resource in the past or first in first served)
5.3 Intergenerational justice (we need to think of the needs of future generations)
5.4 Environmental rights (the environment deserves its own allocation)
5.5 Property rights (individuals should be given rights to amounts of resources based on some contract)
5.6 Economic good (natural resources should be treated as economic goods with prices and markets)
5.7 Utilitarian theories (resources should be managed to maximise community welfare – there are many versions of this)
5.8 Moral imperative (people in one location have a duty to ensure they don't negatively affect people in other locations).

Further, Lukasiewicz and colleagues (Lukasiewicz *et al.* 2013; Lukasiewicz and Baldwin 2014; Chapter 17) have developed a Social Justice Framework using some of these concepts to provide a means by which the inclusion of justice can be evaluated in decision making. These efforts provide a concrete basis in empirical terms as to what justice may look like, and unfortunately to this point have yet to be used systematically in the decision-making process. Much of the reason for this is the dynamics of justice in highly contested decisions and the political confusion that surrounds them.

The dynamics of justice

Relative deprivation, deservingness and entitlement

It has been noted by many authors that there are motivations relating to the feeling of injustice that arise and drive both individual and group emotions. These pertain to the concepts of relative deprivation, deservingness and entitlement (Feather 2015). In simple terms, these concepts relate to the questions of: 'How am I doing in relation to other relevant individuals or groups?'; 'Did the person or group in question do enough to deserve that outcome?'; and 'Is this outcome something that I or they am entitled to because of my social position?' All of these concepts overlap and interact but in their own right can affect the emotions that feed feelings of injustice.

The concept of entitlement is particularly important in natural resource allocations, such as water. Often water rights are granted to irrigators on the basis of their past history of use. To remove such rights can often arouse great dissatisfaction because it is seen as a loss of entitlement (even in Australia where water is 'owned' by the state). As Kahneman and others (Kahneman and Tversky 1979; Vanhoven *et al.* 2000) have demonstrated, the valuation of resources that are perceived as lost is significantly higher than resources gained and thus the perception of injustice can be highly felt. This is not to say that irrigators consider that the first-come first-served principle is just. Those who have missed out also have a motivation to feel injustice, although it is not expressed as vocally. The adoption of the neoliberal concepts of water rights for irrigators and the ability for those who wish to acquire water through markets has been seen as a neat way for government to step out of the conflict. But the justice issues still remain in terms of identifying what the preferred allocation of water should be between uses.

Each of the three concepts have a role in defining and redefining water justice problems over time and all are used by individuals and interest groups in a variety of ways. They emphasise the observation that justice arguments in this area are all about resolving the question: 'Justice for who?' In answering this question, justice firmly enters the area of politics rather than synoptic or rational planning, because the government itself can in many cases be seen as an advocate group for those who have elected them. This makes life difficult for those who wish to advocate justice evaluations of decisions. The frame of the justice question can tend to change as people interpret the issue in terms of relative deprivation, deservingness and entitlement. One often-used method of reframing justice arguments is changing the scale or level at which justice gets expressed (Patrick *et al.* 2014), which often changes the 'who' justice is for.

Scales and levels of justice

Internationally, participative planning has tended to neglect the issue of scale or level of decision making and how it may affect the framing of justice issues and the implementation of change in natural resource management. At the basest level, this concern can be expressed in the highest level of decision making and political power designing participative plans to manipulate outcomes or delay decisions at lower levels. For example, Patrick *et al.* (2014) demonstrated how water policy changes can be implemented by tacit agreement between local, state and national to ascribe justice issues at a national-public-good scale so that local implementation can occur without the opprobrium that lower levels of government were to blame for locally perceived injustice.

During this negotiation, many of the principles in Box 3.1 can be seen to be appropriate from each participant's viewpoint. This makes it difficult for justice evaluations to provide cogent advice for the community, or for that matter the decision makers. Some would say that justice in terms of the overall public good has precedence over localised concerns. But it is at the local or individual project level that most injustices are personally felt and the motivation to act in protest is greatest. How to marry local and higher level frames of justice when neoliberal imperatives are seen to take precedence is a difficult issue that requires further discussion. The failure to marry these levels of justice in the end leads to the chronic environmental justice issues identified by the sociologists from the 1970s.

Evidence-based decision making

Many claims of injustice revolve around issues related to quantifiable amounts of things. For water it is gigalitres. For agriculture it is the need to protect increasingly scarce hectares of arable land. For conservation it is about defining endangerment in terms of numbers of species or numbers of a specific species. The hope is that by objectifying what the management problem is the justice arguments become self-evident.

There are often calls for evidence-based decision making by governments and scientific studies coupled with social impact assessments are used to hopefully provide a defensible and rational debate. In a neoliberal world, by underpinning these decisions with property rights and markets, the political involvement with justice decisions is hoped to be seen to be at arms-length from the government and thus non-controversial.

There are many problems associated with this assumption. Science is not seen to be value free and science itself does not speak with one voice, especially when uncertainty is high (Tuckwell 2012). Social impact assessment, and even economic analyses, contain assumptions that can always be contested. Finally, by framing issues in quantitative terms, when policies change allocations and lessen numbers for some, the powerful psychological phenomena of the endowment effect ('I have had this water for years so this is a personal loss and an attack on my expectations') and social discounting ('I don't identify a responsibility to others who are not close to me') drive perceptions of injustice. When sustainability arguments demanding attention to the rights of future generations are offered, the inevitable tendency of humans to discount the future arises as a significant hurdle. Justification for feelings of injustice can so often be underpinned by the phenomenon of nostalgia for both irrigators and conservationists alike ('we have had droughts before', 'the community was great then'; 'the environment was pristine') (see Syme 2014 for discussion).

Some authors have suggested that, because of the nature of environmental issues, the concept of defining benefits of resource access is a better way forward in dealing with allocation or policy issues (see Syme and Nancarrow 2008 and van den Brink *et al.* 2014 for two contrasting approaches). That is, for renewable resources in particular, the same resource can be used for multiple benefits and negotiations can therefore move from those of a zero sum game. For example, the same body of water moving through a catchment can provide multiple benefits. There is much room for the development of integrated decision making using this framework and incorporating justice criteria. It is possible that some of the effects of the endowment effect can be mitigated and a wider range of negotiation principles used (Hoekstra *et al.* 2001). However, as we progress in this area it is clear the hopes that objective analyses will provide for less confused or emotive decision making are unlikely to be fully realised. We have no alternative but to deal with justice issues more systematically if we are to achieve sustainable outcomes. One contribution may be to lessen the reliance on stakeholder-based decision-making processes.

Group interest

The reference above to social discounting and the tendency to avoid feelings of moral responsibility for those not in the 'in-group' as a major issue is underscored by the role of group identity in defining 'deservingness' and 'entitlement'. Feather (2015) provides an account of how intergroup resentment associated with perceived injustice arises. The feeling of relative deprivation occurs when the out-group achieve positive allocations that they are not perceived to deserve through positive actions. Feather (2015) notes that, on occasion, longer standing agreed forms of entitlement are accepted when cultural norms are complied with. But in the case of positive outcomes awarded to the out-group that are the result in negative actions or perceived self-entitlement, resentment can occur. For example, in the groundwater allocation problem investigated in justice terms by Syme *et al.* (1999) there was no resentment of greater allocations but only so long as responsible best standard water conservation measures were in place. If those conservation measures were not in place, the water was not earned and therefore not deserved. Regardless of the behaviour of other groups, however, there is always a bias towards one's own group's deservingness.

Unfortunately, participative planning in regard to natural resources has tended to exacerbate these within-group biases because of the constant recourse to stakeholder-based advisory groups in Australian natural resources planning (Ellemers 2002; Evans and Pratchett 2013). This tends to promote advocacy rather than transactive planning and win–lose modes of discussion, which enhance the in-group bias. For most individuals, this ends in necessitating a good outcome for their group, or at least a demonstration that they have been a worthy advocate. This participatory approach thus has the potential to nullify broader considerations of justice, especially when amounts of resources are the outcomes of most focus.

There are alternative forums and techniques, especially in the scenario-based futures literature, that can be applied that have a better potential for delivering normative justice. The important thing is that the selection of the methods and processes involved need to go beyond representation by interest groups into representation by the array of justice considerations. Participatory planning processes need to be planned with justice in mind and with the consideration of all levels of impact, thus avoiding cumulative justice impacts.

Improving the influence of justice research in practice

Having started this chapter with a brief summary of the emergence of planning roles in response to the early justice literature, then defined some building blocks and briefly discussed some issues relating to the dynamics of justice, I have concluded that there is no getting away from dealing with these issues methodically, no matter what the challenges seem to be.

Advancing the priority for justice evaluations

In the 1970s when I began justice research I noted that words to the effect that water would be equitably allocated occurred in most state water legislation. In interviews with public servants, I asked what equity really meant. The reply at the time was pretty much 'it's equitable because we make decisions based on the Act'. Justice or equity were neither defined not given priority.

More recently, with the rise of the sustainability concept for most justice researchers, it seems evident that the search for justice is an essential and integral part of achieving sustainability. The implicit hypothesis is that resilience of solutions will be better if not associated with ongoing resentment because of perceived injustice. It has come to some surprise in discussions with planners, economists and politicians that this may be true but that justice is not necessarily the top priority in decision making. Even if it was, justice research is likely to be too theoretical and static to help in the political vagaries that occur. Even in formal planning processes there is a lack of priority, or even inclusion, of justice explicitly in many multi-criteria or systems approaches to decision making.

For justice researchers this should be seen as a challenge rather than an inevitable situation. We need to demonstrate in multi-criteria assessments and other approaches that if the criteria had included justice then the results may have been different. This problems can be overcome with well worked and reported exemplars, as is starting to happen in recent years with several authors providing detailed interpretations of case studies analysing justice, particularly for water resources (e.g. see Gross 2014; Patrick *et al.* 2014).

More importantly, this recent work can provide the basis on which specific exercises in accessible justice analyses can be included in the education of natural resource managers, planners and others interested in integration and systems thinking. These need not be 'state of the art' justice evaluations, but sufficient for professionals for the inclusion of justice and ethical considerations in daily practice (e.g. Groenfeldt 2013; Byrne and Houston 2014) The benefits of this will be felt by politicians, many of whom consider because they have been elected it is their view of what is fair that should prevail, sometimes to the detriment of their constituents.

Creating confidence and a language

Miller (1999) and Montada (1998) described the tacit language and culture that predominated during the 1990s and still holds sway today. That is, there is a culture in which logic and economic jargon predominates at higher level decision making. Solutions are presented in economic terms and in regard to costs and benefits and market goods. Sometimes the terms are misused, but with little accountability. Forecasts are given with great confidence in television programs by 'well-known' economists. There is no effective counter language that enables justice considerations be included in such a coherent way. Fairness issues are left to political debates and interest groups, often with a view to advancing their own agenda.

Such a lack of justice discourse is also evident at local levels where people are coy about pro-social issues such as sharing benefits or losses. A good example from my own practice was in presenting the results of a community survey on coping with reduced groundwater allocations. The most inclusive solution had four components. Two of these favoured economic rationality and two pro-social sharing intent. On presenting the outcomes to a group of farmers with larger allocations, I was accused of making up the results. When asked why, it was an objection to the 'thinky feely' pro-social components. The justification for the accusation was that 'we're all hard bastards up here, farming is a business'. A lively discussion then ensued among the group that did incorporate the justice issues. The point is that it was more acceptable to talk in profits because the language was salient and everyone knows the value of a dollar. Conversations in Australia, at least, don't naturally turn to ethics and the community good. The tacit culture is based in traditional economics.

By attending to the justice priority, easily accessible language will emerge as confidence with these issues will rise among planners and decision makers, especially in relation to procedural justice issues. But justice arguments tend to be expressed tentatively by justice researchers: 'On the one hand there consequences like this on the other it could be that' is the type of phrase used by social scientists. The same caveats could easily be expressed by economists, and often are, though not often in the public arena. Whether we feel comfortable or not with our justice conclusions in the natural resource area when there are likely to be major injustices we should ensure that those issues are confidently pushed in public debate so that a better balance can be achieved in the long term.

Avoiding instrumentalism

Much of the argument about public participation practice had revolved about how to give voice to the public. The reason for this is that, on many occasions, if one is hired to facilitate public programs there is often a predilection of the government or private client for the outcome to support what they considered to be appropriate in the first place. Often if things don't seem to be going in the 'right' direction resources are withdrawn, the project manager disappears and the promised workshop does not eventuate. As a result, the client's motives are not trusted by the participants. Sometimes the minster or CEO cancels the project in mid process. All this leaves a legacy that erodes the ability of participative planning to be successfully conducted.

The only way to prevent this is to spend the time to negotiate an agreement with the agency or client at a high enough level about what power is to be afforded to the public. If this is made clear at the outset, the participants' expectations are appropriate for the influence the client is prepared to offer. This will mean that the ethics and procedural justice issues are clear to begin with. Some of these will be sub-optimal for both the client's and the community's interests. But pragmatic agreements in sharing the uncertainty of outcomes can afford progress in the longer term. There will be a latitude of acceptance of justice considerations and this will affect flexibility in attitude change (Eagly and Teelak 1972). Perfect processes never occur. But it is important that the client at the highest possible level acknowledges it's a partnership and if not does not conduct the public program.

Integrating social justice evaluation in the political process

This is probably the most difficult area for justice researchers and practitioners. As stated earlier, many politicians consider that they personally, and their party generally, have the mandate to make fairness and justice judgements because people have been given a choice at the ballot box. Judgements in this area have been delegated to them and their party colleagues. Such an attitude leads to interest-group-type politics because the party and the individual protect the interests of the groups who align most closely to their values. Since they provide many of the funds for their operation these days, it is not surprising.

The fact is that justice advice can be of great value in avoiding the indignation associated with feelings of relative deprivation as polices are being formulated. Perfect justice cannot be achieved and the dynamics of political debate in the natural resource and environmental area will change with new dynamics as new technologies and knowledge appear. Nevertheless, knowledge of unintended injustice outcomes will improve decision making.

Perhaps we can help improve justice outcomes in this area is by acknowledging that the politician is a client, like many others. There will be different preferences for guiding principles (see Box 3.1) between political parties and groups within parties. If the justice-based practical policy outcomes can be transparent and inclusive, they will be attractive to politicians. But the 'language' of justice is not currently available to policy makers, whether they be politicians or otherwise.

Thus, as suggested before, the justice-oriented planner need not shy off policy debate and can become much more active in a political sense. As the prominence of the styles of justice discourse used becomes greater, there is a greater likelihood of social justice evaluations being included in political circles. At the very least, they may mitigate against unexpected consequences. This may seem a long term goal, and it is. Nevertheless, with the increase of researchers and practitioners who are literate in the justice area there is room for hope.

Conclusions

In this chapter I have attempted to sketch the history of the development of the social justice research and practitioner community and outline the basic concepts involved. In discussing the dynamics of justice I have identified some challenges in a specific sense in terms of issues of practice. Finally, I have outlined the challenges social justice practitioners face in terms of making social justice considerations count in the real world. These challenges are significant and will take time to overcome. With the gradual rise of the justice theory and practice, however, the community dynamics among us will become stronger and the opportunities more significant. In taking these opportunities, I have no doubt that that injustices will be significantly lessened and the public interest in natural resource decision making enhanced. All that is needed is persistence.

References

Arnstein SR (1969) A ladder of citizen participation. *American Institute of Planners Journal* **35**, 216–224. doi:10.1080/01944366908977225

Byrne J, Houston D (2014) Environmental equity/justice. In *Australian Environmental Planning: Challenges and Future Prospects.* (Eds J Byrne, N Sipe and J Dodson) pp 206–217. Routledge, Oxford, UK.

Eagly AH, Teelak K (1972) Width of the latitude of acceptance as a determinant of attitude change. *Journal of Personality and Social Psychology* **23**(3), 388–397. doi:10.1037/h0033161

Ellemers N (2002) Social identity and relative deprivation. In *Relative Deprivation: Specification, Development, and Integration* (Eds I Walker and HJ Smith), (pp. 239–264). Cambridge: University Press, Cambridge, UK.

Evans M, Pratchett L (2013) The localism gap-the CLEAR failing of official consultation in the Murray Darling Basin. *Policy Studies* **34**, 541–558. doi:10.1080/01442872.2013.862448

Feather NT (2015) Analyzing relative deprivation in relation to deservingness, entitlement and resentment. *Social Justice Research* **28**, 7–26. doi:10.1007/s11211-015-0235-9

Feldman S, Turner D (2014) Why not NIMBY? *Ethics. Policy and the Environment* **17**(1), 105–115.

Groenfeldt D (2013) *Water Ethics: a Values Approach to Solving the Water Crisis.* Routledge, Oxford, UK.

Gross C (2014) *Fairness and Justice in Environmental Decision Making: Water under the Bridge.* Routledge, Oxford, UK.

Hoekstra AY, Sevenije HHG, Chapagain AK (2001) An integrated approach towards assessing the value of water: A case study on the Zambesi Basin. *Integrated Assessment* **2**, 199–208. doi:10.1023/A:1013368524528

Hudson BM (1979) Comparison of current planning theories: counterparts and contradictions. *Journal of the American Planning Association* **45**, 387–398. doi:10.1080/01944367908976980

Institute for Participatory Planning (1981) *Citizen Participation Handbook for Public Officials and Others Serving the Public.* IPP, Laramie WY, USA.

Kahneman D, Tversky A (1979) Prospect theory: analysis of decision making under uncertainty. *Econometrica* **47**, 263–291. doi:10.2307/1914185
Lawrence RL, Daniels SE, Stankey GH (1997) Procedural justice and public involvement in natural resources decision making. *Society & Natural Resources* **10**, 577–589. doi:10.1080/08941929709381054
Lukasiewicz A, Baldwin C (2014) Voice, power and history: ensuring social justice for all stakeholders in decision making. *Local Environment: The International Journal of Justice and Sustainability* [online]. doi:10.1080/13549839.2014.942261.
Lukasiewicz A, Bowmer KH, Syme GJ, Davidson P (2013) Assessing government intentions for Australian water reform. *Society & Natural Resources* **26**(11), 1314–1329. doi:10.1080/08941920.2013.791903
Miller D (1999) The norm of self interest. *The American Psychologist* **54**, 1053–1060. doi:10.1037/0003-066X.54.12.1053
Montada L (1998) Justice: Just a rational choice? *Social Justice Research* **11**, 81–101. doi:10.1023/A:1023299119352
Patrick MJ, Syme GJ, Horwitz P (2014) How reframing a water management issue across scales and levels impacts on perceptions of justice and injustice. *Journal of Hydrology* **519**, 2475–2482. doi:10.1016/j.jhydrol.2014.09.002
Rasinski K (1987) What's fair is fair – or is it? Value differences underlying public views about social justice. *Journal of Personality and Social Psychology* **53**, 201–211. doi:10.1037/0022-3514.53.1.201
Schnaiberg A (1975) Social syntheses of the societal-environmental dialectic: the role of distributional impacts. *Social Science Quarterly* **56**(1), 5–20.
Steele W (2009) Australian urban planners: hybrid roles and professional dilemmas? *Urban Policy and Research* **27**(2), 189–203. doi:10.1080/08111140902908873
Syme GJ (2014) Struggling with uncertainty in Australian water allocation. *Stochastic Environmental Research and Risk Assessment* **28**(1), 113–121. doi:10.1007/s00477-013-0694-1
Syme GJ, Fenton DM (1993) Perceptions of equity and procedural preferences for water allocation decisions. *Society & Natural Resources* **6**, 347–360. doi:10.1080/08941929309380833
Syme GJ, Nancarrow BE (2005) Sharing the pain of water reallocation: creating consent by taking fairness and justice seriously. In *Consent and Consensus, Politics, Media and Governance in Twentieth Century Australia* (Eds D Cryle and J Hillier) pp 371–387. API Network, Curtin University of Technology, Perth, WA.
Syme GJ, Nancarrow BE (2008) Justice and the allocation of benefits from water. *Social Alternatives* **27**(3), 21–25.
Syme GJ, Nancarrow BE, McCreddin JA (1999) Defining the components of fairness in the allocation of water to environmental and human uses. *Journal of Environmental Management* **57**, 51–70. doi:10.1006/jema.1999.0282
Syme GJ, Reddy VR, Ranjan R (2015) Justice and equity in watershed development in Andhra Pradesh. In *Integrated Assessment of Impacts of Watershed Intervention: Assessing Hydrological and Biophysical Influences of Livelihoods.* (Eds VR Reddy and GJ Syme) Elsevier, Amsterdam, Netherlands.
Tuckwell E (2012) Science in dispute: debating the authority of knowledge in environmental contestation. *Oceania* **82**, 308–322. doi:10.1002/j.1834-4461.2012.tb00136.x
van den Brink R, Estevez-Fernandez A, van der Laan N, Moes N (2014) Independence of downstream benefits in riverwater allocation problems. *Social Choice and Welfare* **43**(1), 173–194.

Van Boven L, Dunning D, Loewenstein G (2000) Egocentric empathy gaps between owners and buyers. *Journal of Personality and Social Psychology* **79**, 66–76. doi:10.1037/0022-3514.79.1.66

Wenz PS (1988) *Environmental Justice*. State University of New York Press, Albany NY, USA.

4

Justice approaches: methods and methodology in environmental justice research

Sonia Graham, Claudia Baldwin, Jennifer McKay and Sue Jackson

Summary

This chapter explores the diverse methods used by authors of this book in their research on environment and justice in Australia. It explains the reasoning behind choosing these methods and the perceived advantages and disadvantages of using these methods to study questions of justice. In doing so, it identifies ways in which various qualitative and quantitative methods can be used separately, or in combination, to provide a more comprehensive perspective on justice considerations. The chapter also provides examples of the questions posed to participants in interviews, questionnaires, focus groups and visual studies by authors of this book, to illustrate how researchers have employed and operationalised justice concepts.

Introduction

Researchers who study justice employ a wide array of methods in their efforts to tackle inequalities and injustices in society (Mertens 2007). This is reflected in the 14 methods that underpin the work in this book (Fig. 4.1). The methods used by researchers in this book are so diverse that no single social research methods textbook published to date covers them all. Indeed, some methods, such as network analysis, doctrinal analysis, statutory interpretation, spatial mapping and computer simulations, do not appear in key textbooks used to teach social research methods in Australia (e.g. *Applied Social Research* (Hall 2008), *Social Research* (Sarantakos 2013) or *Social Research Methods* (Bryman 2016)).

The researchers in this book employ a wide range of qualitative and quantitative methods across their portfolios and within particular projects. This enables engagement of the diverse stakeholders listed in Fig. 4.2; the majority of lead authors of chapters[1] in this book engage with at least five of the listed stakeholders in their past and current research. Methodological diversity is seen to be essential to understand the plurality of interests and

1 Each of the 15 lead authors of the 19 chapters of this book were asked to specify what methods they use and what stakeholders they engage with in all their environmental justice research. Throughout the chapter the term 'the lead chapter authors' is used to refer to these 15 lead authors.

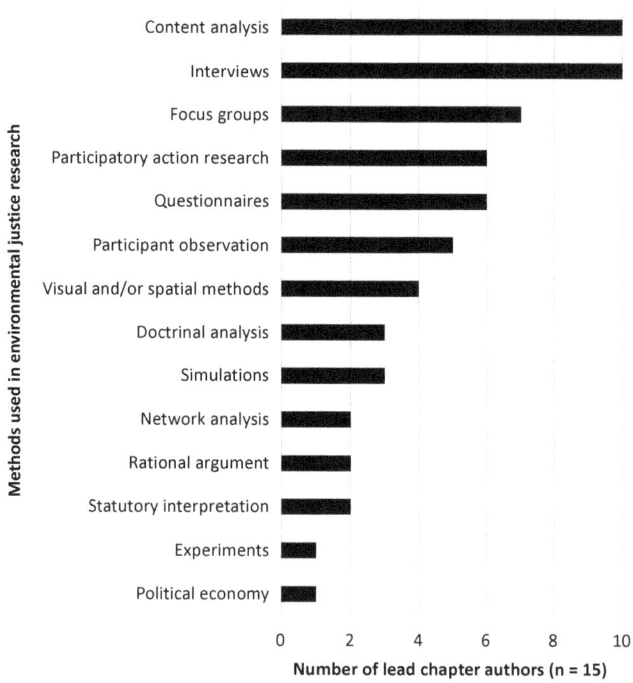

Figure 4.1. Methods used by lead chapter authors in their past and present research on environmental justice.

perspectives involved in natural resource management (Chapter 8), gain descriptive, analytical and normative insight, as well as practical application to improve justice outcomes. The visual and spatial methods also bolster the expression of values in natural resources management by providing non-verbal means to elicit values from respondents (Chapter 11). Ultimately, applied research provides opportunities to involve the public in

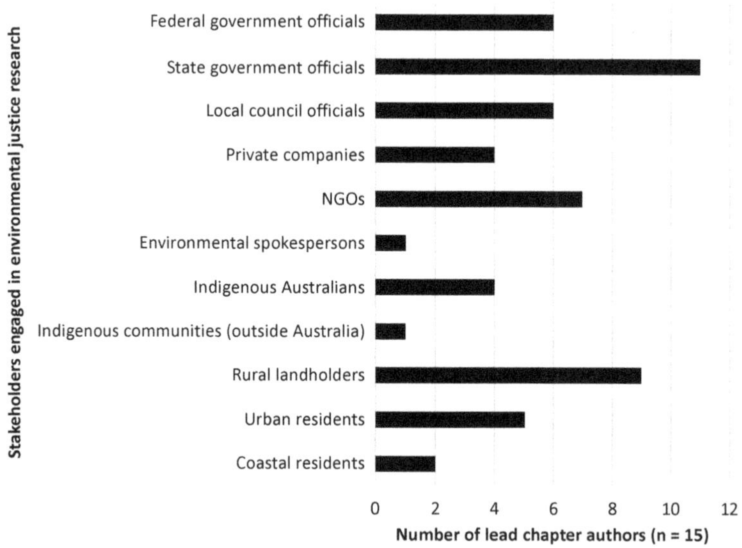

Figure 4.2. Stakeholders engaged in environmental justice research by the lead chapter authors in their past and present research.

dealing with conflict, collaboratively developing and providing input into potential solutions, and influencing decision makers (Chapter 3).

Past reviews of methods and methodological approaches in environmental or justice research have sought to: justify the use of mixed methods approaches (e.g. Mertens 2007, 2013); explore the way in which particular methods are applied, such as multiple criteria analysis (e.g. Romero and Rehman 1987; Mendoza and Martins 2006; Hajkowicz and Collins 2007) or stakeholder analysis (e.g. Reed *et al.* 2009); and evaluate the strengths and weaknesses of various tools (e.g. Lynam *et al.* 2007). The only methodological review that pertained to environmental justice[2] specifically, sought to evaluate the quality of 42 empirical research studies (Bowen 2000 2002) using 'scientific standards' (Bowen 2002: 3). In Bowen's (2000) critique, qualitative research was not seen to be sufficiently rigorous to be used to inform policy decisions. The overview provided in this chapter strongly challenges such a conclusion; almost all the authors here find qualitative research to be highly appropriate and useful for studying and contributing to environmental justice.

There is a gap in our understanding of what methods are used by researchers who study environmental justice in Australia and internationally, and why these methods are chosen. Thus this book provides an opportunity to begin a broad conversation about the perceived validity of different methods in the context of environmental justice and how such research can contribute to improving processes and outcomes for participants. The chapter outlines the rationale for the methods chosen, and the advantages and disadvantages of each. Examples are provided of the questions posed to participants in interview, questionnaire, focus group and visual studies, to illustrate how justice concepts have been operationalised. Such an exploration can provide insights to researchers, who are new to the field of environmental justice, about the range of methods available, how they have been applied, what methods and questions might be most appropriate for their research, and what refinements may need to be made. Such explanations and justifications can also be used to enhance the ethics and quality of justice research; many of the methods discussed in this chapter require approval from human research ethics committee before their implementation and methodological choices will be evaluated by peers before publication in the academic literature.

Methods used in environmental justice research in Australia

The four main methods used by the lead chapter authors presented here and elsewhere are qualitative, namely: content analysis of documents, interviews, focus groups and participatory action research (Fig. 4.1). Other qualitative methods, such as participatory Geographic Information Systems (GIS) and three-dimensional (3D) visualisation, are more recent additions to the suite, yet are proving to be useful for studying environmental justice, particularly for their ability to enable spatial analysis (Chapter 11; Jackson *et al.* 2012).

Quantitative research methods, such as questionnaires, are occasionally used, but are often used in conjunction with qualitative methods (Chapter 3). This is consistent with other justice research, where mixed methods are often used in sequential and/or iterative stages to understand how stakeholders perceive the issue and decision-making processes, determine how representative these concerns are among the wider population affected, and then present research findings back to participants for comment (e.g. Lauber 1999; Liu 2001; Mertens 2013).

2 See the section on 'Justice and the environment – what are we actually talking about' in Chapter 1 for an explanation of the broad field of research we consider to be described by the term 'environmental justice'.

Table 4.1. Methods used and types of justice studied by lead chapter authors

Methods	Distributive	Procedural	Interactional	Spatial	Temporal
Content analysis	5, 9, 11, 12, 14, 17	5, 9, 11, 12, 14, 17	5, 11, 14, 17	5, 14	5, 14
Interviews	3, 7, 11, 13, 17	7, 8, 11, 13, 17	7, 11, 17	7	7
Workshops	7, 11, 17	7, 11, 17	7, 11, 17	7	7
Participatory action research	3, 13	11, 9, 13			
Participant observation	3, 17	9, 17	17		
Questionnaires	3, 5, 11, 13	5, 11	5, 11		
Visual and spatial	11	11	11	9, 11	11
Doctrinal analysis	6, 9, 12, 16	6, 9, 12, 16	6	12	12
Simulations	13	13			
Statutory interpretation	6, 16	6, 16	6		
Rational argument	10, 15				
Experiments	13, 15	13			
Historical analysis	2				
Political economy	13, 15	13			

Both qualitative and quantitative research methods have been used to explore distributive, procedural and interactional dimensions of fairness in this book (Table 4.1). Only qualitative methods of data collection and analysis – interviews, workshops, content analysis of documents, visual and doctrinal analysis – have been used to investigate spatial and temporal dimensions of fairness by these authors. While this clearly reflects the preferences of the authors, it may also reflect the limited research that exists on these types of fairness in the field of environmental management (Graham *et al.* 2015), and the need for exploratory research to better conceptualise and operationalise these types of fairness.

The numbers in Table 4.1 indicate the chapters of the book that draw on each method. The types of justice referred to in this table are based on the definitions in Graham *et al.* (2015). Note that some authors have focused on principles underlying these broader concepts of justice, such as equity (Chapters 13 and 15), which is usually associated with distributive justice. For a conceptual map of the links between the types of justice, see Chapter 1.

The rest of this section explains how each of the methods are used in the context of environment justice research in Australia. The methods are ordered from the most to least most commonly used (Fig. 4.1).

Content analysis of documents

Content analysis of documents is used by many authors of this book in their qualitative research (Fig. 4.1). Content analysis is useful for understanding environmental justice in Australia because of the wealth of documentary material generated in environmental policy and debates. This includes government reports, non-government reports, community submissions, newspapers, internet blogs and audio-visual material. These materials are often available in the public record and are increasingly available on the internet, but documents can also be accessed from private exchanges between parties.

Access to a broad range of documents produced by many stakeholders allows for insights into how justice is conceived and framed—the different values, notions of justice and perspectives of fairness that shape environmental policy, decisions and responses. Indeed, content analysis of documents enables researchers to understand the positions of a wider range of stakeholders than may be possible using other methods, which is particularly important in justice research. Content analysis is also useful for discovering the justice *intentions* of particular stakeholders, which can then be compared with the *implementation* and *perception* of justice revealed through later documents or other methods. For example, Lukasiewicz *et al.* (2013) used a semi-qualitative content analysis to identify the key justice principles in the Murray–Darling reform process, which was then cross-checked through semi-structured interviews.

Because the documents being analysed inevitably reflect stakeholders' societal attitudes and cultural values, content analysis can be used to illuminate some of the forces that enable or prevent justice questions from coming to the fore in public debates. Thus, analysing documents produced over a time period can provide insights into changes in community and government perspectives on environmental justice. For example, content analysis is useful for understanding how environmental policies have previously treated, and continue to treat, Indigenous rights, and the priority given to Indigenous access compared with other groups (Chapter 9).

Content analysis can be done by deriving categories or themes directly from the text data (open coding). It can start with a theory as guidance for initial coding of text or it can include counting the number of times a theme or keyword appears, accompanied by the interpretation of the underlying context. Because the specific term 'justice' or 'fairness' might be used only rarely, content analysis benefits from researchers deconstructing the concept of justice to effectively explore its components. For example, procedural and interactional fairness concepts, such as transparency and being treated with respect, can be identified through a content analysis that is well informed by theory.

A limitation of solely using content analysis of documents is that material accessible and analysed may be produced for a specific purpose, by specific stakeholders and may fail to incorporate a range of perspectives. Omissions in key documents can, however, point to the deliberate or inadvertent exclusion of marginalised or minority groups (e.g. Indigenous communities) or of particular issues.

Note that the term content analysis is similarly used in analysing interviews and focus groups, usually by systematically coding transcripts. Content analysis differs from discourse analysis, which looks beyond the sentence, keywords and themes to examine the way language is used.

Interviews

Semi-structured interviews allow for a detailed investigation into how justice is understood and implemented by a wide variety of stakeholders (Fig. 4.2). Interviewees' responses do not need to be constrained by existing definitions and conceptualisations of justice – interview questions may indirectly or directly ask about justice and other associated concepts (see questions in Box 4.1). An interview guide allows for consistency among interviews and interviewers, with the flexibility to enable views to be queried in-depth.

Interviewees are more likely to share contentious views with the researcher when discussing the issue one-on-one, compared with group settings. While interviews may allow for a rich understanding of individual perspectives on environmental justice, it may also be difficult to access alternative perspectives if the range of stakeholders is unable or unwilling to be interviewed. Further, interviews do not allow for group learning or deliberation, which is increasingly one of the objectives of participatory research methods in natural resource management (Straton *et al.* 2011).

Box 4.1: Sample of semi-structured interview questions asked in the context of water management in Australia

Business owners perspectives
(Chapter 11, case study 4)
In relation to the January 2011 flood, please review this timeline and indicate on it:
- when you first heard about potential flooding
- what the message was about
- who conveyed the message
- how you received it
- your perception of the information.

Did you assess it to be credible? Why?
What would you have done differently?
Can you suggest any guidelines or procedures for protection of your goods, your site, or for emergency planning?

Government perspectives
(Lukasiewicz 2012)
What are the greatest justice issues in water management today?
What is the guiding philosophy for distributing water? (prompts: equity, equality, self-interest, need (whose?), efficiency, fairness)?
What are the essential criteria for a fair process?
How is the participation of Indigenous people assured? Is this adequate?
How can power relations among stakeholders be managed?
How does your department communicate with its stakeholders? Is it adequate?

(Keremane and McKay 2011)
Please describe a fair process for reducing water use/allocations.
Prior to obtaining approval for a desalinisation project, are there any set of principles/guidelines within which the project proposal is analysed and decided upon? If yes, please explain.

Some ethical challenges may arise while conducting interviews on justice, such as the potential disclosure of illegal activities or personal criticism. Interviewees may also have unreasonable or unattainable expectations about what the researcher can achieve for them and their cause, which can create moral dilemmas for researchers who, as mentioned in Chapter 1, see their justice research as an extension of their moral commitments to justice.

Focus groups and workshops

Focus groups are particularly useful for research into justice in environmental policies and practices. They are also advantageous for researchers interested in the subtle ways in which unequal power relations are structured in group settings and their effect in dominating minority or marginalised positions. For example, the method may bring to light or expose assumptions and practices relating to gender and race.

In a focus group setting, the researcher or facilitator leads a discussion around a small number of specific questions and promotes interaction. For example, Dare (2013) asked focus group participants 'Can you give me an example of how [policy/activity] has impacted on you and/or your community?' As people share and elaborate on their issues, attitudes, beliefs and norms, and identify reasons behind their views, they provide rich and cumulative data. Dialogue can facilitate self-disclosure and self-validation. A researcher can use such groups to better understand the dynamics between parties and how individuals are influenced by others in a group (Litosseliti 2003). They enable diverse opinions to be voiced, realistic conversations to be had about potential trade-offs of different policy options, and can result in opportunities for individual and social learning.

One of the challenges of focus groups, in environmental justice contexts and beyond, is that stakeholder willingness to contribute ideas may depend on who else is present or be constrained due to confidentiality concerns. Although a skilled moderator or facilitator may help, it is not always possible to ensure that everyone's views are voiced and heard. One remedy is to complement focus groups with individual interviews to minimise potential for groupthink, false consensus or hesitation at sharing deeply held views with others. Such mixed methods can therefore allow for a comparison of how people talk about justice in front of their peers, other stakeholders or decision makers, compared with conversations in private.

Workshops are often used when engaging people with opposing views, policy decision makers or when using particular group-based decision-making tools, such multiple criteria analysis. Workshops tend to be longer in duration than focus groups and may have larger numbers of participants. For example, the workshop described in Chapter 7 was a full-day event with policy makers from local, regional and state government departments. In comparison, focus groups tend to aim for six to 10 participants, usually from within a similar cohort or belief system (Bryman 2016). Given the larger number of participants and longer duration, workshops are best managed by an independent facilitator.

In both focus groups and workshops, conversations can be derailed by participants who refuse to follow the schedule and seek to promote their own agenda. This is not an insurmountable challenge, but requires deliberate thinking about the purpose of the research and management of stakeholder expectations. Similarly, it is essential that the facilitator does not (intentionally or unintentionally) impose their own view of justice on participants (Stewart *et al.* 2007). In some research contexts, the focus group is not possible or practical so non-invasive methods, such as content analysis, may be preferable.

Participatory action research

Participatory action research (PAR) seeks to go beyond ascertaining views or eliciting values to achieving some degree of procedural justice from within the research process itself. In PAR, participants take an active role in parts or all of the research process, from project design and methods, to facilitation of activities, analysis and evaluation. This approach, which facilitates co-production of knowledge, has gained popularity in environmental justice research as a result of the trend towards decentralised natural resource management governance, giving rise to experimentation in research methods that seek to strengthen citizen engagement in processes that ascertain the views, attitudes and preferences of all affected groups. The strength of PAR is that it fosters inclusivity and empowerment, builds trust in data and crosses the science–practice boundary to make research more useable by stakeholders (Lemos and Morehouse 2005; Vogel *et al.* 2007). The iterative process of creating and 'owning' jointly created data has been found to foster uptake of 'citizen science' or local knowledge (Dilling and Lemos 2011; Jackson *et al.* 2012).

Some participatory methods, such as citizen's juries, introduce evidence or expert opinion to encourage participants to discuss, judiciously argue, deliberate and move towards identifying possible solutions (Delli Carpini 2004; Straton *et al.* 2011). A central aim of these types of effort is to bring about social learning, enhance accountability, reduce or resolve conflict and contribute in the longer run to deliberative democracy. A requirement for the group concerned to reach consensus on procedures that are conducive to effective and equitable participation and deliberation, as well as criteria for evaluating outcomes, can be demanding. In situations in which stakeholder positions were previously very polarised and based on seemingly fixed attitudes or preferences, researchers report positive improvements to the level of debate and willingness to alter opinions and positions in light of discussion (Proctor and Drechsler 2006). As with workshops and focus groups, skilled facilitation is critical in order to minimise conscious manipulation and for all concerned to be alert to unconscious bias.

Other research methods might benefit from PAR principles, even if PAR is not adopted. Communicating with communities early in a research process can assist in assessing suitability and feasibility of potential case studies as well as methods, and ensure a better uptake of research outcomes. Having skilled researchers or facilitators are a clear advantage in bringing together multiple stakeholders in a working relationship. Similarly, experience can assist in moderating the blurring of the distinction between 'researcher' and 'researched' (Jackson *et al.* 2012; Mackenzie *et al.* 2012). In addition, many of these principles are consistent with the principles of Indigenous research methodologies (Gibbs 2001; Louis 2007), particularly the expectation that research should ethically draw on Indigenous knowledge and be of benefit to the affected community (Jackson and Douglas 2015), making it suited to justice research that pertains to Indigenous rights or interests.

Questionnaires

Structured quantitative surveys in environmental justice research are often used because they access a larger number of people (through mail, telephone and the internet) than is often possible through qualitative research and can still provide opportunities for respondents to express themselves through open-ended questions (see survey questions in Box 4.2). Surveys allow for generalisations to be made and for relationships to be drawn between perceptions of fairness and the characteristics of respondents (e.g. their social-economic status). The ability to quantify data and apply statistical analyses can make it easier to communicate the results and recommendations to policy makers, because they may perceive quantitative results to be more valid and valuable (e.g. the critique provided by Bowen 2000).

Surveys can be used for different purposes and can enable evaluation of diverse responses to environmental policies, laws and their administration (Box 4.2). In sequential research, surveys can be used in at least two different ways. Surveys can be used early in the research process to identify areas for the focus of future in-depth inquiry. Later in the research process, surveys allow for the ideas generated through qualitative research to be tested and generalised to a broader population.

One of the key disadvantages of surveys for environmental justice research is that respondents might want to elaborate more on some of the issues raised in the survey than in the space allowed. Even if surveys leave space for 'other comments', by their nature, surveys predetermine the issues or themes on which views are sought, thereby potentially influencing responses or missing valuable data. This is likely if participants have a different conceptualisation of justice than that which guides the researchers. Researchers will want to be sure that their terms are clearly defined and that as far as possible the meaning of justice concepts is conveyed accurately to avoid misinterpretation. Limitations such as these provide further incentives to employ mixed methods in environmental justice research. For example, a focus group process might generate insight into popular understandings and definitions of key terms or concepts.

Box 4.2: Open-ended and closed questions asked in past questionnaires in the context of water management in Australia

Closed questions

Graham (2002)

Please indicate whether you strongly agree, agree, disagree, strongly disagree or are not sure about the following statements:

- The outcomes of the Environmental Flow Rules are fair.
- The Environmental Flow Rules should allocate less water to wetlands.
- There was adequate consultation of rural communities during the development of Environmental Flow Rules.
- Environmental flows occur too frequently.
- Farmers were adequately consulted about their opinions on the Environmental Flow Rules before they were introduced.

(Chapter 5)

Please indicate your level of agreement with the following statements:

- I trust the decision-making processes in water-related matters.
- The Government's water policies and rules are easy to understand.
- Government pays special attention to different groups such as youth, women, Indigenous groups etc. in water policy design and implementation.

Do you think there is a human right to clean fresh water in urban and rural Australia? (Yes/No/I don't know).

Open-ended questions

What do you see as public interest in urban water supply?

Please name policy areas where independence is least/greatest.

Participant observation

Participant observation is used in environmental justice research to observe how decision making takes place and the ways in which stakeholders are engaged in the process. Thus it is more likely to be used to study procedural and interactional, rather than distributive, fairness. Participant observation is useful for evaluating whether people act in accord with their stated beliefs (e.g. as expressed in interviews). For example, policy makers and community members alike often espouse the importance of treating others with respect (an element of interactional justice); participant observation allows researchers to observe if and how respect is demonstrated.

One of the challenges of participant observation in justice research is that it is not always possible to observe a whole process from start to finish. Policy and planning processes can take place over many years. In addition, social relations are strongly influenced by past experiences, which may have occurred before the research started. This may make it challenging to interpret the quality of relationships among stakeholders. For example, one of the lead authors encountered a government advisory group meeting where people were shouting at one another. While this seemed to indicate a lack of respect among participants, it was later revealed that the situation had previously been more acrimonious, and that this meeting was a significant improvement. For reasons such as this, many of the authors of this book use multiple methods to ensure that they understand justice concerns and their contexts as comprehensively as possible.

Visual and spatial methods

A range of visual and/or spatial methods have been used by lead chapter authors in their resources environment and justice research, such as participatory GIS or geospatial mapping, photovoice and 3D visualisations. They range in varying levels of technical complexity, some requiring specific expertise to establish (3D visualisations), some being easy to use once established (participatory GIS) and others being able to be used by participants with little training (photovoice).

Visual methods are effective in engaging participants in research, exchanging understandable information with (to *and* from) participants, enhancing dialogue for perspectives that are difficult to verbalise or that are emotionally charged, and encouraging cooperation and shared understanding (Keremane and McKay 2011). For example, the construction of a participatory 3D groundwater visualisation tool assisted in building trust between representatives of a regional water allocation committee, scientists and water agency staff (Baldwin *et al.* 2012; Jackson *et al.* 2012). Visualising the groundwater resource was found to be an effective means of bringing water users to the table to 'see' the issue more clearly and, with this enhanced transparency, discuss how to reduce the environmental and social impacts of over-allocation.

Photovoice is increasingly used in environmental justice research to empower local residents and to elicit hidden values. It requires participants to take photos that represent their views on certain themes and then discuss them in interviews and/or a focus group to create a common story or shared understanding. In case studies 1 and 2 in Chapter 11, participants are simply asked to 'take photos that illustrate sustainable use of water from your perspective'. The sharing and discussion of photos taken with such open questions represents 'participant-generated' data that minimises researcher bias (Baldwin and Chandler 2010; Baldwin and Ross 2012). The resulting 'voices' captured in captions on the photos, or in stories, can be used to inform decision makers of issues and priorities (McKay *et al.* 2010; Keremane *et al.* 2014; Grant *et al.* 2015).

Visual methods are innovative and novel, different from the 'usual' methods, and therefore often interesting to, and engaging for, participants (Keremane and McKay 2011). For environmental justice research, they can be an effective option for engaging participants from different ethnic backgrounds who have difficulty expressing themselves verbally (e.g. some people with disabilities) or who are unaccustomed to certain terminology. They are often used in combination with an adapted focus group or workshop format. One of the limitations of visual methods is that they often require significant technological input in their development or application. For example, creating 3D visualisations are time-consuming and require specific skills. Moreover they require inter-disciplinary collaboration so that methods, such as possible scenarios or visual stimulus, are designed to ask the appropriate questions. Even simpler methods, such as photovoice, involve logistical complexity in downloading and collecting participants' digital photos and printing them before the focus group. This can reduce the efficiency or increase the cost of these methods.

Doctrinal analysis

Law has a powerful influence on people and place by regulating the use and access of natural resources by social groups. Doctrinal research asks what the law is in a particular area and uses primary sources – cases, legislation, policy documents and other historical materials – to systematically describe and critique a body of law and how it applies. It seeks to understand the interactions between legal rules, highlight problems in the legal framework and may also make suggestions for law reform, taking into account likely social, economic and environmental transformations in society (e.g. McKay 2010).

Environmental justice researchers use doctrinal analysis to interrogate interactions between law and natural resources or the environment as they are revealed in legal determinations. Legal pronouncements make or amend the rules that are intended to regulate human behaviour (Bartel *et al.* 2013; McKay 2015) and so can serve as a window to the powerful societal presumptions, norms, ideologies and socio-spatial processes that produce uneven outcomes in environmental disputes or contests. The explicit attention given to concepts such as procedural fairness and equity in legalistic interpretations makes assessing justice claims within a court case easier to navigate and focus on. Furthermore, doctrinal analysis studies can elucidate several conflict resolution measures (McKay 2011).

Beyond focusing specifically on the law and its implications, doctrinal analysis is useful because litigation can highlight broader aspects of the community debate (see Chapter 12 for an example in the context of coal seam gas and Chapter 16 for an example in relation to water allocation plans). The material required is publicly accessible and are often commented on extensively in the media before and after legal proceedings have concluded.

One of the disadvantages of using doctrinal analysis is that it requires an existing body of statutes, court cases and societal debates on the issue of interest. Where the state of the law is in its infancy, statutory interpretation may be more useful (McKay 2008). Another disadvantage of using doctrinal analysis in justice research is that analysis of the law need not reveal all justice matters pertaining to a court case (Chapter 16). This means that narrowing in on legal interpretations of justice may provide only a partial understanding of the justice concerns that relate to a particular law. Thus doctrinal analysis can be complemented by other methods, such as interviews, that provide insights into perceptions on the implementation of laws. Such multi-method evaluations of the operational aspects of laws are common in sociolegal studies, but they have a limited focus on environmental justice. There is much scope here for further research.

In a settler society such as Australia, where an Indigenous minority has maintained a parallel customary legal system in spite of colonial legal institutions, it is important for researchers to reflect on the capacity of the former colonial legal system to recognise Indigenous concepts that are considered essential to social justice, namely sovereignty and Indigenous systems of property (Strelein and Tran 2013). The doctrines, precepts and norms held by Indigenous or other minority communities can also be a valuable source of information for environmental justice scholarship. These are often documented in court cases, such as Mabo (Mabo v Queensland (No. 2) ('Mabo case') [1992] HCA 23; (1992) 175 CLR 1) and subsequent cases such as Wik (Wik Peoples v Queensland ('Pastoral Leases case') [1996] HCA 40 (1996) 187 CLR 1), and the plethora of other decisions that reveal the content of particular Indigenous doctrines and norms. These cases also show how these Indigenous notions are articulated and represented in a court environment that follows the standards and traditions of the dominant society.

Simulations

Computer simulations of the real-world implications of policies can be quantitative (e.g. Powersim, Vensim and APSIM) or qualitative (e.g. STELLA). They can be used to help affected stakeholders understand the distribution of impacts for themselves, others and the environment. They can also be used to help stakeholders propose alternative distributive outcomes, if they have the opportunity to investigate alternative scenarios or challenge the way the decision rules are being implemented in computer models (Graham 2009). By providing stakeholders with a greater understanding of the implications of decisions, computer simulations and modelling of scenarios can help to tackle power imbalances in decision-making processes that result from decision makers having a greater access to knowledge. As such, they can be used to help communities and policy makers identify the 'fairest' policy or decisions in situations where different stakeholders hold conflicting views on justice (Chapter 15).

Use of experts in participatory modelling in collaboration with the community can increase credibility in options, establish shared meaning and co-generate new knowledge (Henriksen and Barlebo 2008; Hoverman *et al.* 2011). In this way, computer simulations can form part of PAR and facilitate fairer procedural and interactional fairness in practice. Bringing in external expertise needs to be handled sensitively, to ensure experts are seen as credible – whether local or brought in from outside the research setting – and by selecting appropriate persons based not only on their expertise, but their personal qualities and experience in dealing with stakeholders (Mackenzie *et al.* 2012).

A more traditional application of computer simulations involves exploring or mapping out the various options available and comparing the outcomes of each. A notable example is given by Cazorla and Toman (2000), who compare the impacts across countries of different burden-sharing rules for greenhouse gas emission reductions, with each rule implementing a different equity principle. Sensitivity analyses of simulations are used to identify the critical factors that make a difference, as well as conflicts that make no difference to the final outcome (see Chapter 13 for an example).

Network analysis

There has only been limited use of network analysis in environmental justice research in Australia – much of its full potential is yet to be explored. Network analysis is useful for understanding the relational (procedural and interactional) dimensions of justice because it facilitates the exploration of social dynamics and relationships in communities and organisations. Knowing about the quality and nature of social relations allows the researcher to track and unfold the processes of how voice is exercised and how the characteristics of interpersonal relationships affects decision-making processes and outcomes. A refined analysis of existing networks provides insights into who is able to voice their concerns about emerging environmental policies and other relevant decisions, argue for their needs to be accommodated, and thus negotiate and access benefits. This can be particularly important in communities and contexts where decisions are made informally.

To date, network analysis has primarily been used as a qualitative method in environmental justice research in Australia. There is much potential for it also to be used quantitatively, using programs such as UCINET.[3] Such programs can be used to explore the similarities between and relative strengths of social relations (Borgatti *et al.* 2013).

Rational argument and conceptual analysis

Conceptual analysis and rational argument are the standard tools used by philosophers working in the so-called analytic tradition. Roughly speaking, in the analytic tradition,

3 See <https://sites.google.com/site/ucinetsoftware/home>.

philosophy, as a discipline, is akin to science – its primary aim is to produce knowledge and help people understand the world. Just like scientists draw conclusions on the basis of empirical evidence, philosophers likewise draw conclusions after examining evidence. For philosophers, however, the evidence takes the form of premises purportedly supporting the conclusion of an argument. It is the task of the philosopher to ask whether the relationship of support between premises and conclusion is strong enough to support the claim being made.

As well as evaluating arguments, philosophers also construct them. Just as a scientist undertakes research with the aim of testing a hypothesis, a philosopher presents an argument for the purposes of defending or undermining a position. In theory, the philosopher is meant to be as impartial and as committed to the 'experimental' method as a scientist. It is basic practice for a philosopher to test the adequacy of their argument by considering objections to their position. The rationale is that the more objections a philosopher can identify and satisfactorily answer, the stronger their argument becomes. For the philosopher, this process of 'objection and replies' serves to sharpen the argument; just as focusing a microscope serves to bring a scientist's laboratory research findings into view.

One of the advantages of rational argument is that it helps to clarify and, therefore, render transparent, rival positions in a debate. It can also help to reframe orthodox conceptualisations. Debate about environmental justice, in theory and practice, can be emotionally charged: theorists, no less than activists, may allow their personal convictions to inform their arguments. While every researcher needs to be cognisant of their potential biases, it is the philosopher's job to distinguish between claims that support a particular position or theory, and claims that serve only to 'muddy the waters'.

Conceptual analysis likewise helps to bring clarity to a debate. Concepts such as 'justice' and 'fairness' can mean different things to different stakeholders. The concern is that without getting clear about the concepts in question, people will be simply 'talking past one another'. The aim of conceptual analysis is to clarify the terms that feature in the premises of arguments. Roughly speaking, conceptual analysis can be understood as a method of establishing credible definitions or authoritative analyses. It is purportedly useful as an aid to projects that involve identification or categorisation. For example, Hadley (Chapter 10) employs specific analyses of concepts such as 'property', 'ownership' 'justice', 'interests', and 'rights'. A presupposition of his project is that, based upon conceptual analysis of the relevant concepts, it is logically meaningful to, first, consider animals as property owners and, second, to consider animal property rights as a suitable topic for justice research.

A limitation of both rational argument and conceptual analysis is that such methods impose constraints upon debate and, in the process, may foreclose potentially useful lines of enquiry. Some stakeholders may need to make their claims in 'non-rational' ways (e.g. story-telling, humour or activism), and it might be good if some concepts resist precise definition.

Statutory interpretation

Statutory interpretation has become increasingly important in legal environmental justice research over the last decade. There are many statutes covering environmental management, and many of those impose very broad objectives on all stakeholders, including landholders, industry and governments. Parliaments not only develop and enact statutes but also contribute to interpretive processes: legislative provisions now direct courts, tribunals and others as to how to go about interpreting statutes and delegated legislation. Therefore, where once the interpretation of judicial pronouncements and the common law were considered of paramount importance to proper legal practice, today practitioners must also be able to properly read and interpret the myriad of statutory instruments that dominate the legal landscape. As Justice Kirby (2003: 95) aptly notes:

> *the construction of statutes is now, probably, the single most important aspect of legal and judicial work. In Australia, courts have discovered that many lawyers intensely dislike this feature of their lives. They find the obligation to read Acts of Parliament, from beginning to end, so distasteful that they will do almost anything to postpone the labour. The High Court of Australia has been moved to protest at this unwillingness to grapple with the words of the statutory text, instead of returning to the much loved words of judges, written long ago and far away, who uttered them before the legislature's text became the law. Whilst this tribute to the judiciary is touching, it does not represent the law. The world of common law principle is in retreat. It now circles in the orbit of statute. Where statute speaks—and particularly a curious statute such as a Constitution or a Human Rights Act—there is no escaping the duty to give meaning to its words. That is what I, and every other judge in the countries of the world that observe the rule of law, spend most of our time doing.*

Compared with doctrinal analysis, statutory interpretation focuses on the legislative outputs of parliament as they currently stand. Statutory interpretation involves deliberating over the intent of parliamentarians where words in statutes are not obvious and clear, based on their ordinary meaning. This may involve using extrinsic documents, such as parliamentary speeches on a bill or law reform reports to help understand parliament's intent for a piece of legislation. For the justice researcher, statutory interpretation can assist in understanding the factors that influenced a statute's enactment, appeal and/or repeal. An advantage of using statutory interpretation is that it can be used before the first court challenge to a provision of an Act is undertaken, when there is a lack of judicial conclusions for doctrinal analysis.

Experiments

The main disciplines that use controlled (laboratory), partially controlled and field experiments in environmental justice research are behavioural economics and social psychology. The advantage of laboratory experiments involving human participants is that they offer control. They require the researcher to start by developing specific hypotheses as to what factors might influence an individual's or a group's relationship with some aspect of justice. Once the hypotheses are developed, the researcher can systematically control for those factors and study whether, and under what conditions, the hypothesised effects occur (see Chapter 15 for an example).

The disadvantage of laboratory experiments is that they involve decontextualisation: factors are held constant that are not constant outside of the laboratory environment. Field experiments offer one solution to this limitation. Although field experiments allow for less control, they allow for the study of contextual factors not accessible in the laboratory. For example, a field experiment can more easily account for past experience, failures and successes, trust, and expectations about the future.

The use of at least partially controlled experiments holds great potential for a better understanding of the factors underpinning approaches to justice 'in action' – a potential that environmental justice researchers are only beginning to become aware of. In environmental justice research, different kinds of experiments can complement one another, as well as complement non-experimental approaches, such as the other methods discussed in this chapter.

Political economy research

Political economy research lends itself to the study of environmental justice because it involves exploring conflicts of values in decision making, particularly conflicts between social equity and economic efficiency (Chapter 13). The political economy approach embeds justice research, and considerations of justice, within a broader framework that takes account of social and political realities. These typically include the prevalence of budgetary constraints or the need to compromise with potentially conflicting goals, be they geostrategic and military, dealing with pervasive corruption or with strong opposing forces, and existing property rights regimes. As Chapter 15 illustrates, the social reality of conflicting values around justice can create the greatest obstacles to achieving justice. The 'economy' in political economy usually brings into the picture the existence of scarcity of some kind – of resources, but also, and sometimes more importantly, scarcity of time, knowledge, trust or self-confidence. The political economy of justice emphasises the importance of context, which, when very constraining, can sometimes challenge the results of other approaches. Thus political economy research can be usefully applied in environmental justice research in conjunction with other qualitative and quantitative methods.

Conclusions

Justice research is at its best when it reveals and explains the diverse values, interests and perspectives of those who make, and are affected by, decisions. It is for this reason that many of the authors of this book use multiple methods to understand how various stakeholders relate to the environment and believe that justice can be achieved. Quantitative methods provide a broad indication of a sample population's norms about justice, which may be of benefit to decision makers who wish to represent, take into account or even shape broad community sentiment. Qualitative methods, on the other hand, provide an opportunity to explore the 'why', the underlying values and interests that shape community views. Furthermore, qualitative methods can be particularly useful for understanding the views of those language challenges or cultural sensitivities. Because of the logistical necessity of smaller sample population sizes for qualitative research, the results can provide insight to the views of both powerful stakeholders and marginalised groups. Like quantitative methods, this too can assist in decision making. Several of the researchers in this book used mixed methods in a sequential manner. The generation of data from one method may be contingent on analysis of data from another method. Mixed methods also can be used to complement each other to build a composite picture of environmental justice components and/or to validate data gathered through other methods. To date, though, there has been little analysis of the appropriateness of using different methods separately or in conjunction for such research.

The variety of methods used in environmental justice research reflects the diversity of disciplinary backgrounds of contributing scholars. The methods reflect that applying social justice to environmental justice necessitates a crossing of disciplinary boundaries. For example, to evaluate distributive justice effectively, researchers will benefit from an understanding of both physical and social science; that is, not relying solely on perspectives of impacted or benefited stakeholders, but also accessing independent expert views and scientific reports. To understand and influence procedural justice, researchers will benefit from legal, institutional and social science expertise. Interview and document analysis methods will no doubt remain core to in-depth understanding of environmental justice. Just as visual and simulation or modelling methods can play a stronger role, so too can analysis of social media through content analysis and other innovative techniques. This supports an argument for both interdisciplinary teams of researchers and individual researchers being skilled in use of different methods.

Justice is a normative concept and justice research can be value-laden. Researchers need to be aware of this and reflect on the role they play as they apply their methods. Ultimately, fair environmental justice research relies on researchers being competent, rigorous and systematic in research design and analysis. This chapter contributes to one of the aims of this book: to nurture, improve and progress environmental justice research. The chapter represents the beginning of a conversation about the purpose, advantages, limitations and application of different methods for research about environmental justice.

Acknowledgements

We wish to thank our colleagues Mirella Gavidia, John Hadley, Anna Lukasiewicz, Steve Schilizzi and David Turton for their invaluable contributions to this chapter.

The four authors of this chapter have contributed to multiple sections of this book. Each is happy to be contacted for further information on the methods in which they have particular expertise. Jennifer McKay is a specialist in legal research methods, such as doctrinal and socio-legal studies, using empirical methods, such as focus groups, surveys, key actor interviews, content analysis and photo-supported workshops that involve jointly creating the Photostory method. Claudia Baldwin has expertise in visual and spatial methods, participatory action research, focus groups, and content analysis. Sue Jackson is a specialist in participatory action research, visual methods and legal methods, with a particular focus on engaging indigenous communities. Sonia Graham is enthusiastic about simulations, interviews and questionnaires.

References

Baldwin C, Chandler L (2010) At the water's edge: community voices on climate change. *Local Environment* **15**(7), 637–649. doi:10.1080/13549839.2010.498810

Baldwin C, Ross H (2012) bridging troubled waters: applying consensus-building techniques to water planning. *Society & Natural Resources* **25**(3), 217–234. doi:10.1080/08941920.2011.578120

Baldwin C, Tan P-L, White I, Hoverman S, Burry K (2012) How scientific knowledge informs community understanding of groundwater. *Journal of Hydrology* **474**, 74–83. doi:10.1016/j.jhydrol.2012.06.006

Bartel R, Graham N, Jackson S, Prior J, Robinson D, Sherval M, *et al.* (2013) Legal geography: an Australian perspective. *Geographical Research* **51**(4), 1–15.

Borgatti SP, Everett MG, Johnson JC (2013) *Analyzing Social Networks*. Sage Publications, London, UK.

Bowen W (2000) *Environmental Justice Through Research Based Decision-Making*. Garland, New York, USA.

Bowen W (2002) An analytical review of environmental justice: what do we really know? *Environmental Management* **29**, 3–15. doi:10.1007/s00267-001-0037-8
Bryman A (2016) *Social Research Methods.* Oxford University Press, Oxford, UK.
Cazorla MV, Toman MA (2008) International equity and climate change policy. In *Climate Change Economics and Policy: An RFF Anthology.* (Ed Toman MA). pp. 235–247. Resources for the Future, Washington DC, USA.
Dare M (2013) Localism in practice: insights from two Tasmanian case studies. *Policy Studies* **34**(5–6), 592–611. doi:10.1080/01442872.2013.863572
Delli Carpini MX (2004) Mediating democratic engagement: The impact of communications on citizens involvement in political and civic life. In *Handbook of Political Communication.* (Ed LL Kaid) pp. 395–434. Lawrence Erlbaum, Manwah NJ, USA.
Dilling L, Lemos M (2011) Creating usable science: opportunities and constraints for climate knowledge use and their implications for science policy. *Global Environmental Change* **21**(2), 680–689. doi:10.1016/j.gloenvcha.2010.11.006
Gibbs M (2001) Toward a strategy for undertaking cross-cultural collaborative research. *Society & Natural Resources* **14**, 673–687. doi:10.1080/08941920120547
Graham S (2002) Being green when you're in the red: the attitudes of rice and grace growers towards environmental flows for wetlands. BSc (Hons) thesis, University of Sydney, Sydney.
Graham S (2009) Irrigators' attitudes towards environmental flows for wetlands in the Murrumbidgee, Australia. *Wetlands Ecology and Management* **17**(4), 303–316. doi:10.1007/s11273-008-9108-4
Graham S, Barnett J, Fincher R, Mortreux C, Hurlimann A (2015) Towards fair local outcomes in adaptation to sea-level rise. *Climatic Change* **130**, 411–424. doi:10.1007/s10584-014-1171-7
Grant B, Baldwin C, Lieske S, Martin K (2015) Using participatory visual methods for information exchange about climate risk in canal estate communities. *Australian Journal of Maritime and Ocean Affairs* **7**(1), 23–37. doi:10.1080/18366503.2015.1014012
Hajkowicz S, Collins K (2007) A review of multiple criteria analysis for water resource planning and management. *Water Resources Management* **21**, 1553–1566. doi:10.1007/s11269-006-9112-5
Hall R (2008) *Applied Social Research: Planning, Designing and Conducting Real-World Research.* Palgrave Macmillan, Melbourne.
Henriksen H, Barlebo H (2008) Reflections on the use of Bayesian belief networks for adaptive management. *Journal of Environmental Management* **88**, 1025–1036. doi:10.1016/j.jenvman.2007.05.009
Hoverman S, Chan T, Ross H, Powell B (2011) Social learning through participatory integrated catchment risk assessment in the Solomon Islands. *Ecology and Society* **16**(2), 17.
Jackson SE, Douglas M (2015) Indigenous engagement in tropical river research in Australia: The TRaCK Program. *The International Indigenous Policy Journal* **6**(2).
Jackson S, Tan P, Nolan S (2012) Tools to enhance public participation in the development of the Howard East aquifer water plan, Northern Territory. *Journal of Hydrology* **474**, 22–28. doi:10.1016/j.jhydrol.2012.02.007
Keremane G, McKay JM (2011) Using PhotoStory to capture irrigators emotions about water policy and sustainable development objectives: a case study in rural Australia. *Action Research* **9**, 405–425. doi:10.1177/1476750311409598
Keremane G, Wu Z, McKay JM (2014) The emotional connection to urban water through the lens of the water customer. *Water (Basel)* **41**, 61–65.

Kirby M (2003) Towards a grand theory of interpretation: the case of statutes and contracts. *Statute Law Review* **24**(2), 95–111. doi:10.1093/slr/24.2.95

Lauber TB (1999) Measuring fairness in citizen participation: a case study of Moose Management. *Society & Natural Resources* **12**, 19–37. doi:10.1080/089419299279867

Lemos M, Morehouse B (2005) The co-production of science and policy in integrated climate assessments. *Global Environmental Change* **15**(1), 57–68. doi:10.1016/j.gloenvcha.2004.09.004

Litosseliti L (2003) *Using Focus Groups in Research*. Continuum, London, UK.

Liu F (2001) *Environmental Justice Analysis: Theories, Methods, and Practice*. Lewis Publishers, Boca Raton FL, USA.

Louis R (2007) Can you hear us now? Voices from the margin: using indigenous methodologies in geographic research. *Geographical Research* **45**, 130–139. doi:10.1111/j.1745-5871.2007.00443.x

Lukasiewicz A (2012) Lost in translation: where is the social justice in Australian water reform? PhD thesis, Charles Sturt University, Albury, WA.

Lukasiewicz A, Bowmer KH, Syme GJ, Davidson P (2013) Assessing government intentions for Australian water reform using a Social Justice Framework. *Society & Natural Resources* **26**(11), 1314–1329. doi:10.1080/08941920.2013.791903

Lynam T, de Jong W, Sheil D, Kusumanto T, Evans K (2007) A review of the tools for incorporating community knowledge, preferences and values into decision making in natural resources management. *Ecology and Society* **12**(1), 5.

Mackenzie J, Tan P-L, Hoverman S, Baldwin C (2012) The value and limitation of action research methodology. *Journal of Hydrology* **474**, 11–21. doi:10.1016/j.jhydrol.2012.09.008

McKay JM (2008) Insubstantial, tenuous and vague laws – the achievement of ecologically sustainable development by water supply. *Australian Business Law Review* **36**, 432–445.

McKay JM (2010) Some Australian examples of the integration of environmental, economic and social considerations into decision making – the jurisprudence of facts and context. In *Global Justice and Sustainable Development*. (Ed. D French) pp. 327–339. Leiden, Boston MA, USA.

McKay JM (2011) Australian water allocation plans and the sustainability objective-conflicts and conflict-resolution measures. *Hydrological Sciences Journal* **56**, 615–629. doi:10.1080/02626667.2011.580456

McKay JM (2015) Australian water law history: the move from introspective state sovereignty to a national interest approach and the influence of international law. In *A History of Water: Sovereignty and International Water Law* (Eds T Tvedt, O McIntyre and TK Woldesadik). pp. 490–507. Tauris Publishers, New York, USA.

McKay JM, Keremane GB, Gray A (2010) *Picturing Fresh Water Justice in Rural Australia*. Cooperative Research Centre for Irrigation Futures.

Mendoza GA, Martins H (2006) Multi-criteria decision analysis in natural resource management: a critical review of the methods and new modelling paradigms. *Forest Ecology and Management* **230**, 1–22. doi:10.1016/j.foreco.2006.03.023

Mertens DM (2007) Transformative paradigm: mixed methods and social justice. *Journal of Mixed Methods Research* **1**(3), 212–225. doi:10.1177/1558689807302811

Mertens DM (2013) Emerging advances in mixed methods: addressing social justice. *Journal of Mixed Methods Research* **7**(3), 215–218. doi:10.1177/1558689813493994

Proctor W, Drechsler M (2006) Deliberative multicriteria evaluation. *Environment and Planning. C, Government & Policy* **24**(2), 169–190. doi:10.1068/c22s

Reed MS, Graves A, Dandy N, Posthumus H, Hubacek K, Morris J, *et al.* (2009) Who's in and why? A typology of stakeholder analysis methods for natural resource management. *Journal of Environmental Management* **90**, 1933–1949. doi:10.1016/j.jenvman.2009.01.001
Romero C, Rehman T (1987) Natural resource management and the use of multiple criteria decision-making techniques: a review. *European Review of Agriculture Economics* **14**, 61–89. doi:10.1093/erae/14.1.61
Sarantakos S (2013) *Social Research.* Palgrave Macmillan, Melbourne.
Stewart D, Shamdasani P, Rook D (2007) *Focus Groups: Theory and Practice.* Sage Publications, Thousand Oaks CA, USA.
Straton A, Jackson S, Marinoni O, Proctor W, Woodward E (2011) Exploring and evaluating scenarios for a river catchment in northern Australia: using combined scenario development, multi-criteria analysis and a deliberative process as a tool for water planning. *Water Resources Management* **25**(1), 141–164. doi:10.1007/s11269-010-9691-z
Strelein L, Tran T (2013) Building indigenous governance from native title: moving away from 'fitting in' to creating a decolonized space. *Review of Constitutional Studies* **18**, 19–48.
Vogel C, Moser S, Kasperson R, Dabelko G (2007) Linking vulnerability, adaptation, and resilience science to practice: pathways, players and partnerships. *Global Environmental Change* **17**, 349–364. doi:10.1016/j.gloenvcha.2007.05.002

PART 2

FROM THE GROUND UP

5

Concept of public interest in the Adelaide Desalination Plant delivery: lessons for public policy makers on improving social justice

Elnaz Ettehad, Jennifer McKay and Ganesh Keremane

Summary

This chapter examines aspects of public interest and social justice in the delivery of the Adelaide Desalination Plant, one of the major forms of water infrastructure built during Australia's Millennium Drought under a partnership agreement between public representatives (the state government) and the private sector. Under the current neoliberal paradigm and the associated market-based governance reforms, a new wave of discussion has emerged around the increasing role of the private sector in the provision of public infrastructure and services. The question is to what extent the benefits of such partnerships are equally distributed between various parties, which therefore result in outcomes that are perceived to be 'just'. With an intention to provide a view on this issue, this chapter operationalises the concept of 'public interest' using a conceptual framework for 'good governance' involving four important elements: accountability and transparency; public participation and inclusiveness; efficiency; and ownership of public assets and resources. The findings provide useful empirical insights into the notion of the public interest in partnerships between the public and private sectors.

Introduction

In the late 1980s, and particularly since the first Council of Australian Governments (CoAG) reforms in 1994, water institutions in Australia have undergone federally sponsored and fiscally led transformations of associated laws and institutional arrangements (McKay 2005). Issues such as the achievement of ecologically sustainable development, regional water plans and pricing have been raised and, most relevant here, the private sector has been encouraged to participate in the provision of water supply under the assumption that this would promote efficiency. In 2004, these objectives were reinforced and defined in further detail, and the National Water Initiative (NWI) was agreed upon by the CoAG to increase the efficiency of Australia's water use. Under the NWI, the National Water Commission was created to monitor the implementation of the reforms, but this has now been abolished with all its functions being allocated to several government departments (National Water Commission 2016). The next step was in 2007, with Prime Minister Howard expressing frustration with the reforms and passing the *Water Act 2007* to promote these objectives and other policy reforms. Much of this process was driven by the prolonged drought across the country (Heberger 2011; van Dijk *et al.* 2013). The outcome explored in this chapter is that Australian federal and state governments invested around A$10 billion to build desalination plants in five major coastal cities (Productivity Commission 2011). This was undertaken in order to create greater water security for these cities.

However, expensive public infrastructure projects, such as desalination plants, can give rise to controversy among different groups within the community. Heberger (2011) mentioned desalination as one of the most expensive water supply alternatives because the cost of plant construction would result in raising water rates and 'passing on the expense to the customers', despite the Australian Government's financial support. Nevertheless, desalination was seen as being politically acceptable as there were several contemporary studies demonstrating that the Australian community were concerned about using recycled water for drinking and personal purposes (Hurlimann 2007; Hurlimann and McKay 2004; Hurlimann and McKay 2007).

The service delivery method selected by each state government was to partner with various multinational companies. In order to regulate these partnerships, several governments have developed a range of new laws, policies and guidelines, generally with a clause or clauses focusing on protecting the public interest. For example, the Department of Infrastructure and Regional Development (2015), Partnerships Victoria (2001) and the NSW Government (2012) have developed a 'public interest test' to be conducted before any investment and project procurement decisions can be made. The test is designed to ensure that the procurement method remains in the public interest. The notion of public interest has been related to public and constitutional accountability (see e.g. Uhr 1999; Funnell 2003; Barton 2006). Due to the private sector's business-based approaches and objectives, deficiencies in accountability are more observable in this sector than in the public sector (English 2003). In addition, several studies have demonstrated that these partnerships do not necessarily lead to equity or greater public participation (Kamieniecki *et al.* 1999; Miraftab 2004; see also Rosenau 1999; Post and Obirih-Opareh 2003). In Australia, the drought and the consequent severe water restrictions resulted in hurried planning processes with limited transparency, customer engagement, poor coordination and uncertainty about the roles and responsibilities of the various participating parties (National Water Commission 2011).

Therefore, in the context of the need to have water-secure capital cities, the aspects of public interest and social justice explored in this paper relate to the governance of the Adelaide Desalination Plant that began operating in 2012. These aspects are:

- accountability and transparency
- public participation and inclusiveness
- efficiency
- ownership.

The privatisation paradigm in the urban water sector

Water in Australia has traditionally been delivered within a highly regulated environment by the statutory authority and through a monopoly function (Dovers 2001; McKay 2005;

Colebatch 2006; Frontier Economics 2013). Abbott *et al.* (2011) considered Australia's recent water reforms to have taken the form of the 'commercialisation of government owned water utilities, private sector involvement in capital works and other contracting out, and the separation of some regulatory functions'. On a global basis, water services are considered to be one of the main targets of privatisation advocates as a result of their perceived profitability (Walker and Walker 2000: 28). In Australia too, the special features of major water infrastructure projects appear to be encouraging governments to involve skilled private companies in project delivery and management functions. For example, South Australia (SA) has outsourced the management and operation of water and sewerage to a consortium of private firms, known as ALLWATER, under a long-term contract (Keremane *et al.* 2015). As a result of the recent privatisation paradigm, various types of contractual methods have been devised to outline these partnerships between the public and the private sector (see e.g. Sharp and Tinsley 2005; Infrastructure Australia 2008; Keremane *et al.* 2013). While there is a range of context-specific definitions for Public–Private Partnerships (hereafter, PPP/PPPs) in the literature, there is no widely recognised one. The Organisation for Economic Co-operation and Development (OECD) defines a PPP as 'a long-term agreement between the government and a private partner where the service delivery objectives of the government are aligned with the profit objectives of the private partner' (OECD 2011). Ideally, the key objectives of initiating such partnerships are achieving value-for-money and sharing resources on major projects (OECD 2011; Infrastructure Australia 2016). The performance of PPPs is assessed according to their efficiency (the economic benefits) and effectiveness (quality and accessibility of the service) (Post and Obirih-Opareh 2003).

Despite its perceived advantages, criticism exists around private sector involvement in the provision of public services. According to the recent privatisation paradigm, and under the current market-based reforms in most Western countries, including Australia, the ethical principles of public governance – namely citizenship, representation, equality, accountability, neutrality, welfare, integrity and justice – have been undermined in the face of commercial values such as efficiency, competition, profit, productivity, value for money and customer satisfaction (Haque 1996, 1999, 2001). Moreover, Rosenau (1999) argued that '[PPPs] have not necessarily resolved problems with regards to equity, access, participation and democracy'. Investigating a partnership between government and industry in environmental policy implementation in the United States, Kamieniecki *et al.* (1999) posed the question of whether market-based approaches restrict democratic public participation in such policy processes. Indeed, with the limited actual participation of different groups and citizens, concerns have been raised in regards to fulfilling environmental justice (Kamieniecki *et al.* 1999). As Lukasiewicz and Baldwin (2014) have emphasised, public participation in decision making about water is a key factor leading to better decisions and ensuring fairness. The authors also noted that if such processes are 'embraced in a justice culture', then proper public participation can be achieved. Accordingly, the OECD promotes political leadership to clearly define outputs in the PPP contracts and, in so doing, it calls for the involvement of end users to enhance the legitimacy and fairness of PPPs, and to make them easy to be understood (OECD 2012). In relation to PPPs in the Australian water sector, Godden (2009) argued that public and private actors are assigned unequal responsibilities, resulting in negative environmental impacts and public financial obligations. Lack of equality here is mostly related to power issues in such partnerships (Post and Obirih-Opareh 2003).

It is important to note that the principles of fairness, equity, equality, efficiency and participation fall into the three components of distributive, procedural and interactive social justice, according to the Social Justice Framework (see Lukasiewicz *et al.* 2013; Lukasiewicz and Baldwin 2014). In light of the above-mentioned issues, it seems that the extent to which a PPP can, in a practical sense, be fair, democratic and equitable or, alternatively, adhere to the principles of social justice, remains an open question. Indeed, it is pertinent to ask how far the outcome of a PPP can be 'just' to the public, as each partner attempts to achieve the commercial values discussed by Haque (1999). Fulfilment of government responsibilities entails protecting the interests of the public in such processes; however, their commitment and capacity to do so is open to debate (Post and Obirih-Opareh 2003). Thus, this discussion firmly lays the groundwork for evaluating the public interest in delivery of the ADP in a social justice context.

In the next section, the concept of the public interest is further described.

Public interest in public–private partnerships

Consideration of the public interest is an ethical requirement of policy formulation, and the processes of public administration should be 'honest, open, proper and fair' (Niland and Satkunandan 1999). According to the Australian Law Reform Commission (ALRC), public interest matters include freedom of expression, freedom of the media, public health and safety, national security, open justice and the economic wellbeing of the country (ALRC 2016). However, there is no single clear definition for the concept of 'public interest'. According to the ALRC, 'public interest' should not be defined, but instead, there should be a non-exhaustive list of public interest matters to provide the parties and the courts with useful guidance. According to the NSW Ombudsman, the public interest refers to:

> *Considerations affecting the good order and functioning of the community and government affairs, for the wellbeing of citizens or for the common good. (NSW Ombudsman 2012)*

Since the 1970s, following the development of legal aid schemes and community legal centres, public interest law has aimed at improving social justice through a range of differing processes and organisations (Noone 2011). Among these organisations is the Public Interest Advocacy Centre (PIAC) in NSW, an institution 'committed to social justice for all in the community' (PIAC 2016).

On the other hand, water privatisation can impose risks on the public interest as well as the environment (Pacific Institute 2016). Furthermore, PPPs are multifaceted arrangements in nature and can create controversy in a range of ways. While some groups, such as construction companies, directly benefit from these arrangements and support them, opposition to such contracts largely comes from public employee unions and consumer advocacy groups (Ortiz and Buxbaum 2007). Overall, in the presence of economic benefits for the partners, the public interest may be compromised (Rosenau 1999).

As previously noted, various private companies are involved in the provision and operation of Australia's major desalination plants under long-term contracts ranging from 10 to 50 years (WSAA 2013). In this regard, Australian governments have emphasised the protection of the public interest in the procuring of public projects. According to an inquiry report by the NSW Parliament:

> *The Committee believes that public interest in and mistrust of the involvement of the private sector in the provision of public infrastructure and services means that the*

need for transparency and accountability in the process is paramount. (Parliament NSW 2006)

According to the National PPP Guidelines, the public interest is part of the overall investment evaluation and subsequent procurement models (Infrastructure Australia 2008; Department of Infrastructure and Regional Development 2015). Some state governments have developed policies to evaluate public interest matters in the delivery of PPP projects. For example, based on national PPP policies, Partnerships Victoria has set a 'public interest test' for PPP projects. The test includes eight elements: effectiveness, accountability, affected individuals and communities, equity, consumer rights, public access, security and privacy (e.g. Partnerships Victoria 2001). South Australia does not have a specific public interest test, but transparency and accountability are required for all PPPs (Sharp and Tinsley 2005). Despite this, some of Australia's major desalination projects have been examples of inadequate project selection prompting costly outcomes for Australian taxpayers. In a supplementary submission to the Productivity Commission for the Public Infrastructure draft report, it was mentioned that 'modern corporations do not necessarily or inherently work for the public interest' (Productivity Commission 2014). Godden (2009) argued that the clear alignment of PPPs with the public interest in water governance needs further debate because the construction of major water infrastructure, such as desalination plants, has privileged the right to water of certain groups over others (i.e. urban over rural areas). Furthermore, in a national survey on public perceptions towards desalination in Australia, people showed concern about the distribution of costs and benefits in terms of 'fairness' and 'equality' (King *et al.* 2012).

On the other hand, the OECD has developed a set of comprehensive, mutually complementary water governance guidelines to provide a systemic framework to evaluate water governance practices. This framework has three dimensions: effectiveness, efficiency and trust and engagement (OECD 2015). Effectiveness relates to defining and implementing clear and sustainable water-policy goals. The efficiency dimension is about maximising the benefits of sustainable water management and welfare at the least cost to society, while the trust and engagement principle relates to ensuring the inclusiveness of the relevant stakeholders in governance practices through democratic legitimacy and fairness for the whole of society. The OECD principles and the public interest test have overlaps with the value-related issues of water governance arrangements defined by Tortajada (2010), namely responsibility, accountability, transparency, equity and fairness.

In the next section, the theoretical framework for this study will be discussed. The focus will be the Adelaide Desalination Plant (ADP): a state-owned asset, which was delivered under a Design–Build–Operate–Maintain (DBOM) contract. The contract term is 20 years and the private companies – ACCIONA Agua and Trility Pty Ltd – operate and maintain the facility (WSAA 2013; ACCIONA 2016).

Theoretical framework

In order to operationalise the concept of public interest, based on the work of Haque (1999) and Tortajada (2010) and the guidelines developed by Infrastructure Australia (2008), the Department of Infrastructure and Regional Development (2015) and the OECD (2015), a framework for good governance was developed for this study (Fig. 5.1). Commonalities and overlaps exist among the elements of the above-mentioned academic studies and the pragmatic guidelines because they all prescribe criteria for better governance. Therefore,

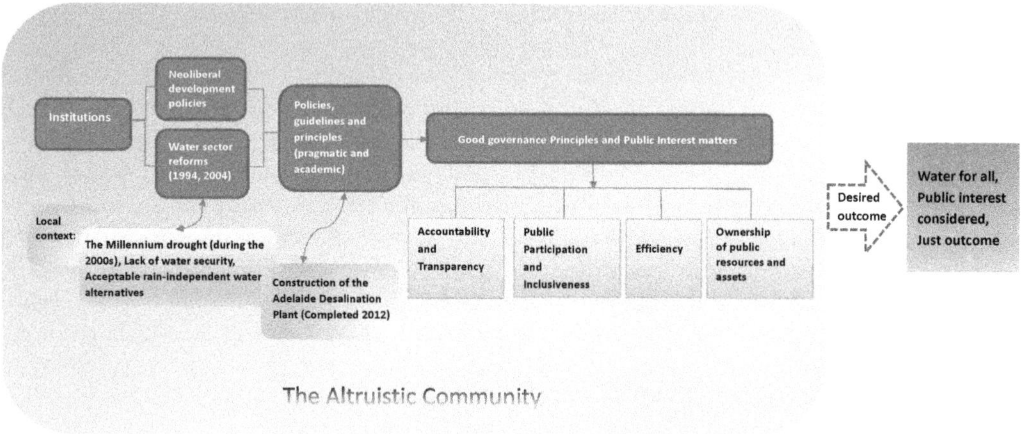

Figure 5.1. The elements of good governance for delivering desalination plants under PPPs in a community with a desire to be altruistic (Sources: adapted from Haque 1999; Partnerships Victoria 2001; Infrastructure Australia 2008; Tortajada 2010; OECD 2015; Department of Infrastructure and Regional Development 2015).

the authors have developed this framework based on what are considered to be four of the more important governance elements, namely accountability and transparency, public participation and inclusiveness, efficiency and ownership. According to Tortajada (2010), participation, accountability and transparency are among the fundamental elements of good governance. Efficiency is considered to be a core objective of involving the private sector in the delivery of public services. The ownership element was added by the authors, because the ownership of public assets and natural resources has grown into a crucial issue under the recent privatisation paradigm, especially in Australia. This is even more important for water resources. Further, this conceptual framework is grounded in an altruistic context within which the community wishes to have water for all. A more detailed explanation of this will be taken up in the results section. In this framework, the 'accountability and transparency' and the 'public participation and inclusiveness' elements correspond with the procedural justice dimension of social justice, while the 'efficiency' element relates to the distributive dimension of social justice (see Lukasiewicz and Baldwin 2014). Therefore, this framework is used to look at the aspects of public interest and social justice in the delivery of the ADP under a PPP.

Research methods

Two methods were applied to evaluate the good governance elements of the ADP delivery. A computer-aided qualitative content analysis of newspapers related to the ADP was used to obtain an insight into the debates leading up to the construction of the plant and those subsequent to its commencement. An online survey of households in the Adelaide metropolitan area was also conducted to illustrate how the community perceived the public interest in the procurement and delivery processes of the ADP.

Computer-aided qualitative content analysis of newspaper articles on the ADP

A qualitative content analysis method was applied to the text of the selected newspaper articles in relation to the ADP. The entire period covered was between 2002 and 2012. The newspapers were obtained from the NewsBank database. As the amount of collected data was quite large (3841 articles), a random sampling approach was applied and 15% of the articles in the peak month of each year were selected, resulting in a total number of 94 articles. The selected articles were from Australian national and state newspapers, weekend newspapers and local newspapers. The articles were from the 'News', 'Editorial', 'Letters to the editor', 'Opinion' and 'State/national edition' sections. The content of the data was analysed using Leximancer, a text analytics software, to explore the main themes and concepts within the textual data. The Leximancer software quantifies and displays the conceptual structure of the text and generates a list of concepts (groups of semantically similar words that generally appear together within the entire text) and themes (the collection of contextually related concepts) (Leximancer 2016).

For the purpose of analysis, the articles were divided into pre-construction and post-construction periods. The findings of the content analysis revealed that in the pre-construction period, the set of concepts within the top themes represented the dominance of debates around the ongoing drought, water shortages, and the existing high-level water restrictions imposed by the state government. Overall, the drought conditions could be seen as the main reason to bring desalination to the forefront of water debates by different politicians (e.g. The Advertiser 2007a,c). The analysis showed that the costs of the state government's water plans, including the construction of a desalination plant, were perceived to be an important reason behind increases in water bills (Nankervis 2007; Owen 2007; The Advertiser 2007c). The text segments from the newspapers revealed both favourable and unfavourable public opinions following the announcement of the desalination plant (The Advertiser 2007a,b). The state government was also considering other supply alternatives, such as wastewater treatment and stormwater harvesting.

In the post-construction period, the initial implication of the most important themes was the dominance of debates around the desalination plant and water-related issues. There was some dissatisfaction regarding the high cost of water infrastructure, including the desalination plant (The Advertiser 2011a). Among the criticisms, *The Sunday Mail* (Crouch and Mennie 2011) mentioned that the plant's principal contractor, AdelaideAqua consortium, would make a A$500 million profit in the face of a 50% rise in the cost of water for consumers 'to pay for the plant'. The government was criticised for choosing desalination over other water supply options (The Advertiser 2011b). There were also reports of project delivery delays, with SA Water's possible penalties on the consortium and the plant's mothballing, and their potential impact on taxpayers (Crouch and Mennie 2011). Following good rainfalls, the plant was perceived to be an expensive 'white elephant' and the government's approach to be a knee-jerk policy (The Advertiser 2012a,b). Eventually, the government announced the plant's mothballing until the advent of the next drought (The Southern Messenger 2012).

The analysis shows that public opinion towards major water infrastructure can rapidly change during and after periods of drought (e.g. The Advertiser 2011a).

Elements of good governance in the selected newspapers

The analysis revealed further issues related to the elements of good governance defined in the theoretical framework.

In regards to the element of accountability and transparency, although SA Water promoted community engagement throughout the project delivery phase (Kumar and Farinola 2012), the government was questioned about why it had never explained that desalination was the preferred option over other cheaper and more environmentally friendly alternatives (The Advertiser 2009).

The South Australian Council of Social Services' (SACOSS) suggestion to the government to offer concessions for water and electricity for low-income households (The Advertiser 2012b) and a possible new pricing structure for those who can least afford the services (Jenkin and Henderson 2007), fall under the public participation and inclusiveness dimension.

The delays, costs and possible mothballing of the ADP fall into the efficiency category. Further, *The Advertiser* reported that from 2011, SA Water was supposed to justify the price rises to the Essential Services Commission of South Australia (ESCOSA). Moreover, a potential review of the investment decisions was also being considered by ESCOSA (Kemp 2010). The efficiency dimension of the OECD (2015) guidelines request that water regulatory frameworks be implemented and enforced in pursuit of the public interest. Hence, the presence of a regulatory body (ESCOSA) is related to the efficiency dimension.

Moreover, the presence of organisations such as ESCOSA, community service agencies and a non-government representative body for health and community services, SACOSS, was noted in the newspapers. These institutions aim at promoting justice and equality in the overall community and to protect the long-term interests of South Australian communities.

An online survey in the Adelaide metropolitan area

After reviewing the Leximancer results, an online survey was designed to gain a deeper understanding of Adelaide residents' perceptions of the public interest, with particular reference to the ADP. The respondents were located in the Adelaide metropolitan area, except for those in the Gawler and Adelaide Hills regions further away from the ADP and, therefore, less likely to know about it.

Qualtrics, an online survey software, hosted the survey. An email list was bought from a marketing company and the participants were randomly selected. The survey comprised mixed question types and was emailed to 1670 people. Overall, 428 respondents completed the survey over a 5-day period.

Survey results related to the human right to water supply and public interest

Part of the survey included general questions to examine participants' understandings of what constitutes the 'public interest'. As well, in order to determine if the community has altruistic attitudes, the survey added a question on human rights and water. A major part of the survey included specific questions related to the good governance elements shown in Fig. 5.1.

The altruistic urban population and their attitude to the human right to water supply

The respondents were asked if there exists a human right to clean fresh water in urban and rural Australia. All of the 428 respondents answered this question and the majority of the participants (90%) thought that there is a human right to clean fresh water in urban and rural areas in Australia. It is important that the community has an altruistic attitude and wants equal access to water for all, because this has implications for governance practices. According to Good (2011), 'at present the human right to water is inadequately formally protected in Australia'.

Participants' perceptions of public interest in urban water supply

In order to prevent any bias, the respondents were asked to provide their personal interpretation of the public interest in urban water supply using an open-ended question, rather than providing them with a range of specific options. The response rate for this question was high (75%).

The text analysis feature of Qualtrics revealed that alongside the word 'water', the words used most frequently by the respondents were clean, supply, cost, reliable, public, good and quality. A manual content analysis of the responses also showed similar results. About 50% of respondents regarded the public interest in water supply as having access to clean water for daily living. Some of the other aspects mentioned by the respondents included reliability and affordability of the water supply, water conservation and consideration of other supply alternatives. According to one participant:

> *The public regards clean potable water as essential but is mindful of the cost. The government should look at wastage such as stormwater runoff and recycling wastewater before expensive desalination plants.*

Survey results related to the elements of good governance

The survey questions and results pertinent to the elements of the good governance framework (shown in Fig. 5.1) are presented in Table 5.1.

Accountability and transparency

The results from the community survey indicate relatively low levels of accountability and transparency throughout the project delivery phase. Under the current market-driven governance reforms, the principle of accountability has been negatively affected because of issues around the openness and accountability of the private sector (Haque 2000; English 2003; Barton 2006). Barton (2006) argued that the commercial-in-confidence information about outsourcing contracts imposes a 'veil of secrecy' on government activities. This was indeed an issue in the earlier stages of this research, as most of the information related to Australia's desalination industry is considered to be commercially sensitive, making it difficult to collect data for research purposes, especially in relation to the contracts for the plants. Similarly, the Productivity Commission report on Australia's urban water sector revealed a lack of transparency and accountability in supply and regulation decisions. The report also mentioned the typical confidentiality associated with the cost and benefit of supply options as a factor leading to less transparency and accountability in the delivery of utilities (Productivity Commission 2011).

Public participation and inclusiveness

The results indicate that, despite the ADP being an exemplary case, the community did not feel consulted. Furthermore, when the respondents were asked about their source of information on the ADP, 61% indicated that they had heard about the ADP through radio and/or television and 38% via newspapers. Word of mouth, the SA Water website, council newsletters and public meetings were the other options. Additionally, around 92% of the respondents said they had not attended any public meetings, while 6% had attended one or two meetings. Perhaps these results indicate that the government needs to adopt alternative information-sharing channels if they wish to reach the public.

Table 5.1. Summary of survey results for the elements of good governance

Good governance elements	Question/statements	Survey participants' responses
Accountability and transparency	Level of accountability of government organisations involved in ADP delivery to the public	49% selected 'very little or not at all' accountable, and 35% selected 'to some or great extent' accountable.
	Level of accountability of private organisations involved in ADP delivery to the public	49% selected 'very little or not at all' accountable, and 28% selected 'to some or great extent' accountable.
	Level of transparency and public reporting around project selection, procuring, and funding processes associated with the ADP	Around 40% selected 'least or less transparent', over 20% were neutral.
Public participation and inclusiveness	Feeling included in government's decision-making processes around major water infrastructure such as the desalination plant	61% disagreed with the statement, 27% were neutral.
	Level of involvement throughout government's decision-making process throughout construction, operation, and maintenance of the ADP	Around 77% of the respondents said they were 'very little or not at all involved' at these stages.
	Had received sufficient information before/during the plant's construction	More than 50% of the respondents disagreed with the statement.
	There was accessible information throughout the plant's planning and construction processes.	48% were neutral and 35% disagreed with the statement.
	Always trust the information that I have received from the government.	52% disagreed with the statement and 32% were neutral.
	Did you receive any information regarding the procurement process for the ADPt?	94% answered 'No'.
	Knowledge of the ADP contract type	84% had 'no or little knowledge'.
	Knowledge about the background to the project and a summary of the key contractual terms and conditions	83% had 'no or little knowledge'.

Efficiency	Government should have first considered other less costly supply alternatives, such as wastewater recycling and stormwater harvesting, before building the seawater desalination plant	63% agreed with the statement, 26% were neutral.
	Community satisfaction over the costs associated with delivering the ADP	60% selected 'dissatisfied' and 32% were neutral.
	To what extent did the cost of the plant's construction, operation and maintenance impact upon the household's water bills?	65% selected 'to some or a great extent', and 16% selected 'very little or not all'.
	Private finance should play a bigger role in the Australian water sector to increase economic efficiency	47% of the respondents held a neutral opinion, 27% agreed with the statement.
Ownership	Which of the following organisations do you prefer to own the ADP?	The state government received the highest rating (38%), followed by the federal government (19%) and SA Water (17%). PPPs, the general public, the private sector, local councils, and regional authorities received lower ratings.

A relative lack of trust in the information received from the government confirms earlier results from King *et al.*'s (2012) study on public perceptions of desalination, suggesting that governments are among the least trusted sources of information.

Moreover, regarding the procurement of the ADP, the survey results show that there has been a relative deficiency in actual public engagement in decision-making processes around the ADP.

Efficiency

Concerning the efficiency dimension, the survey results indicate a lack of consideration of other water supply alternatives, before the decision was made to build the ADP. This is in line with the Productivity Commission report (2014), which states:

> *The SA Government opted to invest in a desalination plant, when it appears that an alternative that would have been much cheaper was available. The Australian Government's actions in insisting on a doubling of the plant's capacity as a condition of providing additional funding made matters worse.*

Furthermore, the results revealed that there was some level of dissatisfaction over the costs of the ADP, which is compatible with the results from the newspaper analysis. This is also in line with the Productivity Commission report about some of Australia's major desalination projects having inadequate project selection and costly outcomes for Australian taxpayers (Productivity Commission 2014).

Respondents' neutral attitudes towards the role of private finance in increasing economic efficiency may be because of their lack of information on this matter. It may also reflect the findings from previous studies showing that private sector participation (PSP) might not necessarily equate to greater efficiency. Prasad (2006) linked increasing feelings of discontent and resistance against privatisation to the fact that, although sometimes perceived as useful, the economic benefits have not actually been realised, and the social impacts have not been thoroughly analysed. According to Worthington (2011), research

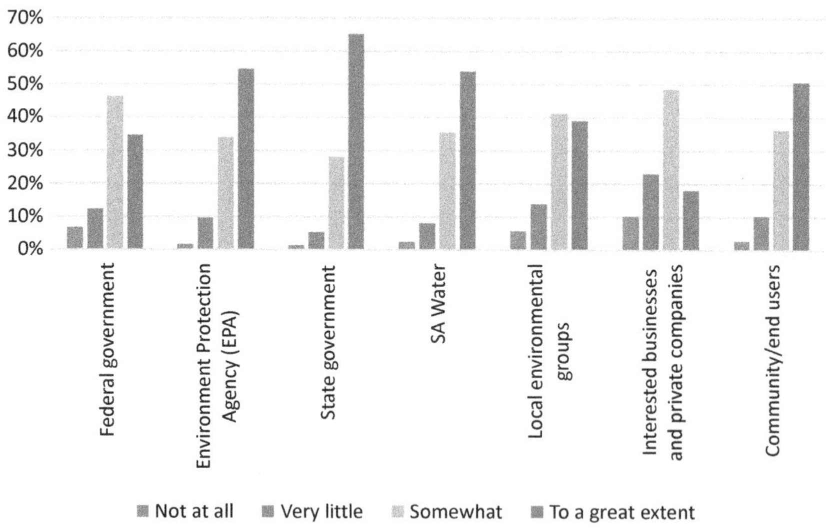

Figure 5.2. Community's preference for various authorities to be responsible for selecting and planning desalination projects.

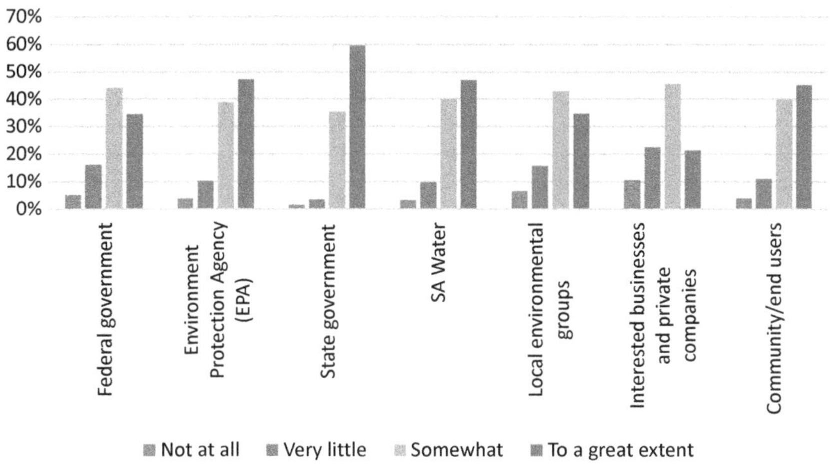

Figure 5.3. Community's preference for various authorities to be responsible for procuring desalination projects.

into the efficiency of urban water utilities, and the role of ownership and regulation, suggest that there is 'no clear consensus emerging on the relative efficiency of private over public water utilities.' Nevertheless, according to the Productivity Commission (2014), private involvement in infrastructure delivery can bring efficiency gains in some sectors, depending on the implementation of a proper policy and regulatory framework.

Ownership of the ADP

As previously noted, the ownership of public water infrastructure has been the subject of growing debate in academic and policy-making spheres.

Regarding the ownership of the ADP, the results confirm that the Australian urban community prefers government ownership of water assets. 'Remunicipalisation', or transfer of water services from private companies to municipal authorities, is a growing trend, especially in the water sector (McDonald 2012; Pigeon *et al.* 2012). In the remunicipalised cases, there has been more equitability, transparency and efficiency compared with when the private sector provides or operates the services (McDonald 2012).

Further to the governance of the ADP, respondents were asked to rank their preferences on two questions in relation to various authorities taking responsibility for the planning and the procurement processes for desalination projects. The participants regarded the state government to be the body with the highest level of involvement in such processes, followed by the Environment Protection Authority (EPA), SA Water and end users. The private sector, the federal government and local environmental groups were selected as being moderately involved in such processes (see Figs. 5.2 and 5.3). This indicates community support for a stronger role for state government in project planning and procurement. Moreover, the relatively high ranking of the EPA in both questions may be an illustration of community concern about the environmental impacts of desalination.

The policy implications of these findings will be discussed in the next section.

Conclusion and policy implications

Under the recent water privatisation paradigm, despite the fact that the private sector must have a place at the table with respect to public–private partnership policies, the principal consideration of decision makers must be that of the public interest. This study has revealed insufficient accountability, transparency and actual public participation in the ADP delivery process (despite government efforts), and a lack of public knowledge about the advantages of PSP in terms of economic efficiency. These issues might be the reason behind the dissatisfaction over different aspects of the ADP such as cost, and in a more general sense, can lead to a lack of community trust and support for PSP and privatisation. Effective governance and accountability arrangements are vital in order to reap the potential benefits of PSP (Productivity Commission 2014). More effective and appropriate information-sharing methods throughout the planning, selection and delivery stages of major water infrastructure need to be considered.

Further, the survey results show that the community interprets public interest in urban water supply as having access to clean water, reliability and affordability of water supply, and consideration of water conservation measures. Additionally, the Adelaide community considers access to water to be a human right, and wants 'water for all'. This imposes obligations on governments to provide equality and access to water for everyone and to protect this right. This research also indicates that the community wants greater government control and responsibility over public assets and decision-making processes in the water sector. Overall, these findings call for a greater role for government in ensuring that PSP in the provision of water will protect the human right to this vital commodity.

South Australia does not yet have a public interest framework for project delivery. Setting realistic, context-specific and comprehensive public interest guidelines and devising thorough monitoring mechanisms could be beneficial for government agents and private parties involved in public infrastructure delivery. Being more transparent, accountable and inclusive throughout the decision-making process in relation to public–private partnerships can positively influence public perceptions of such processes and contribute to 'just' outcomes that are 'in the public interest'.

Acknowledgements

The authors acknowledge the financial support of the National Centre of Excellence in Desalination Australia, which is funded by the Australian Government through the Water for the Future Initiative. The authors thank the Adelaide community and several key public servants who assisted us in coming to grips with the issues. As always, Neil Palmer made many suggestions of people in the desalination industry to interview.

References

Abbott M, Wang WC, Cohen B (2011) The long-term reform of the water and wastewater industry: the case of Melbourne in Australia. *Utilities Policy* **19**, 115–122. doi:10.1016/j.jup.2010.08.002

ACCIONA (2016) *Adelaide Desalination Plant.* ACCIONA, Adelaide, <http://www.acciona.com.au/projects/water/desalination-plants/adelaide-desalination-plant/>.

ALRC (2016) *Balancing Privacy with Other Interests.* Australian Law Reform Commission, Sydney, <*https://www.alrc.gov.au/publications/8-balancing-privacy-other-interests/meaning-public-interest.*

Barton AD (2006) Public sector accountability and commercial-in-confidence outsourcing contracts. *Accounting, Auditing & Accountability Journal* **19**, 256–271. doi:10.1108/09513570610656114

Colebatch HK (2006) Governing the use of water: the institutional context. *Desalination* **187**, 17–27. doi:10.1016/j.desal.2005.04.064

Crouch B, Mennie S (2011) It's all systems flow – desal water in July (or maybe later). *The Advertiser*, 29 May, p. 4.

Department of Infrastructure and Regional Development (2015) *National Public Private Partnership: Policy Framework*. Department of Infrastructure and Regional Development, Canberra.

Dovers S (2001) Institutional barriers and opportunities: processes and arrangements for natural resource management in Australia. *Water Science and Technology* **43**, 215–226.

English L (2003) Emasculating public accountability in the name of competition: transformation of state audit in Victoria. *Critical Perspectives on Accounting* **14**, 51–76. doi:10.1006/cpac.2002.0523

Frontier Economics (2013) *Review of Urban Water Customer Choice Options, Policy Drivers and Regulatory Instruments: a Report for the National Water Commission*. Frontier Economics, Melbourne.

Funnell W (2003) Enduring fundamentals: constitutional accountability and auditors-general in the reluctant state. *Critical Perspectives on Accounting* **14**, 107–132. doi:10.1006/cpac.2002.0519

Godden L (2009) Towards a new ethic in Australian water law and policy. In *Climate Change on for Young & Old*. (Ed. H Sykes) pp. 46–60. Future Leaders, Melbourne.

Good M (2011) Implementing the human right to water in Australia. *University of Tasmania Law Review* **30**, 107-143.

Haque MS (1996) The intellectual crisis in public administration in the current epoch of privatization. *Administration & Society* **27**, 510–536. doi:10.1177/009539979602700403

Haque MS (1999) Ethical tension in public governance: critical impacts on theory-building. *Administrative Theory & Praxis* **21**(4), 468–473. doi:10.1080/10841806.1999.11643404

Haque MS (2000) Significance of accountability under the new approach to public governance. *International Review of Administrative Sciences* **66**, 599–618. doi:10.1177/0020852300664004

Haque MS (2001) The diminishing publicness of public service under the current mode of governance. *Public Administration Review* **61**, 65–82. doi:10.1111/0033-3352.00006

Heberger M (2011) Australia's millennium drought: impacts and responses. In *The World's Water Volume 7*. (Ed. PH Gleick) pp. 97–125. Center for Resource Economics, Island Press, Washington DC, USA.

Hurlimann AC (2007) Is recycled water use risky? An urban Australian community's perspective. *The Environmentalist* **27**, 83–94. doi:10.1007/s10669-007-9019-6

Hurlimann A, McKay J (2004) Attitudes to reclaimed water for domestic use: part 2. Trust. Water: *Journal of the Australian Water Association* **31**(5), 40–45.

Hurlimann A, McKay J (2007) Urban Australians using recycled water for domestic non-potable use – an evaluation of the attributes price, saltiness, colour and odour using conjoint analysis. *Journal of Environmental Management* **83**, 93–104. doi:10.1016/j.jenvman.2006.02.008

Infrastructure Australia (2008) *National Public Private Partnership Guidelines*. Infrastructure Australia, Sydney.

Infrastructure Australia (2016) *National Public Private Partnership Policy Framework*. Infrastructure Australia, Sydney, <https://infrastructure.gov.au/infrastructure/ngpd/files/National-PPP-Policy-Framework-Oct-2015.pdf>.

Jenkin C, Henderson N (2007) Water bills to rise by at least $80 a year. *The Advertiser*, 18 September, p. 1.

Kamieniecki S, Shafie D, Silvers J (1999) Forming partnerships in environmental policy the business of emissions trading in clean air management. *The American Behavioral Scientist* **43**, 107–123. doi:10.1177/00027649921955182

Keremane G, McKay J, Ettehad E, Wu Z (2013) The evolution of business models for non-major desalination plants in Australia – issues for governance. In *Proceedings of the International Desalination Association World Congress on Desalination and Water Reuse.* 20–25 October, Tianjin, China. (Ed. L Henthorne). International Desalination Association, Topsfield MA, USA.

Kemp M (2010) SA Water warned on prices. *The Advertiser*, 23 December, p. 4.

Keremane G, McKay J, Ettehad E (2015) Hearing from the people on the collective or private governance of urban freshwater in Australia. *Water Utility Management International* **10**, 17–21.

King TJ, Ooi D, Cary J, Fisher A, Schibeci R, Murphy K, *et al.* (2012) 'Public perceptions of, and responses to, desalination in Australia: a report on findings'. Working paper. Series two. (Ed. P Kelly). Alfred Deakin Research Institute, Deakin University, Melbourne.

Kumar M, Farinola M (2012) 'Adelaide Desalination Project – final project report, report #4'. SA WATER, Adelaide.

Leximancer (2016) *Text in. Insight out.* Leximancer Pty Ltd, Brisbane, <http://info.leximancer.com/>.

Lukasiewicz A, Baldwin C (2014) Voice, power, and history: ensuring social justice for all stakeholders in water decision-making. *Local Environment* [online]. doi:10.1080/13549839.2014.942261

Lukasiewicz A, Bowmer K, Syme GJ, Davidson P (2013) Assessing government intentions for Australian water reform using a social justice framework. *Society & Natural Resources* **26**, 1314–1329. doi:10.1080/08941920.2013.791903

McDonald DA (2012) Remunicipalisation works! In *Remunicipalisation: Putting Water Back into Public Hands.* (Eds M Pigeon, DA McDonald, O Hoedeman and S Kishimoto) pp. 8–22. Transnational Institute, Amsterdam, Netherlands.

McKay J (2005) Water institutional reforms in Australia. *Water Policy* **7**, 35–52.

Miraftab F (2004) Public-private partnerships the trojan horse of neoliberal development? *Journal of Planning Education and Research* **24**, 89–101. doi:10.1177/0739456X04267173

Nankervis D (2007) Great debate – desal decision hinges on report. *The Advertiser*, 9 September, p. 15.

National Water Commission (2011) *Urban Water in Australia: Future Directions.* National Water Commission, Canberra.

National Water Commission (2016) *Closure in 2014.* National Water Commission, Canberra, <http://www.nwc.gov.au/organisation/closure-in-2014>.

Niland C, Satkunandan S (1999) The ethical value of red tape! The role of specialised institutions in ensuring accountability (the NSW experience). *Australian Journal of Public Administration* **58**, 80–86. doi:10.1111/1467-8500.00075

Noone MA (2011) ADR, public interest law and access to justice: the need for vigilance. *Monash University Law Review. Monash University. Faculty of Law* **37**, 57–80.

NSW Government (2012) *NSW Public Private Partnerships Guidelines.* NSW Treasury, Sydney.

NSW Ombudsman (2012) *Public Interest.* NSW Ombudsman, Sydney.

OECD (2011) From lessons to principles for the use of public-private partnerships. 32nd Annual meeting of Working Party of Senior Budget Officials, Public Governance and Territorial Development Public Management Committee, Luxemburg, <http://www.oecd.org/gov/budgeting/workingpartyofseniorbudgetofficialssbo.htm>.

OECD (2012) *Recommendation of the Council On Principles for Public Governance of Public-Private Partnerships*. OECD Publishing, Paris, France.

OECD (2015) *OECD Principles on Water Governance*. OECD Publishing, Paris, France.

Ortiz IN, Buxbaum JN (2007) *Protecting the Public Interest: the Role of Long-Term Concession Agreements for Providing Transportation Infrastructure*. Keston Institute for Public Finance Policy and Infrastructure Policy, University of Southern California, Los Angeles CA, USA.

Owen M (2007) Desal plant on the way – Maywald seeks federal funds. *The Advertiser* **8**(September), 1.

Pacific Institute (2016) *Water Privatization*. Pacific Institute, Oakland, CA, USA, <http://pacinst.org/issues/sustainable-water-management-local-to-global/water-privatization/>.

Parliament NSW (2006) 'Second report: the Cross City Tunnel and public private partnerships'. Joint Select Committee on the Cross City Tunnel, Sydney.

Partnerships Victoria (2001) *Partnerships Victoria Practitioners' Guide*. The Department of Treasury and Finance, Melbourne.

PIAC (2016) *PIAC Strategic Plan*. Public Interest Advocacy Centre, Sydney, <http://www.piac.asn.au/sites/default/files/publications/extras/15.08_piac_plan_2015-18_-_summary.pdf>.

Pigeon M, McDonald DA, Hoedeman O, Kishimoto S (2012) *Remunicipalisation: Putting Water Back into Public Hands,* Transnational Institute, Amsterdam, Netherlands.

Post J, Obirih-Opareh N (2003) Partnerships and the public interest: assessing the performance of public-private collaboration in solid waste collection in Accra. *Space and Polity* **7**, 45–63. doi:10.1080/13562570309244

Prasad N (2006) Privatisation results: private sector participation in water services after 15 years. *Development Policy Review* **24**, 669–692. doi:10.1111/j.1467-7679.2006.00353.x

Productivity Commission (2011) 'Australia's urban water sector'. Final inquiry report. Productivity Commission, Canberra.

Productivity Commission (2014) 'Public infrastructure inquiry report'. Productivity Commission, Canberra.

Rosenau PV (1999) The strengths and weaknesses of public-private policy partnerships. *The American Behavioral Scientist* **43**, 10–34.

Sharp L, Tinsley F (2005) PPP policies throughout Australia: a comparative analysis of public private partnerships. *Public Infrastructure Bulletin* **1**(5), 21–34.

The Advertiser (2007a) Opinion. *The Advertiser,* 3 September, p. 17.

The Advertiser (2007b) Go slow on desalination. *The Advertiser,* 10 September, p. 17.

The Advertiser (2007c) No word on federal funding. *The Advertiser,* 13 September, p. 12.

The Advertiser (2009) Growth priority. *The Advertiser,* 22 May, p. 17.

The Advertiser (2011a) Incentive to save water is lost in Labor's pricing logic. *The Advertiser,* 13 May, p. 20.

The Advertiser (2011b) Letters: But not a drop to drink. *The Advertiser,* 17 May.

The Advertiser (2012a) Don't rub debt into the desal wound. The Advertiser, 5 October, p. 22.

The Advertiser (2012b) $30m desal pipeline pain. *The Advertiser* 6 October, p. 8.

The Southern Messenger (2012) Not a 'disaster'. *The Southern Messenger,* 24 October.

Tortajada C (2010) Water governance: some critical issues. *Water Resources Development* **26**, 297–307.

Uhr J (1999) Three accountability anxieties: a conclusion to the symposium. *Australian Journal of Public Administration* **58**, 98–101. doi:10.1111/1467-8500.00079

van Dijk AIJM, Beck HE, Crosbie RS, De Jeu RAM, Liu YY, Podger GM, *et al.* (2013) The Millennium Drought in southeast Australia (2001–2009): natural and human causes and implications for water resources, ecosystems, economy, and society. *Water Resources Research* **49**, 1040–1057. doi:10.1002/wrcr.20123

Walker B, Walker BC (2000) *Privatisation: Sell off or Sell out*. ABC Books, Sydney.

Worthington AC (2011) 'Productivity, efficiency and technological progress in Australia's urban water utilities'. Waterlines Report Series No 62, National Water Commission, Canberra.

WSAA (2013) *Seawater Desalination Information: Pack Two*. Water Services Association of Australia, Melbourne.

6

Company–community relations in the mining context: a relational justice perspective

Mirella Cobeleanschi Gavidia and Deanna Kemp

Summary

This chapter explores issues of justice and fairness in relationships between mining companies and locally affected communities. The focus is on relational justice, defined as the way interactions between mining companies and communities are formed and managed, both in terms of how parties communicate their interests and the processes through which decisions are made. Using a case study in the Brazilian Amazon, relational justice is explored from the perspective of 'voice', 'capabilities' and 'trust' – collectively these are the elements of justice. Ethnographic methods were applied to conduct this research, and four key findings are presented. First, community–company relations are affected by insufficient understanding among local communities about mining and its social and environmental impacts. Second, elements of justice are impacted by communication processes established by the mining company to engage with local communities. Third, the participation of external actors to support communities has the potential to maximise justice by improving local community capabilities, especially in terms of improving communication and negotiation. The fourth finding was that relational justice is affected by internal organisation and community representation. We conclude by arguing that justice in the mining context can be maximised by improving access to information, and establishing participatory and inclusive process for decision making.

Introduction: why relational justice matters in the mining context

Issues of 'justice' and 'fairness' lie at the very heart of almost every mining-related conflict or controversy. Although mining generates significant revenue for host nations through taxes, royalties and employment, it can also cause great disruption to local people, their land and their livelihoods. These disruptive impacts include pollution and contamination of soil and water; dust, noise and vibration. Further stresses are placed on local communities through in-migration or involuntary resettlement, and an increase in violence. The broad range and uneven distribution of mining-related impacts and benefits can trigger concerns and grievances at the local level.

In the absence of adequate regulations and state dispute resolution bodies many mining companies and host communities are left to resolve issues and negotiate directly (IIED 2002). Many local communities lack the means to effectively relate to and negotiate with the corporate entity. In the resulting power imbalance, mining activities and mining companies themselves are often associated with a range of injustices (Hamann and Kapelus 2004). Allegations of injustices appear to be most audible when operations are located in politically, culturally and ecologically vulnerable contexts. The complex set of dynamics associated with mining negotiations is readily characterised as a 'battleground' of rights and interests (Ballard and Banks 2003).

Applied research in mining and community relations is largely concerned with the distributional elements of justice (e.g. Bebbington *et al.* 2009; Romero *et al.* 2012; Morrice and Colagiuri 2013). Researchers and advocates focus on the outcome of conflicts and contestations over mining and the allocation of impacts and benefits across different groups (Kemp *et al.* 2011). While a distributive perspective is vitally important for understanding the dynamics of justice and injustice, there is an increasing recognition that in the context of mining, relational dynamics – that is, the way different parties relate to and engage with each other – is a significant determinant of justice. Until recently, the relational dynamics of justice have been overlooked in scholarly debates on the mining sector in its negotiations with local people (Whiteman and Mamen 2002).

This chapter fills this gap by focusing on relationship dynamics between global mining companies and local communities from a relational justice perspective. First, the chapter presents definitions. Relational justice is defined as the way the interactions between mining companies and locally affected communities are formed and managed throughout time, both in terms of how parties communicate their interests and the processes through which decisions are made. In this chapter, relational justice is used as a term to encompass the concepts of procedural and interactional justice. These concepts are largely discussed in the field of social psychology (Colquitt *et al.* 2001) and increasingly adopted in the fields of natural resources management (Smith and McDonough 2001). Second, the chapter presents the interpretation of justice used in this research, and the theoretically grounded framework proposed for examining factors that enhance or hinder relational justice between mining companies and affected communities. Finally, the framework is applied to a case study in the Brazilian Amazon, involving a large-scale bauxite mine located in the Juruti municipality and local communities. The mine is owned by global mining giant, Alcoa.

Elements of justice in company-community relations in mining

Justice is a relative concept that is interpreted differently, depending on the position and perspective taken by the observer. For this reason, justice can be a challenging concept to deal with in scientific research, unless a framework of analysis is established (Charmaz 2005). Here we use a theoretically grounded framework that draws on the work of Nobel Laureate, Professor Amartya Sen, and in particular his work, *The Idea of Justice* (Sen 2009), which summarises and extends his previous work on justice and fairness. The framework developed in this chapter is also informed by a multidisciplinary literature review across cognate fields such as public participation, negotiation and social psychology, and community development (e.g. Hollander-Blumoff and Tyler 2008; Deutsch *et al.* 2011; Liss 2011; Eversole 2012).

Following Sen, we interpret justice as an ideal concept, and a practical impossibility for society. From this perspective, society (and this includes relationships between mining companies and affected communities) is permeated by injustice that cannot be fully eliminated, only minimised. Although the world we live in may never be fully just, it can certainly be less unjust than it currently is. In this context, social justice is maximised when social injustices are identified and minimised.

A pragmatic way to minimise social injustice is to identify 'elements' of justice, and the factors that hinder justice, so that these factors can be minimised. We propose that this approach is more effective in terms of advancing justice in mining than engaging in a discussion about what an ideal set of 'just relations' should look like. We therefore focused on identifying those elements that are relevant for justice in the way mining companies and affected communities relate to each other. In other words, the elements that affect relational justice in the context of company–community negotiations, decision making, community engagement and public participation practice. Researchers in these fields agree that, for relational justice, elements of 'voice' and 'capabilities' of parties themselves, and the levels of 'trust' between them, are central determinants of relational justice (Lewicki *et al.* 2007; Hollander-Blumoff and Tyler 2008; Couldry 2010).

By 'voice', we mean the opportunities and the available channels that enable communication and the flow of information between members of the community and company representatives. We define 'capabilities' as the level of understanding about issues relevant to the relationship, and for decision making based on that understanding. It also includes community capacity of political organisation and to engage in negotiation processes. 'Trust' refers to the perception that the other party is acting in good faith, and seeking mutually beneficial relationships. When these elements are maximised, justice itself is maximised because stakeholders are better able to articulate and negotiate their interests. Ultimately, the maximisation of relational justice between community and mining companies supports a better management of mining impacts, and may increase the likelihood that the community benefits.

Methodology and sample

The relational justice framework is applied to the relationship between people living in the municipality of Juruti, located in the Brazilian Amazon (Fig. 6.1) and Alcoa, a multinational mining company and the largest aluminium producer in the world. The company operates a mine and a processing plant in the Juruti Velho Lake region, a train line to transport the bauxite and a road (paved by the mine and also used by local population) that connects the mine to the port and to the town. Bauxite is loaded in ships and transported to Maranhão, where Alcoa owns a refinery where that bauxite is used to produce aluminium. The company's corporate functions are located at the port, including the community relations team. Alcoa positions its Juruti mine as a flagship site in terms of corporate social responsibility (CSR) and sustainable development (Abdala 2010).

This case study exemplifies the social justice challenges embedded in the relationship between large multinationals and local communities located in socially and environmentally vulnerable contexts. The ethnographic methods used to collect data involved *in situ* interviews and participant observation during two extended field visits in 2012 and 2014. This qualitative case study describes the field site, and the relationship dynamics between the mining company and local people.

More than 100 people participated in the research. The sample was gender balanced and included adults of different ages living in communities close to Alcoa operations. Alcoa community relations staff and other company employees were also interviewed.

Figure 6.1. Brazilian Amazon map showing the location of Juruti.

Data collection sought to describe the characteristics of community–company patterns of communication, and the structures of community organisation. Mapping these patterns and structures is important considering that many individuals interact with the company yet have their interests voiced and articulated only via community leaders and local associations. Data were analysed using thematic analysis. Themes discussed in this chapter are selected on the basis of prominence and high frequency.

Mines and communities: relationships in the Brazilian Amazon

This section provides a brief overview of communities in Juruti, and describes some of the key characteristics of the relationship between the company and local population in terms of communication channels and representation structures.

Juruti is a municipality located in the heart of the Brazilian Amazon, on the banks of the Amazon River. It has a population of 53 000 people distributed in more than 200 rural communities and a town defined as an 'urban area' (IBGE 2015). Juruti can only be accessed by boat, and the local economy is based on cassava flour production, fishing, local commerce, subsistence and small-scale agriculture, and tourism. In 2000, the local economy was boosted by direct and indirect employment created by the arrival of the mine. Juruti has a literacy rate of 15%, high levels of poverty, deficient infrastructure and insufficient basic services for the population (FGV 2011).

Similar to many other places in the Amazon, Juruti is characterised by large families, extended social networks, and an informal approach to business and negotiations. Local government buildings and most public services are located in the town. Rural communities vary in size, from only a few nuclear families, up to more than 200 families. The size and distance of these communities to the town determines the level of remoteness from services, infrastructure and information. Travel to town from some communities takes up to 2 days by boat. The level of access to relevant information about Juruti and the mine is therefore strongly affected by physical location.

Traditional communication structures are based on verbal forms, with 'gossip' providing a powerful channel for information flow. Although literacy levels are low, people in Juruti and surrounding areas value written communication because they perceived it to be more reliable than the spoken word, including 'gossip'. Nevertheless, written information is scarce and Juruti has no newspapers or magazines, and the majority of population has limited access to the internet and the company's website.

The relationships between Juruti people and Alcoa began when the company began prospecting in 2000 (Bartolini *et al.* 2010). The company was granted preliminary and installation licenses in 2005, and construction of the mine began in 2006. In 2007, as a requirement for obtaining its legal licence to operate, Alcoa organised a public hearing with local populations and other key stakeholders in Juruti to discuss the project's impacts and potential benefits. Before this hearing, Alcoa conducted a series of 'pre-hearings', or smaller meetings, to prepare the population for the official event. These hearings were important to the Juruti–Alcoa relationship because they influenced people's expectations of the project, impacts and potential benefits. Many informants from the communities indicated that during these pre-hearings, promises were made by the company and gifts were distributed such as t-shirts, bicycles, caps and soccer gear. Alcoa was granted final licenses, and operations started in 2009.

Despite a high level of engagement in the early stages of the project, the relational dynamics between the company and the community changed at the different stages of the project. These changes had significant impacts on the elements of justice: voice, capabilities and the trust of community people. There are several key points to be considered in terms of maximising relational justice in the mining context. The following two sub-sections provides a brief overview of channels of communication and community organisation and local representative structures in Juriti before findings are presented in the section that follows.

Channels of company–community relations in Juriti

Patterns and channels of communication between a community and a company provide an important domain of analysis for the identification of factors that affect relational justice (Whiteman and Mamen 2002). Structures of communication shape the elements of justice and help determine factors that enhance or hinder relational justice. Data collected in Juruti demonstrate that verbal and written forms of communication have different impacts on the elements of justice.

Verbal communication occurs through formal community visits and informal interactions between company representatives and members of the community. Research participants explained that the frequency of meetings decreased as mining activities progressed. During the construction stage, local people could visit the Alcoa community relations office in the town (called *Centro de Referência*) to engage with company representatives, ask questions and make requests. The office was closed once Alcoa commenced operations. Currently, locals have to travel to the port to meet with the company. Community representatives and Alcoa employees also engage during the meetings of the Sustainable Juruti Council, which is a formal structure established by Alcoa once mining operations commenced. The focus of these meetings is to discuss possibilities for a sustainable future, rather than engaging community concerns about mining impacts, complaints and requests.

While there are different ways through which company and community communicate verbally, the company places more emphasis on written communication. Alcoa representatives explained that all complaints and requests must be presented in writing. Formal correspondence establishes a record of communication and is said to improve corporate accountability. In this context, communities have come to express their voice, and communicate their interests, concerns and requests through written requests, called *ofício*.

Alcoa also publishes a newsletter, which at the time of the field work, was produced every 3 months. According to participants, this same newsletter was published monthly during the construction period. Several local people noted that the newsletter is a one-way communication and does not lend itself to shared dialogue.

Community organisation and local representative structures

Not all mine-affected local people engage and communicate with the company in the same way and it is impossible for every individual to participate directly in the relationship. In addition, not everyone develops the same level of interest for mining-related topics. In engagements with the company, local people (and their interests) are mostly represented by community leaders and local associations. In this context, understanding the representative structures of communities, and how representatives engage with the company is relevant for understanding the dynamics of voice, capabilities and trust of the communities.

Communities in Juruti had a functional degree of political organisation before the arrival of Alcoa. An employee working in Alcoa's community relations team mentioned that Alcoa was surprised by the level of social organisation in the Juruti community. The presence of structured associations and cooperatives is a characteristic of Amazonian communities, although they are not always functional. At the time of the research, Juruti had more than 60 associations with different degrees of formalisation and engagement with Alcoa.

The Association of the Communities in the Juruti Velho Region, for example, is a formal organisation that has had extensive engagement with the company. The organisation represents more than 2500 families and negotiated the payment of royalties to the community for use of their land. This was the first case in Brazilian history where a mining company was legally requested to pay royalties to a community association, rather than only paying royalties to the government.

The association is recognised by local people as the main interlocutor between locals and the company. Most participants in the research acknowledged that communication with the company occurs via the association. For a member or group within the community to voice a request or a complaint through the organisation, the standard process is to inform the Coordinator or President of the community who communicates with the leadership of the association. The association then takes ownership of the matter and communicates with Alcoa, in written form. Feedback is then directed to those who raised the matter. Communication and articulation of community interests is centralised through leaders. The findings section, presented below, discusses those factors that influence relational justice, including: community understanding about mining; the scope of interaction between company and community; the role of external actors; and internal community dynamics.

Findings: Factors that influence relational justice

Insufficient understanding about mining, its impacts and areas of interest

We were surely not prepared to host such a huge mining project. We had no idea about what would happen to us and to Juruti, and; they [the company] have explained in the public hearings but still we couldn't build the picture in our minds, that the Project would be like that, big and transformative like that. (Female, 25 years old, from Juruti town)

In retrospect, the Juruti population perceived that they were not prepared to receive a mining project, and to negotiate their interests with Alcoa. There was a strong feeling among research participants that local people had an insufficient understanding about topics that are relevant for community decision-making processes. This included access to basic information about mining activities, social and environmental impacts, negotiation of compensation, and the rights and responsibilities of the company and the community.

In the context, where communities feel that they lack an understanding about the situation, relational justice is hindered because people's capabilities to develop a critical opinion about the company, the mine, and their relationship with it, is compromised. Lack of knowledge and understanding also affects the ability of communities to perform strategically when it comes to articulation and negotiation of their interests. The element of voice is also affected, because without information and knowledge to discuss issues, people in Juruti signalled that they tend to remain quiet and passive, instead of engaging in dialogue with Alcoa.

Lack of accurate, accessible and timely information is a factor that minimises relational justice. It also increases the chances of misunderstanding between the parties. In Juruti, there were cases where local people struggled to understand what company people were saying, prejudicing their ability to engage and exercise their voice in community meetings. The use of technical and legal language became a barrier for communication, and some local people described instances where people were 'puzzled' in such events. Although some people indicated that Alcoa sought to explain mining, the main impacts and potential benefits using a simplified language, these initiatives were not enough to generate sufficient knowledge for local people to critically analyse the situation and to make informed decisions about the project. In addition, lack of understanding also affects trust. Without knowledge people said they find it hard to believe what the company is saying and that their rights are being addressed.

Transitioning to a narrowed and formalised relational structure

Before there were no weekends or holidays as soon as we called, they [company employees] would come. Now it's a huge bureaucracy. We have to write to them and wait for them to schedule it to when they can do it. They just come when they want now. For us nights are better because people are back from work, so they can attend, but they just come in business hours. (Female, 32 years old, from a rural community next to the train line)

At the point when Alcoa moved from construction to operations, structural changes to the relationship began to take place. These changes related to the regularity and nature of communication, and to the perceived willingness of the company to engage with communities and to fulfil promises made in the past. Once Alcoa began to operate, the company sought to formalise its interaction with the community, and establish a more 'accountable' mode of engagement. The company's preference was for formal and written communication. In contrast, during the previous permitting and construction periods, interactional processes were more individualised, informal and verbal.

Research participants, both from community and company, also explained that during this period of transition, direct communication with a large cross-section of the community diminished. Communication became focused on engagement with collectives and representative bodies. Company employees explained that once they were in operation, the company revised and streamlined its mode of community engagement to enable better tracking and fulfilment of promises.

Local people perceived that the company also changed its approach to social investment and benefit sharing. Research participants said that during the permitting and construction period, the company seemed to be more willing to satisfy community and individual requests for goods such as water tanks, schools, medicine and others. The permitting and construction stages were also characterised by promises of future benefits as a result of mining such as economic development and jobs. Once operations began, research participants explained that these promises were not fulfilled, leading to community frustration.

These changes in the company's approach to community engagement mid-project affect the elements of justice in many ways. In terms of voice, the changes resulted in a narrowing of opportunities for people to voice concerns and aspirations, to build dialogue and mutual understanding with the company. The formalisation of communication processes also required communities to develop a new set of capabilities in order to use formal channels for communication. Levels of internal social organisation may have been high, but this did not automatically translate to a capability in representing local interests to a corporate entity. While formalisation may have functioned as a positive driver for corporate accountability, it limited the ability of some people to communicate. Further research would be needed to evaluate whether new capabilities of communities to perform within such structures emerged over time. Finally, the withdrawal of the company from regular and informal communication correlates with a reduced level of willingness to engage with communities, reducing the flow of information, trust, and finally hindering relational justice.

The pivotal role of external actors in enhancing relational justice

> *The nuns are the soul of the association. Without them, we wouldn't have won the battle to receive royalties from the company. (Male, 41 years old, from a rural community in the Juruti Velho Lake region)*

The presence and participation of external actors can often work to resolve power imbalances in company–community engagement processes (O'Faircheallaigh 2015). Communities in Juruti received assistance from a group of German nuns and a lawyer. These actors were present in the community before the arrival of the mine. Research participants said that the involvement of these actors improved internal community organisation, enabling the formation of the association of communities from the Juruti Velho Lake. They also assisted those communities during negotiations and engagement processes with the company. In fact, the presence and participation of these external actors in the Juruti Velho Lake region was perceived to have enhanced interactions between affected communities and the company in terms of improving the voice and capabilities of the affected population.

These external actors were able to gain access to information and acquire knowledge about issues relevant to the community–company relationship more easily than local people (e.g. information related to mining-related impacts, relevant legislation and models for community organising). The capabilities of these actors were shared with community members, and improved the voice of affected people in their communications with the company and their ability to use formal channels of communication more effectively. These actors improved the community's capability to articulate and negotiate their interests with Alcoa, and to adapt to the more formalised processes of engagement after the transition from construction to operations.

People from Juruti (both from the Lake region and from other regions of Juruti) perceived that the participation of the external actors resulted in greater benefits received by the communities. There is a belief that, without the participation of these actors, these communities would not have successfully negotiated the royalty agreement with Alcoa. This example confirms that, in cases where communities lack inherent capabilities, the participation of external actors with their own capabilities rectifies capability gaps, which can help to maximise relational justice.

Internal community dynamics and challenges of representation

> *Before, the leader organised community meetings all the time. Nowadays, to talk to him is a nightmare, he is never available, never interested in what we have to say. (Male, 53 years old, from a rural community in the Juruti Velho Lake region)*

Factors enabling or hindering relational justice were not solely related to the relational dynamics between company and community and how they relate to and negotiate with each other. Voice, capabilities and trust were affected by the way community organises internally, including the internal structures of communication and the decision-making process.

For example, the association in the Juruti Velho Lake changed the way it engaged its members. Research participants explained that over time, power became more centralised among the leadership. The number of community meetings were said to have decreased, and with fewer opportunities for people's participation in decision-making processes. Changes in the statute of the association transferred decision-making power to leaders, without having to consult with members. These internal changes had significant impacts on the elements of relational justice within the context of a mining operation.

As opportunities for participation in the association decreased, the voice of local people was hindered. Individuals and community groups lost opportunities to express their opinions and to collectively find the best solution for the issues that they had with the company. Capabilities were also limited due to poor internal feedback from the association's leaders about the relationship with Alcoa, limiting the awareness of locals about issues of interest (e.g. royalty payments and social investments). The relationship between the company and the community was damaged by a lack of transparency within the association, which lead to negative gossip, and claims of corruption and manipulation by leaders.

Conclusion

Relational dynamics are important for establishing a comprehensive understanding of how injustice can be minimised in the context of mineral exploration. Relational dynamics affect the ability of affected populations to exercise their voice, to access and understand information, to articulate their interests, and to make informed decisions about the mining project and their future. This case has highlighted a variety of factors that maximise or minimise the voice, capabilities and trust of affected communities. Although most of these factors are directly related to the forms in which companies and communities communicate with each other, internal organisational aspects of the communities were also found to have a strong impact on relational justice.

The Juruti case also demonstrates that the characteristics of engagement and the communication processes established by companies are important for relational justice. The way companies position themselves in the relationship, and the changes in a company's behaviour at different stages of the mine life cycle affect the elements of justice and the levels of trust and opportunities for engaging in meaningful dialogue with affected communities.

Research findings suggest that in order to maximise justice in the mining context, more attention must be given to initiatives that aim to: (a) increase the amount, quality and accessibility of information available to the affected population; (b) enhance understanding within that group; and (c) establish locally appropriate communication channels that foster community participation and engagement in the relationship. These recommendations are to be considered both in the relationship between local people and company representatives, as well as between local people and their representatives.

Although the development of the analytical framework was informed by theory, the case study contributes to a better understanding of how to explore relational justice empirically from a practical and applied perspective. Further research could explore the connections between different types of justice, including the extent to which improvements in relational justice might improve distributive and procedural justice in the context of mining company–community relationships. Additional elements of justice could also be incorporated into the framework, to build an even more comprehensive analysis of relational justice in the mining context.

References

Abdala F (2010) *Sustainable Juruti: A proposed model for mining and local development.* Paper presented at the 6th Brazilian Mining Congress (CBMINA2010), 2–3 August, Belo Horizonte, Brazil.

Ballard C, Banks G (2003) The anthropology of mining. *Annual Review of Anthropology* **32**, 287–313. doi:10.1146/annurev.anthro.32.061002.093116

Bartolini J, Biddle J, Ekiyor T, Kang E, Da Rocha Oliveira M, Serotta R, *et al.* (2010) *Alcoa Brazil's New Sustainability Model for Mining.* Columbia University, New York, USA.

Bebbington A, Bebbington DH, Hinojosa L (2009) *Extraction, Inequalities and Territories in Bolivia.* International Congress of the Latin American Studies Association, Rio de Janeiro, Brazil.

Charmaz K (2005) Grounded theory in the 21st century: applications for advancing social justice studies. In *The Sage Handbook of Qualitative Research* (Eds NK Denzin and YS Lincoln) Sage Publications, London, UK.
Colquitt JA, Conlon DE, Wesson MJ, Porter C, Ng KY (2001) Justice at the millennium: a meta-analytic review of 25 years of organizational justice research. *The Journal of Applied Psychology* **86**, 425–445. doi:10.1037/0021-9010.86.3.425
Couldry N (2010) *Why Voice Matters: Culture and Politics After Neoliberalism.* Sage Publications, London, UK.
Deutsch M, Coleman PT, Marcus EC (2011) *The Handbook of Conflict Resolution: Theory and Practice.* John Wiley & Sons, San Francisco CA, USA.
Eversole R (2012) Remaking participation: challenges for community development practice. *Community Development Journal: An International Forum* **47**, 29–41. doi:10.1093/cdj/bsq033
FGV (2011) *Projeto Indicadores de Juruti – 2a. Edição.* Fundação Getúlio, Vargas, São Paulo, Brazil.
Hamann R, Kapelus P (2004) Corporate social responsibility in mining in Southern Africa: fair accountability or just greenwash? *Development* **47**, 85–92. doi:10.1057/palgrave.development.1100056
Hollander-Blumoff R, Tyler TR (2008) Procedural justice in negotiation: procedural fairness, outcome acceptance, and integrative potential. *Law & Social Inquiry* **33**, 473–500. doi:10.1111/j.1747-4469.2008.00110.x
IBGE (2015) *The Municipality of Juruti.* The Brazilian Institute of Geography and Statistics, Brasília, Brazil. <http://cod.ibge.gov.br/58Y accessed on 04.12.15>.
IIED (2002) *Breaking New Ground: Mining, Minerals and Sustainable Development.* International Institute for Environment and Development. Earthscan, London, UK.
Kemp D, Owen JR, Gotzmann N, Bond CJ (2011) Just relations and company–community conflict in mining. *Journal of Business Ethics* **101**(1), 93–109. doi:10.1007/s10551-010-0711-y
Lewicki RJ, Barry B, Saunders DM (2007) *Essentials of Negotiation.* McGraw-Hill/Irwin, Boston MA, USA.
Liss J (2011) Negotiating the Marcellus: the role of information in building trust in extractive deals. *Negotiation Journal* **27**, 419–446. doi:10.1111/j.1571-9979.2011.00317.x
Morrice E, Colagiuri R (2013) Coal mining, social injustice and health: a universal conflict of power and priorities. *Health & Place* **19**, 74–79. doi:10.1016/j.healthplace.2012.10.006
O'Faircheallaigh C (2015) *Negotiations in the Indigenous World: Aboriginal Peoples and the Extractive Industry in Australia and Canada.* Routledge, New York, USA.
Romero H, Méndez M, Smith P (2012) Mining development and environmental injustice in the Atacama Desert of Northern Chile. *Environmental Justice* **5**(2), 70–76. doi:10.1089/env.2011.0017
Sen A (2009) *The Idea of Justice.* Allen Lane/Penguin Books, New York, USA.
Smith PD, McDonough MH (2001) Beyond public participation: fairness in natural resource decision making. *Society & Natural Resources* **14**, 239–249. doi:10.1080/089419201750111056
Whiteman G, Mamen K (2002) Examining justice and conflict between mining companies and indigenous peoples: Cerro Colorado and the Ngabe-Bugle in Panama. *Journal of Business and Management* **8**, 293–329.

7

Accounting for justice in local government responses to sea-level rise: evidence from two local councils in Victoria, Australia

Sonia Graham and Jon Barnett

Summary

This chapter explains the way justice is being considered by two local governments tasked with the responsibility of facilitating adaptation to sea-level rise. It compares the ways in which the East Gippsland Shire Council and Wellington Shire Council in Victoria have approached adaptation to sea-level rise, how they have engaged their communities on the issue, and the way staff in these two local governments think about the justice dimensions of adaptation.

Introduction

Successful adaptation to climate change requires local-level responses that account for place-specific risks and community needs (Adger and Nelson 2010; Brisley *et al.* 2012). Local governments are argued to be the most appropriate entity for delivering responsible and legitimate adaptation (Measham *et al.* 2011; Bradley *et al.* 2015), because they are well placed to evaluate the likely distribution of climate impacts (Brisley *et al.* 2012), consider the vulnerabilities and adaptive capacity of local communities, engage with local residents (Mukheibir and Ziervogel 2007) and incorporate local knowledge (Dow *et al.* 2006). Therefore it is assumed – if not well demonstrated – that local governments can develop adaptation policies that are perceived to be fair by their local communities.

The growing body of research on climate justice provides few insights into fairness considerations in local government adaptation decision making. Most of this research focuses on: mitigation rather than adaptation (Schlosberg 2012); the global rather than the local scale (Fritze and Wiseman 2009); developing rather than developed countries (e.g. Huq and Khan 2006; Barnett 2009); local community rather than local government perspectives; and distributive and procedural fairness at the expense of other types of fairness, such as interactional, spatial and temporal (Graham *et al.* 2015). While there is an emerging body of literature surrounding just adaptation (e.g. Grasso 2007; Brisley *et al.* 2012), there is much work to be done to understand how local-level decisions on adaptive responses are made (Paavola and Adger 2006).

Box 7.1: Five types of justice in climate adaptation

Graham *et al.* (2015) explored how residents of East Gippsland Shire Council and Wellington Shire Council perceived fairness in the context of adaptation to sea-level rise. Below is an overview of five types, and two sub-types, of justice that emerged from the analysis.

1. *Distributive*: the effects climate change and adaptation have on residents' everyday lives, and how these impacts vary across the community.
2. *Procedural*: the process through which decisions on adaptation are made and by whom, with a particular emphasis on the degree to which local residents are included in the process.
3. *Interactional*: how individuals are treated by government over time. This includes:
   (i) *Interpersonal*: the extent to which residents feel they are treated with dignity and respect in their interactions with government.
   (ii) *Informational*: the extent to which residents feel that governments have provided them with information about sea-level rise projections, the risks sea-level rise poses to their community and the possible adaptation options being considered.
4. *Spatial*: being given access to decision-making forums and not being required to adapt to a greater extent than other communities, especially in other jurisdictions.
5. *Temporal*: the speed of adaptation decision making, which provides timely information to residents, allows them enough time to adapt and occurs close to the impacts of sea-level rise.

There is research that seeks to understand the constraints local governments face in mainstreaming climate change into local government decision making (e.g. Mukheibir and Ziervogel 2007; Measham *et al.* 2011; MacCallum *et al.* 2014; Bradley *et al.* 2015; Pasquini *et al.* 2015). Most of these studies focus on internal constraints, such as local government decision-making processes, resources, leadership, competing priorities and informational constraints, with some recognition of constraints arising from local government interactions with higher levels of government (Measham *et al.* 2011; Pasquini *et al.* 2015). In this body of research, the focus is on the development and implementation of adaptation policies, rather than the array of fairness concerns that local governments are challenged with when engaging with their communities.

The aim of this chapter is therefore to explain the way justice is being considered by two local governments tasked with the responsibility of facilitating adaptation to sea-level rise in Australia. It builds on the work of Graham *et al.* (2015), who examined five types of justice (Box 7.1) from the perspectives of five coastal communities within these two jurisdictions. This chapter examines the way each of these five types of justice is being considered in the work of the two local governments.

Two local coastal councils in eastern Victoria

East Gippsland Shire Council (land area 2.1 million ha; population 43 413) and Wellington Shire Council (land area 1.1 million ha; population 42 319) are the second and third largest councils, by land area, in Victoria. They have similar sized populations and their largest towns, Bairnsdale and Sale, respectively, are located in non-coastal areas. The coastal settlements in both councils are destinations for tourists, and for migrants from metropolitan areas seeking a 'sea change' (McKenzie and Frieden 2010).

Together, the East Gippsland Shire Council (EGSC) and Wellington Shire Council (WSC) span the length of the East Gippsland coast, which has been identified as being one of the most vulnerable to sea-level rise in Australia (DCC 2009). At present, many of the coastal settlements, such as Manns Beach, McLoughlins Beach, Port Albert and Lakes Entrance, experience coastal flooding when there is a confluence of king tides and catchment flooding. The frequency and duration of such flood events are predicted to increase under climate change and associated sea-level rise (GCB 2008), providing an impetus for the two councils to develop adaptation plans.

There are two state government planning decisions that provide a policy imperative for the two local governments to implement sea-level rise adaptation plans. First, the 2008 Victorian Coastal Strategy required authorities to plan for no less than 80 cm of sea-level rise by 2100 (for a detailed explanation of the Victorian coastal planning process see Macintosh (2013) and Hurlimann *et al.* (2014)). This requirement has been maintained in the more recent 2014 Victorian Coastal Strategy. Second, both councils have been at the forefront of controversial planning decisions in the state of Victoria. Specifically, the Victorian Civil and Administrative Tribunal (VCAT) did not grant permits for particular developments in East Gippsland (*Taip v East Gippsland SC [2010] VCAT 1222 (28 July 2010)*) or Wellington (*Ronchi v Wellington Shire Council [2009] VCAT 1206 (16 July 2009)*) based on the sites' vulnerability to sea-level rise and coastal flooding. In the Lakes Entrance case, VCAT raised concerns regarding whether the development constituted '*intergenerational equity*, sustainability, *fair* and socially responsible development' (italics added). These decisions have applied pressure on both councils to plan for sea-level rise to a degree not seen in other parts of Victoria or Australia.

Exploring local government perspectives on fair adaptation to sea-level rise

In September 2010, both councils became involved in an ARC Linkage project with the University of Melbourne, seeking to develop an approach for achieving equitable local outcomes for adaptation to sea-level rise. Other project partners were the Gippsland Coastal Board and the Department of Sustainability and Environment (now the Department of Environment, Land, Water and Planning) and the Department of Planning and Community Development (now the Department of Economic Development, Jobs, Transport and Resources). The project ended in June 2014.

As part of the project, seven members of staff across both councils participated in:

- six steering committee meetings
- five formal and two informal semi-structured interviews that asked them about their experiences planning for adaptation to sea-level rise and their views on vulnerable groups within their communities
- a day-long adaptation options workshop that sought to understand potential future adaptation strategies for the two councils (a detailed explanation of the full suite of methods used is provided in Barnett *et al.* 2014).

The formal interviews (EGSC1, EGSC2, WSC1, WSC2, WSC3) and workshop (EGSC3, WSC4) were audio recorded and transcribed. Thematic analysis was used to explore how the concepts of distributive, procedural, interactional, spatial and temporal fairness emerged in the councils' sea-level rise adaptation deliberations.

The project also provided opportunities for the researchers to attend and participate in internal local government forums, which enabled observation of adaptation planning in action. Meetings attended by the authors included the Lakes Entrance Inundation Steering Committee (December 2010), Regional Councils Victoria Annual Summit (May 2013), Municipal Association of Victoria's Coastal Adaptation Forum (November 2013) and a meeting of the Board of the East Gippsland Catchment Management Authority and the Mayor and Councillors of East Gippsland Shire Council (April 2014).

The results presented in this chapter are primarily based on the formal interviews and adaptation options workshop, but also take into account the additional information obtained from the working relationships established with these two councils throughout the duration of the project. The ARC project focused on sea-level rise adaptation in one coastal community in EGSC – Lakes Entrance – and four small communities in WSC – Seaspray, Port Albert, Manns Beach and McLoughlins Beach. Therefore much of the discussion with the local governments focused on these communities.

Six dimensions of fairness in sea-level rise adaptation in Gippsland East

Local government staff were most concerned about the fair distribution of outcomes and sharing of information in planning for sea-level rise. Staff were not only concerned about achieving distributive and informational fairness for communities, they also expected the state government to treat their local councils fairly. Procedural and interpersonal concerns were also raised, and these reflected broader concerns about community engagement. Some limited concerns about achieving fairness across space and time were also raised. It is important to note that the informational and interpersonal dimensions of fairness were sufficiently different that, in this analysis, they are treated as two separate dimensions of fairness, rather than as two subsets of interactional fairness, as they have previously been treated in the literature (Usmani and Jamal 2013; Graham *et al.* 2015). In drawing this distinction, we recognise that there is considerable overlap across the six dimensions of fairness, reflecting the fluidness of fairness concerns in adaptation decision-making practice.

Distributive fairness

Protecting land values and property prices was a key distributive concern for both councils; they saw it as their responsibility to protect residents' 'investment into their little piece of Australia' (WSC1). Thus councils were keen to ensure that their climate adaptation plans had minimal impacts on the value of properties within their jurisdictions.

Although staff were genuinely concerned about the impacts of changes in property prices for their residents, they also discussed the impacts of adaptation planning for their own organisations' incomes. Thus distributive fairness concerns were directed both within and beyond the council:

> *[Places] affected by coastal inundation ... if we were to treat them all the same and say righto by 2015 you have to be off the coast, potentially a lot of those people might relocate outside the Wellington Shire ... to lose 10, 15, 20% of those people through relocation, significantly reduces our bottom line. (WSC1)*

Staff from both local governments were keen to explain that, although they do their best to protect property prices in adaptation planning, some key planning and policy decisions are made at the state level and thus are beyond their control:

If you had a house on the Esplanade... with a stroke of the pen they've [VCAT] virtually made it worthless or at least de-valued it considerably. Depending how we go with the work on the committees ... will determine the value again. (EGSC1)

[T]his State planning policy that says you've got to plan for sea-level rise trumps it all as being the be-all and end-all things is going to influence the outcomes for your town. (EGSC3)

Beyond property, staff were concerned about 'people's day-to-day lives' (WSC1) and how they could enable fair outcomes for 'people's jobs, creating the liveability and wealth factors that lead to the life people lead' (WSC2), recognising that there is considerable heterogeneity in residents' values:

Deciding where and what we're protecting and keeping people adapting is going to be one of the hardest set of decisions... it's about the sense of community probably more than the built environment... how do you maintain that sense of community? (EGSC3)

Up until now we've been worrying about the buildings and the infrastructure... I just think the people component and looking at the impact on people and trying to get an equitable outcome is part of what we've got to aim to do... this whole social science aspect to the whole adaptation process is really important in terms of understanding the people involved and [that] they respond differently to different things. (EGSC2)

The long timeframes associated with adaptation planning makes achieving distributive fairness challenging because of the uncertainty about the speed at which environmental changes will unfold and what the nature of impacts will be. To achieve fairness in the face of such uncertainty, the local governments suggested scoping what community wants, rather than simply being dedicated to the principle of distributive fairness:

The concept of tolerance to more frequent events needs to be tested ... if we don't know exactly what it means, what do the community think they can live with? (EGSC3)

[W]hat is the preferred outcome of this? What do we want people to be doing? Or what do we want to be happening in our communities in the various futures? And work backwards from there to a point where we can actually put – and what we do may be completely different to what the policy response might have been. (WSC4)

In these last two quotes, distributive fairness blurs into concerns about procedural fairness.

Informational fairness

Staff of both local governments placed importance on having the best available information on which to make adaptation decisions that are distributively fair:

If people are going to be pointing the finger at us about getting it right then I think we need to be sure ... that we've got it right, that we're making good decisions ... there's some detailed information gaps that would be filled over time but at the moment we just need to demonstrate that we've used the best information that we've got ... we've just got to make the best decisions that we can now using the information that we can and document what it is that we're doing. (EGSC2)

Although staff aspired to use the best available information to inform their decisions, they had two key concerns about accessing and using such information. First, staff of both local governments spent considerable time talking about the sensitivity of information regarding climate change. Their primary concern was about providing information to communities in a timely fashion without causing unnecessary concern or unintended or unfair outcomes, such as increased insurance premiums, decreased property prices or increased home renovation costs:

> *[W]e have to be very careful and conscious about when this information's released, how else it will be used? ... So it's not just good information for future planning, it has an impact now ... we're actually trying to deal with the impact of this [climate change] information in our communities right now. (EGSC2)*

> *Right now we are hamstrung in terms of what we can tell someone. An instruction or advice or assistance we can give people who want to develop, who want to do the things that realise their dreams and aspirations. (WSC2)*

Second, staff expressed frustration in gaining access to information about the potential impacts of climate change that is available to the state government:

> *There is one level of government that is working with a particular set of information and a particular set of knowledge that another level of government is not privy to and that level of government is at the front line of the adaptation discussion with the local community and therefore we are kind of having this argument with both our hands tied behind our back and hopping on one leg at the same time ... it is an area that is preventing that really clear dialogue and discussion. (WSC2)*

It is clear from all of these quotes that staff saw informational fairness as being essential for achieving procedural and interactional fairness; staff see transparency and prompt sharing of information with communities as the foundation for building working relationships with their communities, yet they felt that their ability to do this was constrained by the state government.

Procedural fairness

One of the biggest challenges both local governments faced in engaging with their communities about sea-level rise came from the divergent views that members of their communities have about the dynamic nature of the coastal environment and whether climate change will cause significant impacts:

> *[T]here's a lot of people in a lot of denial around sea-level rise right now. If you scratch any of the communities, Lakes Entrance, Port Albert, there'd be a lot of people saying it's not going to happen ... In terms of getting an effective adaptation outcome when you've got a community that actually doesn't believe or doesn't understand the dynamics of the coast environment. (WSC4)*

> *There's a lot of confusion at community level around climate change and climate change adaptation. (WSC1)*

> *I think for our purpose it's about making sure that we can retain that stuff for the community, whether they are thinking about it [climate change] or not. (EGSC3)*

Denial about climate change within the communities makes it difficult for local governments to engage a broad range of interests in meaningful discussions about the future that, as the above quotes indicate, also has repercussions for distributive fairness.

Interactional fairness

Staff were keenly aware that their ability to engage with communities about a sensitive issue such as sea-level rise was hampered by the awkward relationship that exists between local governments and their communities, because staff live and work in these communities:

> *It's almost a sport to target local government as ineffective and slow ... There is that level of cynicism and almost resentment about having to not only pay a certain amount of money to this local agency, that is then charged to tell you what you can and can't do ... it muddies the water when you come to putting in place some sensible contingencies to assist people to make choices around where and how they will develop their land, build their house ... because we are such a close level of government to the people, it's that thing about well you're not the boss of me ... you're just my next door neighbour or the people down the street, who are you to tell me what I can and can't do? (WSC2)*
>
> *[T]he community is pre-disposed to have some degree of contempt for local government. (WSC4)*

These challenges to the local governments' authority make it difficult to establish trust and respect as well as achieve widespread participation and representativeness.

Temporal fairness

To some extent, concerns about temporal fairness were closely associated with concerns about informational and interactional fairness. Staff were keenly aware that information needs to be provided in a timely manner and that the community often perceives the local government to take too long to make decisions, which compromises the amount of trust and authority that the community invests in them. However, staff were also aware that if they proceed too quickly then meaningful community engagement would be compromised, and there may be greater resistance to proposed changes:

> *If you've got the time to transition that's fine but if the impact of decision-making is that the impact is now then it becomes a much harder sort of exercise because we are going to try and rush to fill the gap now ... you need time to bring the community along, to actually understand the impacts ... I'm a bit worried at the moment that we're really going to rush that because everybody's anxious to get the answer. (EGSC2)*

The other dimension of temporal fairness raised by staff pertained to the achievement of fair outcomes in a changing physical and social environment: outcomes that are considered fair today may not be considered fair in the future as the environment and community values change. For staff, this meant that adaptation plans not only need to be informed by current shared values, but that adaptation plans need to be revisited over time to ensure that they continue to deliver fair outcomes:

> *It is an environment that is in transition ... we proceed on that basis that it is a transitioning environment, and accept it for what it is ... People's values change ... and the State Government put a lot of money into coming up with a solution. They've come up with a solution, they've come up with an approach that is far more effective [and] they're still in the bad books. (WSC2)*

Overall, concerns about time relate to both procedural and distributive fairness.

Spatial fairness

Some concerns about spatial fairness were identified in local government staff accounts of adapting to sea-level rise. Staff were keenly aware of the diverse needs of their residents, both within and across their jurisdictions. Thus sea-level adaptation planning cannot be a one-size-fits-all policy. It requires catering to the specific needs of communities:

> *It'll be hard to get policy that suits lots of areas, unless it's well thought out and fluid enough to be delivered at a local level. (WSC1)*

Staff were also concerned about their ability to deliver adaptation solutions that do not exacerbate existing vulnerabilities or that are vastly different to adaptation solutions that are implemented in other parts of the state:

> *Given the lower, poorer access to community infrastructure in the rural areas it will continue to be a challenge to the state and federal governments, how they can deliver equitable solutions that will not be seen as an abuse of the coastal community ... because they're [local communities] going to suffer financial loss if, when they try to sell their block, they can't get the price they want ... the city [coastal] councils there [in Melbourne] would have a little bit more money to develop asset solutions but we don't have the same level of privilege ... if the Federal government expects us to provide the same level of protection or even to deliver equitable solutions they need to understand that we need help ... there is a difference between different regions and if government is continuing to do anything else they need to address that disparity. (WSC3)*

Developing adaptation plans that are locally relevant and do not disadvantage regional coastal communities requires time and genuine community engagement, as well as support from higher levels of government; thus spatial fairness is intimately intertwined with the other dimensions of fairness.

Fairness is in the eye of the beholder: comparing interpretations of fairness from above and below

The emphasis staff placed on informational, temporal and spatial dimensions of fairness in adaptation planning is largely consistent with concerns expressed by their local communities (reported in Graham *et al.* 2015). In the discussion that follows, consideration is given to the extent to which local government staff perspectives of fairness were consistent with community expectations and existing conceptualisations in the justice literatures. From this, we highlight how the informational, temporal and procedural dimensions of fairness require further deliberation by local governments to ensure they meet community expectations.

Local communities' articulated concerns about distributive fairness were largely focused on costs and compensation; they were concerned that adaptation policies were placing an unfair burden on local residents through increased construction costs and reduced expenditure on maintaining key infrastructure (Graham *et al.* 2015). While both local governments were concerned about the economic impacts of sea-level rise and adaptation on their communities, they were also concerned with non-material social impacts. To this end they were keen to understand the range of lived values – valuations that individuals make about what is important in their lives and the places in which they live (Graham *et al.* 2013) – of their residents, how these differ within and across communities, and how these values can be accommodated by adaptation policies. This reflects recent turns in climate justice more broadly that focus on how climate change makes human lives more vulnerable (Schlosberg 2012) and affects the social cohesion and functioning of communities (Schlosberg 2013). Such concern with non-material values is rarely mentioned in research on local governments' adaptation planning processes. Instead, local governments are often characterised as prioritising rates, roads and rubbish (Measham *et al.* 2011). This suggests the need for future research to broaden what is considered within the realm of local governments' climate adaptation planning and to go beyond material concerns regarding changes to property and infrastructure.

Differences between local community and local government perspectives on informational fairness were the most striking. For communities, informational fairness was about receiving information about adaptation planning while it was underway – they wanted to be told what is happening and how decisions are changing, rather than finding out about decisions once they have been made (Graham *et al.* 2015). Thus, for residents, informational fairness was closely associated with procedural fairness concerns. However, both local governments were reluctant to include communities in their adaptation planning deliberations. Such reluctance to create a dialogue with communities about climate change planning is evident elsewhere in Australia (MacCallum *et al.* 2014) and abroad (Marino and Ribot 2012). Clearly, this is a key area where local governments could achieve greater informational fairness in the eyes of the community.

Both local governments were concerned with informational fairness but their focus was on demonstrating that their decisions were made on the best available information and that care was taken in sharing information pertaining to climate predictions. While local communities were concerned about the stigma attached to their communities as a result of climate predictions, this informational concern was not attributed to local governments, but to the media (Fincher *et al.* 2015). Considerable attention has been given to understanding the challenges local governments face in accessing local-level information on the vulnerability of their municipalities to climate images (Measham *et al.* 2011) and how they can source such information through modelling and hazard mapping (Mukheibir and Ziervogel 2007). Given that this does not align with the informational fairness concerns of local communities, future research may benefit from not only considering informational limitations facing local governments, but also informational fairness challenges.

With regard to procedural fairness, local communities were concerned about how adaptation decisions were being made and by whom. They believed that they should be given the opportunity to participate and have their local knowledge incorporated into climate decisions (Graham *et al.* 2015). Despite this expressed willingness on the part of residents to participate, councils indicated that they found it challenging to talk to communities about adaptation because a large proportion of the community does not believe climate change is happening. Our own research indicated that many local residents questioned the validity of sea-level rise predictions (Barnett *et al.* 2014a). To overcome this barrier to engagement, we have proposed elsewhere (see Barnett *et al.* 2014b) that a staged approach to adaptation is required, where temporary plans are put in place until such time as a social trigger occurs in which communities can see that changes need to be made and will agree to initiating the next plan and continuing to prepare for the future.

Local governments and their communities agreed that more could be done to achieve a working relationship between them. Local government perceptions of the lack of respect that their communities hold for them was confirmed by local residents' accounts of their councils. Local residents believed that they were treated with 'contempt' by local government and that an 'authentic' relationship was lacking (Graham *et al.* 2015). Although local governments explained that this unworkable relationship existed because they were too institutionally close to the residents, the residents believed that local governments created too much distance between them; despite living in the region, staff do not really understand the nuances between the different communities and what they need. This community perspective fits with broader concerns in the climate justice literature regarding the lack of recognition of a link between environmental conditions and everyday life (Schlosberg 2013). Local governments need to find a way of demonstrating that communities are being heard, not ignored, are recognised (Schlosberg 2012), and are treated with dignity and respect (Chapter 3).

Perceptions of temporal fairness differed significantly between communities and their local governments. Communities were focused on making sure the timing of decisions reflects the rate of environmental change, so that they are not forced to bear the burden of adapting to future climate change impacts now (Graham *et al.* 2015) that is, they were concerned about intergenerational equity. Staff, on the other hand, were concerned about the pace of decision making to ensure that they maintained legitimacy with their communities and so that communities would appreciate the reason why decisions were being made – that is, that they made proportional decisions and did not overreact (Driessen and van Rijswick 2011). Local governments were also concerned about changing community values and the effect this would have on long-term planning. Communities, on the other hand, tended to emphasise how little things changed over time where they lived (Fincher *et al.* 2014). Communities' desire for stability and reluctance to acknowledge the need to begin adapting to sea-level rise now poses a significant challenge to local governments who wish to be seen to be making decisions that are temporally fair in the eyes of their communities. However, we believe the staged approach to adaptation (Barnett *et al.* 2014b) will not only be beneficial for achieving procedural fairness, but temporal fairness as well.

The scale of spatial fairness concerns differed slightly for communities and local governments. Although communities were primarily concerned with not being treated differently from communities beyond their region, such as Melbourne (Graham *et al.* 2015), local governments were also concerned about fairness within their jurisdictions, recognising that what may work for one place may be maladaptive for another, which is consistent with concerns raised in the broader adaptation literature (e.g. Barnett and O'Neill 2010). To some extent, local communities perceived that the local governments treated them similarly to other coastal towns within each municipality, which they perceived to be fair. Thus if local governments implement appreciably different adaptation policies across their jurisdictions they will need to carefully justify how they are enabling each place to meet the values of each community – that just adaptation is achieved by being context specific (Brisley *et al.* 2012).

Local communities expressed concerns about being excluded from decision-making spaces. Although this could be considered to be a spatial fairness issue, it is also closely related to procedural, interpersonal and informational fairness. In the climate justice literature, consideration is given to the negative consequences of non-recognition (Schlosberg 2012) that arises from being excluded from decision making (Schlosberg 2013). Thomas and Twyman (2005) discuss the need for 'head room' – that is, the right kind of space to enable fair adaptation – if communities are involved in decision-making spaces then they are more likely to have the information required to make autonomous decisions (Brisley *et al.* 2012). Thus, local governments need to consider the ways in which communities are provided opportunities to genuinely engage in the development and implementation of adaptation plans, so that they not only perceive the process to be fair but can also create fair outcomes for themselves.

Conclusion

Local government staff perspectives on what constitutes fairness in adaptation planning were largely consistent with what is expected of them from their communities when it comes to developing working relationships (interpersonal fairness) to deliver fair outcomes (distributive fairness) across their jurisdictions (spatial fairness). Yet, despite considerable agreement about these three of the dimensions of fairness, our analysis points to some mismatch between the theory and practice of informational, temporal and procedural fairness in adaptation planning. In Gippsland East, the communities attached greater significance to the prompt sharing of information than their local governments. This not only had ramifications for informational fairness but also procedural and temporal fairness. Clearly such misalignment needs to be rectified if adaptation to sea-level rise is to be accepted, or at best embraced, by local communities.

Of the six dimensions of fairness considered here, informational fairness is rarely considered in the climate, or even environmental, justice literature, yet it was of key importance to the two councils studied here. Given the extent of climate change denial in regional Australia, and the significant impacts that climate predictions and adaptation policy can have for current generations, we argue that informational fairness merits further consideration and theorisation.

Acknowledgements

The research outlined in this chapter was funded through Australian Research Council Linkage Grant LP100100586, with support from the East Gippsland Shire Council, the Gippsland Coastal Board, the Victorian Department of Planning and Community Development, the Victorian Department of Sustainability and Environment and Wellington Shire Council. We acknowledge also the work of colleagues Professor Ruth Fincher, Dr Anna Hurlimann, Dr Colette Mortreux and Dr Nick Osbaldiston on the broader project of which this chapter is part.

References

Adger WN, Nelson DR (2010) Fair decision making in a new climate of risk. In *Climate Change, Ethics and Human Security.* (Eds KL O'Brien, AL St. Clair and B Kristoffersen) pp. 83–94. Cambridge University Press, Cambridge, UK.

Barnett J (2009) Justice and adaptation to climate change. In *Climate Change and Social Justice.* (Ed J Moss) pp. 131–143. Melbourne University Press, Melbourne.
Barnett J, O'Neill S (2010) Maladaptation. *Global Environmental Change* **20**, 211–213. doi:10.1016/j.gloenvcha.2009.11.004
Barnett J, Graham S, Mortreux C, Fincher R, Waters E, Hurlimann A (2014a) A local coastal adaptation pathway. *Nature Climate Change* **4**, 1103–1108. doi:10.1038/nclimate2383
Barnett J, Fincher R, Hurlimann A, Graham S, Mortreux C (2014b) 'Equitable local outcomes in adaptation to sea-level rise: final project report'. The University of Melbourne, Melbourne.
Bradley M, van Putten I, Sheaves M (2015) The pace and progress of adaptation: marine climate change preparedness in Australia's coastal communities. *Marine Policy* **53**, 13–20. doi:10.1016/j.marpol.2014.11.004
Brisley R, Welstead J, Hindle R, Paavola J (2012) *Socially Just Adaptation to Climate Change.* Joseph Roundtree Foundation, York, UK.
DCC (2009) *Climate Change Risks to Australia's Coast.* Department of Climate Change, Canberra.
Dow K, Kasperson RE, Bohn M (2006) Exploring the social justice implications of adaptation and vulnerability. In *Fairness in Adaptation to Climate Change.* (Eds WN Adger, J Paavola, S Huq and MJ Mace) pp. 79–96. The MIT Press, Cambridge MA, USA.
Driessen PP, van Rijswick HF (2011) Normative aspects of climate adaptation policies. *Climate Law* **2**, 559–581.
Fincher R, Barnett J, Graham S, Hurlimann A (2014) Time stories: making sense of futures in anticipation of sea-level rise. *Geoforum* **56**, 201–210. doi:10.1016/j.geoforum.2014.07.010
Fincher R, Barnett J, Graham S (2015) Temporalities in adaptation to sea-level rise. *Annals of the Association of American Geographers* **105**, 263–273. doi:10.1080/00045608.2014.988101
Fritze J, Wiseman J (2009) Climate justice: key debates, goals and strategies. In *Climate Change and Social Justice.* (Ed J Moss) pp. 187–211. Melbourne University Press, Melbourne.
GCB (2008) 'Climate change, sea level rise and coastal subsidence along the Gippsland Coast: implications for geomorphological features'. Gippsland Coastal Board, Bairnsdale.
Graham S, Barnett J, Fincher R, Hurlimann A, Mortreux C, Waters E (2013) The social values at risk from sea-level rise. *Environmental Impact Assessment Review* **41**, 45–52. doi:10.1016/j.eiar.2013.02.002
Graham S, Barnett J, Fincher R, Mortreux C, Hurlimann A (2015) Towards fair local outcomes in adaptation to sea-level rise. *Climatic Change* **130**, 411–424. doi:10.1007/s10584-014-1171-7
Grasso M (2007) A normative ethical framework in climate change. *Climatic Change* **81**, 223–246. doi:10.1007/s10584-006-9158-7
Huq S, Khan MR (2006) Equity in national adaptation programs of action (NAPAs): the case of Bangladesh. In *Fairness in Adaptation to Climate Change.* (Eds WN Adger, J Paavola, S Huq and MJ Mace) pp. 181–200. The MIT Press, Cambridge MA, USA.
Hurlimann A, Barnett J, Fincher R, Osbaldiston N, Mortreux C, Graham S (2014) Urban planning and sustainable adaptation to sea level rise. *Landscape and Urban Planning* **126**, 84–93. doi:10.1016/j.landurbplan.2013.12.013
MacCallum D, Byrne J, Steele W (2014) Whither justice? An analysis of local climate change responses from South East Queensland, Australia. *Environment and Planning. C, Government & Policy* **32**, 70–92. doi:10.1068/c11295
Macintosh A (2013) Coastal climate hazards and urban planning: how planning responses can lead to maladaptation. *Mitigation and Adaptation Strategies for Global Change* **18**, 1035–1055. doi:10.1007/s11027-012-9406-2

Marino E, Ribot J (2012) Special issue introduction: adding insult to injury: climate change and the inequities of climate intervention. *Global Environmental Change* **22**, 323–328. doi:10.1016/j.gloenvcha.2012.03.001

McKenzie F, Frieden J (2010) *Regional Victoria: Trends and Prospects*. Department of Planning and Community Development, Melbourne.

Measham TG, Preston BL, Smith TF, Brooke C, Gorddard R, Withycombe G, *et al.* (2011) Adapting to climate change through local municipal planning: barriers and challenges. *Mitigation and Adaptation Strategies for Global Change* **16**, 889–909. doi:10.1007/s11027-011-9301-2

Mukheibir P, Ziervogel G (2007) Developing a municipal adaptation plan (MAP) for climate change: the city of Cape Town. *Environment and Urbanization* **19**, 143–158. doi:10.1177/0956247807076912

Paavola J, Adger N (2006) Fair adaptation to climate change. *Ecological Economics* **56**, 594–609. doi:10.1016/j.ecolecon.2005.03.015

Pasquini L, Ziervogel G, Cowling RM, Shearing C (2015) What enables local governments to mainstream climate change adaptation? Lessons learned from two municipal case studies in the Western Cape, South Africa. *Climate and Development* **7**, 60–70. doi:10.1080/17565529.2014.886994

Schlosberg D (2012) Climate justice and capabilities: a framework for adaptation policy. *Ethics & International Affairs* **26**, 445–461. doi:10.1017/S0892679412000615

Schlosberg D (2013) Theorising environmental justice: the expanding sphere of a discourse. *Environmental Politics* **22**, 37–55. doi:10.1080/09644016.2013.755387

Thomas DSG, Twyman C (2005) Equity and justice in climate change adaptation amongst natural-resource-dependent societies. *Global Environmental Change* **15**, 115–124. doi:10.1016/j.gloenvcha.2004.10.001

Usmani S, Jamal S (2013) Impact of distributive justice, procedural justice, interactional justice, temporal justice, spatial justice on job satisfaction of banking employees. *Review of Integrative Business and Economics Research* **2**, 351–383.

8

Exploring the justice in forestry negotiations: trading justice for politics

Lain Dare and Jacki Schirmer

Summary

In this chapter we evaluate the 'Tasmanian forest peace process' from the perspective of procedural justice, focusing on the process rather than on its outcomes (distributive justice). This allows for an exploration of how the negotiation process influenced perceptions of the justice process, providing insights to the trade-offs involved between procedural justice and achieving desired outcomes. Following a brief overview of the study methods, we present key principles of procedural justice using the Social Justice Framework (Lukasiewicz and Baldwin 2014), and identify nine normative indicators of procedural justice that guide our subsequent analysis. We then discuss lessons from the Tasmanian forest peace process for justice, focusing on the challenge of satisfying multiple procedural justice objectives within a highly contested political environment.

Introduction

Environmental justice is an important political challenge, with impacts that extend beyond the environment to affect social and economic interests. Environmental justice proponents call for a greater voice for the environment (see Lukasiewicz *et al.* 2013; Neal *et al.* 2014), and many significant policy reforms have been implemented in response to such calls, including the Tasmanian Forest Agreement. Contention about the impact of forest harvesting on environmental forest values drove decades of conflict, despite several attempts at resolution (see Webb *et al.* 2008; Schirmer 2013; Gale 2013). A new resolution effort, the 'Tasmanian forest peace process' (TFPP), was initiated in 2009 and culminated with the 2012 Tasmanian Forest Agreement (TFA). The TFA introduced significant policy changes, including new conservation reserves and a reduced timber harvest volume, but critics felt it lacked democratic legitimacy and had poor social justice outcomes (Schirmer *et al.* 2016). As a consequence the TFA was short lived, repealed in August 2014 after being legislated in 2013, although many of its provisions remain in place. Despite the repeal, the peace process and resulting TFA represent an important contemporary political process in Australian environmental governance from which many justice lessons can be learnt.

Methods

The idea of justice is a contentious concept that is based on normative ideals set by an individual's beliefs and the prevailing context (see Clayton and Opotow 2003; Skitka 2003). As such, any exploration of justice needs to allow space for such preferences to be exposed and hence provide a baseline of what [in]justice outcomes entail. In this study, we provided such space using a qualitative research methodology. Research was undertaken via 16 semi-structured interviews conducted from December 2012 to December 2013 with key actors directly involved in the TFPP, together with a small number of close observers of the process. This included representatives of environmental, forest industry and community organisations directly engaged in the negotiations, actors who aimed to facilitate or broker aspects of the process, and representatives of organisations who felt excluded from the process. Government stakeholders were involved principally as observers, and largely declined to be interviewed, with government views instead drawn from public documentation and government records. Interviews lasted from 45 min to 3 h, with interviewees asked to: (i) describe their involvement in, or observation of, negotiations; (ii) identify what worked well and poorly during the negotiations; (iii) reflect on what would represent a 'successful' process outcome; and (iv) discuss what factors most influenced the likelihood of achieving this outcome. Ethical approval was obtained from the University of Canberra Human Research Ethics Committee (HREC 13–156). For the purposes of anonymity, we do not name interviewees but instead indicate whether they were representatives of environmental non-government organisations (ENGOs), forest industry representatives (industry), or other interests.

Interview data were analysed using an adaptive theory approach (Layder 2006) where pre-existing theory informs initial analysis, with subsequent analysis contributing to a process in which the theory is adapted based on the empirical findings. Using core principles of the Social Justice Framework, data were analysed to seek evidence of the presence of each principle within the TFPP, and to identify what factors aided or hindered achieving each justice principle. By grounding the study in justice theory and testing justice principles based on the self-reported empirical experiences of the actors involved, a well-established theoretical framework could both inform and be informed by on-ground experience. This ensured a critical approach that drew on justice theory while allowing the nuances of the case and the individual actors to have voice. Analysis took the form of thematic coding with the Social Justice Framework guiding codes, with new and emerging concepts additionally coded where they were present. This was followed by reorganising codes into thematic networks that identified relationships between themes. This approach deliberately blended theory and practice, and was well suited to the exploration of justice, which itself is a theoretical concept that is applied to specific situations (Clayton and Opotow 2003; Skitka 2003).

Assessing procedural justice

Justice and fairness are important considerations in decision making (Syme and Nancarrow 2005), and are particularly critical when making decisions regarding contested resources such as water, energy and forests. Based on initial underlying premises of equity and fairness, justice scholars explore the impact of policy and its implementation on stakeholders. Integral to many of these studies is an understanding of the different dimensions of justice, predominantly described as distributive, procedural and interactive (or interactional) justice (see Cropanzano *et al.* 2002). Distributive justice refers to how resources are distributed within a group or across society, procedural justice is concerned with the processes by which decisions are made, and interactive justice relates to how stakeholders are treated by decision makers. Interactive justice is highly related to procedural justice and some researchers do not distinguish between these two dimensions (Neal *et al.* 2014).

There are blurred boundaries between each of the justice dimensions, with interactions and dependencies stretching between and across dimensions (see Brockner and Wiesenfeld 1996). Although we focus on procedural aspects of the TFPP, elements of other justice dimensions are also incorporated within the analysis where necessary. Schirmer *et al.* (2016) also provide some perspectives on stakeholder views of the interactive and distributive justice of the process.

Our analysis drew on the guiding justice principles provided in the Social Justice Framework developed by Lukasiewicz and colleagues (see Lukasiewicz and Baldwin 2014). Developed from a comprehensive literature review, the framework is an evolving conceptual model useful for justice-based analysis of decision-making processes. In the framework, procedural justice refers to the structure of decision-making processes. Following Paavola's (2007) model, this structure is divided into three distinct elements: (i) representation, (ii) power within decision making, and (iii) process rules, each of which have a series of guiding principles (see Lukasiewicz and Baldwin 2014). The procedural justice principles selected for this analysis were representativeness and the related principle of accessibility, level of power, transparency, accuracy, neutrality, ethics, timeliness, and accountability, each of which is described as it is analysed with regard to the TFPP. While Lukasiewicz and Baldwin (2014) also describe other principles within the procedural justice dimension, two other principles (consistency and correctability of errors) were less directly relevant due to the short-lived nature of the agreement.

The Tasmanian forest peace process

Following decades of conflict about the harvesting of native forests in Tasmania, a new opportunity for negotiating peace in Tasmania's forests arose in 2009 due to a combination of factors including: the rapid economic decline of the industry due to poor market conditions; public debate surrounding Gunns Limited's proposed pulp mill; and Gunns Limited's decision to exit native forest harvesting which created opportunities for new forest reserves (Schirmer *et al.* 2011; Schirmer *et al.* 2016). These conditions resulted in incentives to negotiate with the industry needing to achieve both a certainty of wood supply and publicly acknowledged sustainable forest management practices. Environmental groups had an opportunity to increase the area of forest conversation reserves, and organisations representing workers and communities sought stability in the industry to stem decline in employment (see Schirmer *et al.* 2016). The peace negotiations that emerged to capitalise on this opportunity had five phases of incremental agreements and negotiations (see Lester and Hutchins 2012; Dare 2013; Schirmer *et al.* 2016), and a later sixth phase in which the TFA was repealed, as outlined in Table 8.1.

Phase 1 lasted a year, and included establishment of a core negotiating group that included representatives of the forest industry, environmental groups, community-based groups and unions.[1] Given the challenges of achieving consensus after decades of conflict, initial discussions sought to agree to principles to guide further negotiations, with the

1 Tasmanian forest peace process participants included: Forest Industries Association Tasmania (FIAT); the National Association of Forest Industries (NAFI), which later became the Australian Forest Products Association (AFPA); the Australian Forest Contractors' Association (AFCA); the Tasmanian Forest Contractors' Association (TFCA); Timber Communities Australia (TCA); the Tasmanian Country Sawmillers' Federation (TCSF); the Australian Conservation Foundation (ACF); The Wilderness Society (TWS); Environment Tasmania Inc (ET); and the Construction, Forestry, Mining and Energy Union (CFMEU).

Table 8.1. The phases of the Tasmanian forest peace process (Source: adapted from Schirmer *et al.* 2016)

Phase	Date	Description
1. Tasmanian Forests Statement of Principles	Late 2009 to October 2010	Process stakeholders met to negotiate towards a peace agreement, with a focus on identifying shared principles. The 'Statement of Principles' was presented to the Australian and Tasmanian Governments on 18 October 2010.
2. Tasmanian Forests Intergovernmental Agreement (IGA)	August 2011	Independent facilitator appointed to assess feasibility of developing an implementable agreement. This resulted in governments agreeing to support ongoing negotiations, with the 'Tasmanian Forests Intergovernmental Agreement' (IGA), signed in August 2011, providing support for continued negotiations.
3. Independent Verification Group (IVG)	November 2011 to March 2012	In response to a lack of agreement on environmental values, wood supply volumes and socio-economic impacts, independent experts appointed to assess environmental, economic and social implications of changing forest harvest volumes and reserve areas.
4. Tasmanian Forests Agreement 2012 (TFA)	November 2012	Negotiations culminated in the 'Tasmanian Forests Agreement 2012' on 22 November 2012 and a draft 'Tasmanian Forests Agreement Bill 2012'. The agreement included expanding areas of forest reserves, reducing the volume of timber harvested, and providing support for transition of forest industry businesses and workers.
5. Legislative Council Select Committee on the Tasmanian Forests Agreement Bill 2012	December 2012 to March 2013	Following public backlash, the Legislative Council referred the Bill to a Select Committee. Following an inquiry by the Committee, including public submissions and hearings, substantial modifications were made. The amended Bill passed on 30 April 2013.
6. Legislation of TFA and subsequent Repeal	March 2013 to August 2014	In election campaigns, both state and federal Liberal parties promised to repeal the parts of the TFA their governments were responsible for. Subsequently, the newly elected Liberal state governments repealed the TFA in August 2014, while the federal government unsuccessfully attempted to have a World Heritage Area declaration, made as part of the TFA, repealed.

'Tasmanian Forests Statement of Principles to Lead to an Agreement' announced October 2010. Neither the Tasmanian nor Australian Government was formally involved in this first phase of the peace process.

The Statement of Principles requested that state and federal governments provide support to the process. In response, the governments appointed independent facilitator Bill Kelty to work with 'the Signatories and other stakeholder groups' to determine the feasibility of an agreement (Kelty 2011).[2] Together with lobbying by organisations involved

2 In Phase 2, those involved in process negotiations were referred to as the 'signatories' and we adopt this term from this point.

in the negotiations, Kelty's assessment resulted in the 'Tasmanian Forests Intergovernmental Agreement' (IGA) between the Tasmanian and federal government, where funding was committed to support further negotiations and to support forest industry workers affected by the exit of Gunns Limited from the native forest sector. The government remained observers, rather than active participants, in negotiations.

Negotiation among the signatories was hampered by a lack of agreement on scientific evidence around issues such as forest conservation values and sustainable wood supply volumes. In Phase 3 independent experts (the Independent Verification Group, IVG) were appointed to examine these issues and report to the signatories, with the report handed to the federal and state governments and signatories in March 2012 (West 2012).

In Phase 4, the signatories meet regularly, but negotiations broke down over trade-offs between the desired area of new conservation reserves and volume of continued wood supply (Schirmer *et al.* 2016). After several failed negotiation attempts, Minister Tony Burke publicly announced an intention to unwind the committed public funds (The Examiner 2012), which spurred a shift in positions by the signatories. A peace deal in the form of the TFA was announced on 22 November 2012. However, this agreement was between negotiators, not elected representatives of government. As such, the Agreement needed to be passed by the Tasmanian government to be enacted. Legislation based on the deal was presented to the Tasmanian lower house of parliament within days of the deal being announced.

Although the Bill was quickly passed through the lower house, the Legislative Council sent the Bill to a Select Committee in response to intense public debate regarding the lack of citizen input into the Agreement (see Dare 2013; Schirmer *et al.* 2016). Many felt that the Select Committee provided a critical opportunity for the public to have a role in decision making (Schirmer *et al.* 2016) via the processes of representative government. In April 2013, the Lower House passed a Bill that included several amendments to the original agreement, and the first stages of the peace agreement were implemented, including: the establishment of a Special Council overseeing implementation; a successful nomination by the Australian Government for the first tranche of forest reserves to be declared a World Heritage Area; and establishment of a process for sawmillers to apply for structural adjustment funding.

The peace agreement was made when a Labour–Green government was in power in both the state and national governments. During the 2013 election campaigns, the Tasmanian and Australian Liberal parties both included a platform of repealing the TFA if they were elected, citing a view that the TFA was not in the interests of the economy, and that they felt it had been negotiated by 'unelected groups' who did not represent the interests of the broader community. When elected, the incoming governments held to these promises. Despite the repeal of the TFA in August 2014, the peace process achieved more than previous attempts to resolve Tasmania's forest conflicts, with the signatories remaining committed to working together towards peace in Tasmania's forests; and most aspects agreed to in the TFA remained in place, including lower timber harvest volumes, forest reserves, and a commitment to the state forestry agency (Forestry Tasmania) seeking Forest Stewardship Council certification (Schirmer *et al.* 2016).

Justice within the Tasmanian forest peace process

Analysing the TFPP using procedural justice principles highlights critical challenges and trade-offs made throughout the process. Table 8.2 summarises how well each of the nine procedural justice principles were met, with each described in more detail below.

Table 8.2. Procedural justice within the Tasmanian forest peace process (Source: adapted from Lukasiewicz and Baldwin 2014)

Justice principle	What it means	Assessment of the TFPP
Representativeness	The stakeholder groups that are allowed to participate and whose interests are recognised	*Limited.* Representation was better than for many previous attempts at resolution, but the exclusion of some views – particularly those of the broader general community and opposition political parties – meant there were significant gaps in representation. This, in turn, contributed to eventual repeal of the TFA.
Accessibility	The engagement of stakeholders must be done in a way that enables the stakeholders to participate	*Deliberately low.* Accessibility was restricted in order to provide a safe negotiating space in which trust could develop between negotiating actors. In later phases, access improved.
Transparency	Making all information available to all stakeholders – including feedback and justification of decisions made	*Low.* Little information was available to the public throughout much of the negotiations. This led to high levels of suspicion of the process by many stakeholders with an interest in it.
Accountability	Representatives must be accountable to their constituents	*Varied.* ENGOs and unions had good network communication that ensured accountability to their constituents. Industry had good accountability to key industry members, but not to all due to lack of communication between some industry members. Timber community networks were poor, which limited accountability. There was poor accountability to general citizenry.
Level of power	How much influence stakeholders have over the final decision	*Distributed power.* Negotiating parties created a powerful role, although the agreement they reached was subject to government consideration and approval. Perceptions that negotiating parties, rather than government, held decision-making power contributed to contestation of TFA and its eventual repeal.
Neutrality	Suppression of the decision-maker bias	*Deliberately low.* The TFPP was predicated on the idea of a fair outcome achieved through agreement between parties with 'skin in the game' – in other words, only those without neutrality participated. Thus the negotiations included no neutral parties, instead seeking to include opposing views to ensure outcomes were fair to all. The governments acted as the only 'neutral' decision makers. However, strong lobbying by those vested in the process, together with governments having a strong interest in supporting any agreement reached, limited neutrality of government.

Table 8.2. (Continued)

Justice principle	What it means	Assessment of the TFPP
Ethicality	Decisions being compatible with fundamental moral and ethical values	*Variable.* The Statement of Principles set out by negotiating parties sought to set out key values by which negotiations would proceed. However, most parties felt the final agreement represented a compromise, in which not all their values were met. Ethicality is challenging in a conflict involving contestation of some fundamental values: specifically contestation of what actions are necessary to achieve these values in practice. Parts of the process were viewed as incompatible with the fundamental democratic processes valued by many Tasmanians.
Accuracy	Decisions based on informed opinion with minimum processing error	*Good.* The IVG process enabled contemporary, relevant information to be available to negotiating parties. However, there was limited use of some of this information in the ultimate decision making.
Timeliness	Adequate time being given for the process without undue delays	*Poor.* Although negotiations took considerable time, at various points strong pressure was placed on negotiating parties to achieve an outcome in a short period; each time this pressure was unsuccessful. Strong pressure to finalise the agreement by passing legislation rapidly through the Tasmanian Government contributed to poor perceptions of the eventual agreement.

Representation, accessibility, transparency and accountability

The ability to have your voice heard in any form of decision making is an important justice principle. During the peace process negotiations access was deliberately limited to a narrow range of interests: environmental groups, unions and forest industry, with some timber community interests. However, stakeholders such as local government, farming groups and broader local community interests were not represented. This restriction of interests was due to low levels of existing trust and associated difficulty to negotiate across the large group of stakeholders initially involved in the process, which inhibited effective interest mediation:

> *Probably fair to comment that nobody much trusted anybody, and it took quite a long time to actually develop even sufficient trust in the other people's articulation of their positions and the rationale behind those positions, to even get to the Statement of Principles. We, as an industry we'd spent 35 years chucking rocks at these blokes, and they'd spent 35 years chucking rocks at us. We were all a little bit battered and bruised from the experience, so we needed to get over that ... (Industry representative #1)*

> *You can't always do something around the whole bloody table. Sometimes you got to have a circuit breaker. A circuit breaker can be various things, but one of the things you can use is just having a trusted smaller group that doesn't make a decision but just explores all options. It just expands the field of discussion and therefore expands the options for a settlement. (Industry representative #2)*

Consequently the stakeholders agreed to limit participation to a select group of representatives, and to have closed negotiations and limited announcements to the media (see also Lester and Hutchins 2012):

> *To a large degree during the process we've either agreed on a collective approach in the case of joint media release, or just a general approach that we would each individually take ... it wouldn't have been tenable otherwise ... prosecuting your negotiating position through the media's generally, you know, more likely to lead to it falling over. (ENGO representative #1)*

The small group that subsequently engaged in negotiations did consult their broader constituents regularly. Despite this, limiting the size of the group, and limiting the extent to which it communicated about its negotiations, did reduce the representativeness, accessibility, transparency, and to some extent the accountability of the process. These were traded off in order to improve the ability of those involved to negotiate effectively. Thus procedural justice was sacrificed to enable improved interactional justice, acknowledging the contested environment and the need to overcome high levels of mistrust between stakeholders. However, this trade-off later contributed to failure of the process, because stakeholders who felt excluded questioned its legitimacy based on concerns about representation, transparency and accountability. It also created space for external attacks on the process by those who had no vested interest in it:

> *... there were critical elements outside that process or partly in it, who systematically tried to derail it or destroy it ranging from fanatics to some whom were taken very seriously and got themselves into positions as representatives of whole sections of either side. (Observer #1)*

Level of power

Democratic representation in the form of direct participation in negotiations by the Tasmanian and Australian Governments was excluded from the negotiating process. This was due to poor outcomes of previous forest conflict resolution attempts, and prevailing political stances. In some ways, removing themselves from the negotiations enhanced the power of the government to reject agreement if it proved politically untenable, with several interviewees reporting an active desire by both governments to be formally recorded as being observers only of negotiations:

> *One agency that never was willing to try to come forward with any great solution was government and the politicians. They loved the idea that they could say, 'It's up to them ... we don't have to have a policy. Our policy is to support them'. Everyone is trying to handball it to someone else to actually come up with the solution that integrated the irreconcilable claims. (Observer #1)*

This exclusion, however, was argued by some to represent an illegitimate concentration of power in the hands of non-elected representatives:

Who elected these people [the signatories]? Who gave them the right to deal with public land and make decisions that will impact on all of us? (Observer #2)

However, while perceived by many to have uncontested power to reach and implement an agreement for forest peace, in reality the power of the negotiating parties was circumscribed: any agreement reached was subject to scrutiny by the governments who were asked to implement it. The Tasmanian Legislative Council, who chose not to support the agreement in its initial form, highlighted the limits of the power of the negotiating parties, amending the TFA substantially through their Select Committee review:

I think the Legislative Council acted professionally. I don't say I necessarily agree with every decision they made. But it's not for me to agree or disagree with the Parliament of the land. They are the final arbiters. (Industry representative #3)

So we're working hard still to try and, as a group, to try and pull, to you know, get the Legislative Council into a place where they are comfortable. (Industry representative #4)

However, distributed power was enacted to place substantial pressure on the governments to pass the agreement reached by negotiating parties. Perceptions of illegitimate power being given to negotiating parties by the governments of the day ultimately were a key factor in opposition parties repealing the agreement when they were elected:

The Liberal Government will act on the overwhelming mandate delivered to us by the Tasmanian people and move immediately to tear-up the job-destroying forest deal ... unlike the former government, we won't be held to ransom by unelected groups. (Harriss 2012)

This perception that negotiating parties held complete power over the agreement that enabled them to 'bypass' normal processes of government was not accurate. The substantial amendments made by the Tasmanian Legislative Council to the initial Agreement are testament to the limits of the influence of the negotiating parties. However, the exclusion of government representatives from active participation in negotiations represented a higher degree of power for the negotiators than typically exists in processes involving government-managed assets.

Neutrality and ethicality

Neutrality refers to efforts to suppress bias in decision making, while ethicality refers to approaches used to ensure outcomes are compliant with fundamental societal morals and ethics. Tasmanian forest management, like most conflicts over management of forests, is a deeply contested issue with entrenched polarisation of values and interests across the community (Schirmer 2013). The negotiations were predicated on the idea of negotiation between vested interests – those with what most interviewees described as 'skin in the game', rather than on the justice principle of ensuring neutral decision makers. In fact, negotiating parties deliberately rejected having a neutral facilitator for almost the entire process, instead focusing on ensuring inclusion of opposing viewpoints that, through forcing consideration of all points of views in negotiations, aimed to reach a 'neutral' outcome that met all interests to the greatest extent the negotiating parties felt possible. Although the scientific advice sought in the IVG process could be viewed as neutral, the advice requested was strongly circumscribed by negotiating parties and experts selected by the parties, limiting perceptions of neutrality of the IVG group.

Similarly, meeting fundamental societal morals and ethics was challenging. Although the TFPP sought to provide public benefit through tackling a conflict that was damaging to Tasmanian social, environmental and economic interests, the extent to which fundamental morals and ethics were complied with beyond this is more challenging to assess. This is because, while negotiators agreed on key values, as stated in their Statement of Principles, they did not necessarily agree with each other, or with other members of the public, on how to best meet those values and ethics. In the end, the TFA was viewed as a compromise by all involved, which meant key values of each group were not always met:

> *We all realised that one; we all weren't going to win… No one was going to win by just staying and throwing arrows at each other … we had to get a compromise. And that a compromise could see the thing go forward, so everyone just rolled up their sleeves and basically got into it. (Industry representative #5)*

> *In terms of kind of rank and file conservationists out there in the streets, most people see this as a comprehensive outcome that's riddled with compromise which is inevitable. (ENGO representative #2)*

Accuracy and timeliness

Accuracy was dealt with during Phase 3, where the aim was to resolve data gaps and contested information through the IVG. Although some information generated was contested by some negotiating parties, the process did help negotiators agree on scientific evidence around areas such as wood supply volumes – and helped to identify realities around which they had to negotiate:

> *The [IVG] work not only helped focus people on what was really important, it helped people understand the process of giving up which is hard in a negotiation. Now, look, I know there's different views on different bits and pieces … [but] through that signatories process, it's given us an independent touch stone to refer to. (ENGO representative #3)*

A key challenge to accuracy was the short timeframes set for the IVG scientists to complete their work: accuracy and timeliness were to some extent traded off to enable negotiations to progress as the timeliness of the overall process was problematic. A long period was needed to develop trust among negotiating parties, yet once government funding was provided, there was significant pressure for the parties to reach outcomes, with several government-imposed deadlines being set and passing without being met. The lengthy process needed for negotiations was both understandable and necessary, given the 30-year conflict that had led to negotiations, but also challenging as areas of forest remained in temporary reserves awaiting final outcomes, and concerns about the ongoing loss of employment in the industry continued. It also led to pessimism from key actors, including the then Federal Minister for Sustainability, Environment, Water, Population and Communities (Mr Bourke):

> *I've got to be quite frank, I'm now quite pessimistic that they will be able to bridge the needs that industry say they have and the needs that the conservation movement say that they would require. (The Examiner 2012)*

Federal government pressure, including the setting of publicly stated deadlines, and threats to withdraw federal government funding was sometimes counterproductive. Imposing deadlines did force negotiators to continue attempting to reach agreement, but at the same time reduced their ability to effectively participate, with many negotiators reporting being physically and mentally exhausted. These time pressures inhibited procedural justice outcomes, reducing the time available for consultation with members of the organisations they represented, and with the broader community, which ultimately contributed to the demise of the agreement (Schirmer *et al.* 2016).

Trade-offs between procedural justice and other outcomes

Several key trade-offs were made between meeting key procedural justice principles, and achieving other important outcomes, as part of the TFPP. Any negotiating process is influenced by those processes that preceded it, and the prevailing socio-political context (see Singleton 2002; Cornwall 2004; Luning 2012); this was particularly crucial for the TFPP, where a long history of failed prior processes, and of low trust and conflict between those now seeking to negotiate, presented perhaps the biggest threat to achieving an outcome. Procedural justice principles of representation, accessibility, transparency and consequently accountability were traded off to create space more conductive to the relationship development and deliberation processes necessary for interactive justice.

Power trade-offs were more complex: the power of having the government of the day involved in negotiations was traded off to enable negotiators more freedom to explore options for outcomes that might not be considered by governments. The need to achieve an agreement meant that pressure was placed on achieving ethicality, accuracy and timeliness. Some justice values were compromised and there were time limits placed on analysis of scientific data, which limited consultation with some groups. Neutrality was deliberately rejected as a principle: the goal being to ensure that those who had a stake in the outcomes were putting themselves 'on the line' to achieve a negotiating outcome. These trade-offs were all critical to the process achieving the outcome it did: the first agreement in the long history of Tasmanian forest conflict that was supported by multiple ENGOs and industry representatives. However, the trade-offs made also contributed to its repeal. The rejection of neutrality was a key contributor to widespread community concern about the level of power held by the negotiators, as was lack of time spent consulting broader community interests. Reduced representation, accessibility and transparency helped those involved in the negotiations build trust and interact successfully with each other, but at the same time reduced trust in the process by many of those not directly engaged in it.

The procedural justice trade-offs involved in the TFPP provide insights into the complexity of political processes, and the difficulty in ensuring procedural justice simultaneously with meeting political pressures and achieving a 'safe space' for negotiations. A key question is whether these trade-offs could be handled differently, to simultaneously achieve desired outcomes and procedural justice.

A key consideration is the timing of meeting different procedural justice principles. Although there were reasons for not meeting some principles at particular points in time, a deliberate attempt to reintroduce them at subsequent stages of negotiations may have improved public response to the TFA. In particular, improved public consultation, public reporting and transparency of decision making across Phases 2–4 could have improved procedural justice issues of representation, accessibility, transparency and accountability. Once trust was better established between negotiating parties in Phase 1, from Phase 2 there was greater scope to improve representation, transparency and to some extent accessibility. In Phase 3, more space for dialogue between experts and the public may have improved trust in the process: although the IVG report was made available online, the sheer volume of data and technical nature of the reports made this inaccessible for many stakeholders. Throughout, outcomes of ongoing negotiations were not made available in a consistent and transparent manner, further isolating those not directly engaged in negotiations. A more specific and transparent role for government beyond observing negotiations, particularly in Phases 3 and 4, would have reduced the need for substantial amendment after legislation was tabled in Parliament.

Conclusions

Justice is a manifestation of politics: a contextually based construct that attempts to balance risk, values and impact across society. Politics, however, is more than justice: politics includes prevailing economic and legislative constraints, and includes the voice of people, their interests and their power. As such, the nexus of politics and justice is a complex space that evolves and changes as a process such as the TFPP evolves over time. While acknowledging that achieving all justice principles simultaneously is often unrealistic, the failure of the TFPP provides important lessons about the need to ensure justice principles are incrementally achieved over time, if they cannot all be met all the time. This emphasises the importance of understanding and prioritising justice principles early in a policy development process, planning how to increase procedural justice over time as the process evolves, while also providing space for regular evaluation of justice outcomes to enable the balancing of principles throughout the process. Without a considered approach to justice implications, the policy process may be overshadowed by claims of injustice that risk achieving the desired policy objectives and outcomes: the ultimate repeal of the TFPP highlights the consequences of failing to meet social justice principles considered essential by many in the community. That said, trade-offs can, and will, be needed between achieving procedural justice and other outcomes at given points in time. Pragmatic recognition of the needs and opportunities presented by the prevailing socio-political context is essential to achieve any outcome. Negotiating parties and policy makers need to be clear in their rationale for trade-offs in justice principles, aware of the implications of such trade-offs, and to be prepared to be flexible in their policy-making approach to tackle procedural injustices that negatively affect policy outcomes. Without such a considered approach to issues of justice, policy-making processes can be ineffective, and indeed detrimental, to the public issues at hand.

References

Brockner J, Wiesenfeld B (1996) An integrative framework for explaining reactions to decisions: Interactive effects of outcomes and procedures. *Psychological Bulletin* **120**, 189–208. doi:10.1037/0033-2909.120.2.189

Clayton S, Opotow S (2003) Justice and identity: changing perspectives on what is fair. *Personality and Social Psychology Review* **7**(4), 298–310. doi:10.1207/S15327957PSPR0704_03

Cornwall A (2004) Spaces for transformation? Reflections on issues of power and difference in participation in development. In *Participation: from Tyranny to Transformation? Exploring New Approaches to Participation in Development.* (Eds S Hickey and G Mohan) pp. 75–91. Zed Books, London, UK.

Cropanzano R, Prehar C, Chen P (2002) Using social exchange theory to distinguish procedural from interactional justice. *Group & Organization Management* **27**, 324–351. doi:10.1177/1059601102027003002

Dare M (2013) Localism in practice: insights from two Tasmanian case studies. *Policy Studies* **34**(5–6), 592–611. doi:10.1080/01442872.2013.863572

Gale F (2013) When interests trump institutions: Tasmania's forest policy network and the Bell Bay pulp mill. *Environmental Politics* **22**(2), 274–292. doi:10.1080/09644016.2012.683150

Harriss P (2012) Tasmanian Forests Agreement – Legislative Council Inquiry. Media release, 20 December. Legislative Council, Hobart, <http://www.parliament.tas.gov.au/ctee/Council/Media/TFA_Media%20release_20%20December%202012.pdf>.

Kelty W (2011) *Tasmanian Forests – Interim Report for Consideration.* Department of the Environment, Hobart, <https://www.environment.gov.au/resource/tasmanian-forests-interim-report-consideration>.

Layder D (2006) *Understanding Social Theory.* Sage, London, UK.

Lester L, Hutchins B (2012) The power of the unseen: environmental conflict, the media and invisibility. *Media Culture & Society* **34**(7), 847–863. doi:10.1177/0163443712452772

Lukasiewicz A, Baldwin C (2014) Voice, power, and history: ensuring social justice for all stakeholders in water decision-making. *Local Environment.* doi:10.1080/13549839.2014.942261.

Lukasiewicz A, Syme G, Bowmer K, Davidson P (2013) Is the environment getting its fair share? an analysis of the Australian Water Reform Process USING A Social Justice Framework. *Social Justice Research* **26**, 231 252. doi:10.1007/s11211 013 0186 y

Luning S (2012) Corporate social responsibility (CSR) for exploration: consultants, companies and communities in processes of engagement. *Resources Policy* **37**, 205–211. doi:10.1016/j.resourpol.2011.02.004

Neal M, Lukasiewicz A, Syme G (2014) Why justice matters in water governance: some ideas for a 'water justice framework'. *Water Policy* **16**, 1–18. doi:10.2166/wp.2014.109

Paavola J (2007) Institutions and environmental governance: a reconceptualization. *Ecological Economics* **63**(1), 93–103. doi:10.1016/j.ecolecon.2006.09.026

Schirmer J (2013) Environmental activism and the global forest sector. In *The Global Forest Sector: Changes, Practices, and Prospects* (Eds E Hansen, E Panwar and R Vlosk) pp. 203–235. Taylor & Francis, New York, USA.

Schirmer J, Dunn C, Loxton E, Dare M (2011) 'Socio-economic impacts of forest industry change: a baseline study. Tech.Rep 214: Interim Report'. CRC for Forestry, Hobart.

Schirmer J, Dare M, Ercan SA (2016) Deliberative democracy and the Tasmanian forest peace process. *Australian Journal of Political Science.* doi:10.1080/10361146.2015.1123673.

Singleton S (2002) Collaborative environmental planning in the American West: the good the bad and the ugly. *Environmental Politics* **11**(3), 54–75. doi:10.1080/714000626

Skitka L (2003) Of different minds: an accessible identity of justice reasoning. *Personality and Social Psychology Review* 7(4), 286–297. doi:10.1207/S15327957PSPR0704_02

Syme G, Nancarrow B (2005) Creating community consent through fairness and judgements. In *Consent and Consensus: Politics, Media and Governance in Twentieth Century Australia* (Eds D Cryle and J Hillier) pp. 371–387. API Network, Perth WA.

The Examiner (2012) Forestry peace talks collapse. *The Examiner*, 27 October, <http://www.examiner.com. au/story/425317/forestry-peace-talks-collapse/>.

Webb TJ, Bengston DN, Fan DP (2008) Forest value orientations in Australia: an application of computer content analysis. *Environmental Management* **41**(1), 52–63. doi:10.1007/s00267-007-9011-4

West J (2012) *Final Report on the Work of the Independent Verification Group for the Tasmanian Forests Intergovernmental Agreement*. Department of the Environment, Hobart, <https://www.environment.gov.au/resource/independent-verification-group-report>.

PART 3

LENSES INTO JUSTICE

9

Enduring and persistent injustices in water access in Australia

Sue Jackson

Summary

Indigenous-specific water entitlements are currently estimated at less than one hundredth of 1% of Australian water allocations. This minute figure contrasts starkly with the extent of Indigenous land ownership that is currently in excess of one-fifth of Australia's land mass. Indigenous Australians see clear connections between the past and present in accounting for this skewed distribution of entitlements, and their testimonies indicate an awareness of the enduring effect of historical injustices in the development of Australia water law and policy. This chapter applies concepts derived from political theory to examine this significant case of historical injustice. It argues that there is a poor appreciation within the water sector, and among wider society, of the importance of historical events and structural processes in explaining today's pattern of access to water. By revealing the dynamics of water allocation over time through the trajectory of water law and policy, the chapter shows how injustices were produced and reproduced during the colonial and state administration eras and the ramifications for achieving justice in the recent and ongoing neoliberal reform era. Indigenous peoples' rights and interests in water have been systematically and persistently marginalised because reparative or restorative mechanisms have not been established in any of the processes that have made and re-made water law and policy over the course of Australian history.

Introduction

Indigenous demands for justice present far-reaching challenges for a liberal democracy such as Australia (Waldron 1992; Ivison *et al.* 2000; Gibbs 2009). As a former colonial country that forcibly dispossessed the Indigenous population, the social and political relationships between Indigenous Australians and other citizens are matters of immense significance, as are the ways in which these relationships affect access to land and water and natural resource management more generally. The historical record of colonial events matters not only in accounting for the quality and strength of these intercultural relationships, but also in shaping them as they unfold into the future.

As seen in appeals for acknowledgement and redress and acts of symbolic remembrance, such as the apology to the Stolen Generations, history is present in Indigenous perspectives on the processes and forces that have structured their quality of life, identity, political standing and future prospects as distinct peoples. In other settler countries, connections are also made between past and present policies and injustice, especially when Indigenous representatives 'try to use the past to critique present policy and conditions' (Jung 2009: 1; see also Gibbs 2009). So it should be of little surprise that in their advocacy for water rights Indigenous Australians are making such connections and drawing attention to the 'presence of an unsettled past' (Lu 2011: 263) in the unfinished business of water policy.

Water has been commodified over the past two decades through a series of reforms that entailed the separation of land and water titles, implemented in combination with measures to stimulate water trading, including pricing. In addition, water laws and planning processes were reformed to reframe the environment as a water 'user', legally entitled to water allocations as an equal with other consumers. Australia is now regarded a world leader in water management for having created the world's biggest water market (Nikolakis *et al.* 2013), established a legal requirement to provide environmental flows and set a sustainable diversion limit on water use in the Murray–Darling Basin (MDB). Economic benefits accruing to buyers and sellers of water rights, almost all of whom are irrigators, were in 2011 estimated to be worth A$2.4 billion in annual turnover of trade (Nikolakis and Grafton 2014).

Since reforms began in earnest in the mid-1990s, there have been many appraisals of the condition of Australian water resources and their development trajectory. The most recent accounts have brought Indigenous rights and interests within their scope (Jackson *et al.* 2012), a feature that can be attributed to their inclusion in national policy: the National Water Initiative (NWI) in 2004. The National Water Commission's 2011 biennial assessment of progress on water reform is exemplary. Having acknowledged the improvement in Indigenous consultation and participation in preparing water plans, the Commission concluded that

> *... the full intent of the NWI parties' commitments on Indigenous interests in water has not yet been achieved. Many water plans do not consider Indigenous cultural values and economic development, leaving the cultural and economic expectations of Indigenous Australians as* an unmet demand on the water system. *(NWC 2011: 9; emphasis added)*

The review offers an understated appraisal of a failure of intent, with no acknowledgement that Indigenous Australians could have unique and historical claims to natural resources. Economic framing such as this presents an Indigenous water subject or citizen as a 'new' user, rendering Indigenous claims for water as an afterthought in the bigger Australian story of re-allocation. In that process, in which the contest is largely between non-Indigenous farmers and the environment, distribution based on need holds currency as a justice principle (see Lukasiewicz *et al.* 2013), but in the NWC review, Indigenous Australians have 'expectations', not needs. Moreover, framing Indigenous claims in the market logic of demand and supply puts the onus on Indigenous communities to demonstrate an economic application for water to which they may lay claim – a capacity that, because of the consequences of colonisation, is greatly constrained by access to financial capital, let alone land.

Within the neoliberal formulation of this problem, Indigenous people are conceived as 'late entrants' in what has been described as a 'mature' water economy (Smith 1998). The concept of 'maturity' in this context refers to the stage at which water resources are fully developed (Smith 1998): one reached in parts of southern Australia in the 1990s. This metaphor of growth and development of the water economy naturalises the exclusion of pre-existing Indigenous custom and rights to water from the modes of distribution that have presided over the creation and expansion of this economy, precisely the processes that have structured the accumulation of water rights.

However, the water economy did not mature naturally. Underlying and unjust socio-historical processes established during the colonial period of Australian settlement were entrenched, rather than redressed, during the 'modern' phase of water reform that commenced in the 1990s. Land and water ownership were decoupled, with significant implications for the restitution of land to Indigenous Australians via claims processes or direct acquisition. Although it heralded major changes in access to, and control of, water, this restructuring of water governance took no account of Indigenous needs and aspirations for a full 10 years of institutional change. When Indigenous interests were taken into account in water policy in the second period of reform around 2004, they were narrowly prescribed in an ahistorical and weak attempt to accommodate native title (Jackson and Morrison 2007; Tan and Jackson 2013).

Statements like those made by the NWC reflect an over-riding concern with economic efficiency and competitiveness (Hussey and Dovers 2006). Through this lens, wants or needs that manifest as 'unmet demand' do not constitute a water justice problem: one that is reflected in, for example, the skewed distribution of water resources. Such statements make no reference to special claims founded on historical entitlement or on aspirations for reparation or redistribution, nor do they question how present water users came to acquire their water use rights. Indeed, they are made intelligible by bracketing the historical and geographical factors that shaped today's system of entitlements and its patterns of access (cf. Dempsey *et al.* 2011).

With southern Australia's water resources fully allocated to water users with a history of access and entitlement (deriving from land titles), Indigenous communities in those regions, by virtue of their relative landless condition, are water poor, prompting Sheehan and Small (2007: 1) to ask 'Is a new form of Aboriginal dispossession now subtly occurring?' Preliminary estimates of Indigenous access lend support to this assertion. Jackson and Langton (2012) compared Indigenous land ownership and water entitlements, demonstrating the extent of the inequity: Indigenous Australians own more than 20% of the country's land mass,[1] while Indigenous-specific water entitlements are at present estimated at <0.01% of Australian water diversions. Based on figures relating to the MDB,[2] Darren Perry (*pers. comm.*) found a similarly minute proportion of 0.08% of the Basin's Sustainable Diversion Limit in the hands of Indigenous organisations.

Re-allocation measures (such as a water trust facility, buy-backs or special purpose licences) would be required to remove or reduce inequity in the distribution of water access

1 The estimate of land ownership is drawn from work by Altman and Markham for a parliamentary submission that is cited in Altman and Jackson (2014). Altman and Jackson state that, as of 2014, just over 20% of the land mass is held under exclusive possession Native Title or land rights regimes, while non-exclusive Native Title (shared) has been determined over a further 10%. There are still over 300 Native Title claims to be heard indicating that current figures will grow.

2 Indigenous peoples of the MDB (who comprise approximately 3.4% of the Basin's population) currently hold <0.2% of land (Arthur 2010).

and, although these have been recommended at key points during the past fifteen years (see McAvoy 2006), they have not eventuated on any meaningful scale. Indigenous people remain greatly constrained in their ability to shape the use and management of water, and negligible attention has been given to the unjust social arrangements that continue to generate benefits for non-Indigenous water users. This failure to understand and act on the 'intertwined and co-implicated' (Balint *et al.* 2014) character of past and present water policy reproduces injustice in water access.

Following this introduction, I first provide a brief description of Indigenous perspectives on water justice (or more precisely, injustice), arguing that the positions advanced emphasise historical reflection, recognition and reparation. I then turn to reveal the dynamics of water allocation over time through the trajectory of water policy, showing how injustices were produced and reproduced during the colonial era and the ramifications for achieving justice in the recent and ongoing reform era. The next section discusses several theoretical concepts and ideas that can be usefully applied to the analysis of these forms of water injustice before a final concluding section.

Before turning to Indigenous perspectives, it is necessary to define key terms. Water access is the object of justice in this chapter, where access is defined as 'the ability to benefit from things – including material objects, persons, institutions, and symbols' (Ribot and Peluso 2003: 153). The term is appropriate because it encompasses matters that cannot neatly be contained within the justice categories of distribution or procedure. Property is one set of access relationships, but there are others. 'By focusing on ability, rather than rights as in property theory, this formulation brings attention to a wider range of social relationships that can constrain or enable people to benefit from resources without focusing on property relations alone' (Ribot and Peluso 2003: 154). In this formulation, structural mechanisms are brought into play as a mediating constraint and, as I will argue below, structural-based political theories can help us to see the dynamic nature of historical injustice.

I rely on Meyer and Roser (2010) for a definition of historical injustice. It arises in a situation where 'earlier generations of one community wronged earlier generations of another community and today's generations of both communities are now looking for an adequate way of responding to this historical fact' (Meyer and Roser 2010: 230). Historical injustice calls for kinds of justice that are different to distributional or procedural justice and draw on principles of need, fairness, equality, and so on. Such types of justice respond to a particular historical injustice and might include those that are reparative and/or restorative. Reparative justice, for example, is directed towards 'putting right a past wrong' and can include a range of forms of reparation including apology, compensation, restitution, and guarantees of non-repetition (Gibbs 2009: 45). Reparative justice is largely historical in focus, whereas restorative justice is usually applied in cases where the identities of the 'offender' and 'victim' are known and both are alive. Gibbs argues that these related types of justice are both suited to settler society contexts.

Indigenous perspectives and ideas on what is right or just in water access

Through numerous water governance institutions and negotiating arenas Indigenous Australians are seeking to establish the legitimacy of their water-related values, ethics and practices and to increase their access to water (Weir 2009; Tan and Jackson 2013). Notwithstanding the active role recently played by Indigenous representatives in water forums, as a society we have a poor appreciation of their ideas on what is right or just in water governance. Contemporary Indigenous aspirations for the return of traditional lands, preservation of culture and the right to self-governance are better understood as key concerns for Australian NRM, but what would a just level of access to water look like and how far are we from developing a conception of justice that can be shared across cultures?

All human societies have norms about the kind and amount of access to water that their members should have. We have only started to explore these matters in Australia (see Jackson and Altman 2009; Weir 2009; Jackson and Barber 2013; Lukasiewicz *et al.* 2013; Nikolakis and Grafton 2014). In work in the Pilbara conducted by Barber and Jackson (2011), free access to water was, for some Indigenous respondents, a primary principle of access, based on the widespread tradition of sharing fresh water. Additional perspectives offered in that study indicate that there is an obligation for the Indigenous resource owners to provide the necessities of life for those living in the area. In a north Australian study, Nikolakis *et al.* (2013) found that more than 50% of the Indigenous people they surveyed considered water management regimes to be inequitable and 73% believed that their rights and interests are not reflected in water management policies.

Weir (2009, 2011) has closely examined Indigenous relationships with water in the MDB, as well as associated norms and ethics of responsibility and care. Weir argues that ecological loss and decline represents a form of dispossession because they disrupt, and in places sever, Indigenous relationships with place. This and much of the work on Indigenous cosmology, as well as environmental governance, reminds us that Indigenous notions of justice encompass the field of non-human life and that restoration of socio-ecological relationships under Indigenous institutions are likely to be sought by Indigenous peoples as vigorously as is the restitution of property.

The complexities and any variations in Indigenous notions of water resource ownership and access, and attendant rights and responsibilities, need to be more fully explored in future work. From what has been documented to date, there appears to be a consistently held perspective that emphasises an ongoing sense of Indigenous custodial responsibility and ownership. This is despite more than two centuries of laws that have ignored or actively attempted to remove this sense of entitlement. Moreover, Indigenous Australians see clear connections between the past and present historical injustice in the skewed distribution of entitlements, and their testimonies suggest an awareness of its endurance, as well as limiting effect on the conditions of Indigenous life.

Submissions from Indigenous organisations to the 2014 review of the *Water Act 2007*, for instance, reveal a perception that the current state of water governance is grossly unjust. For example, the submission of the Murray Lower Darling Rivers Indigenous Nations (MLDRIN) states:

> *Our First Nations remain, with minor exceptions, locked out of meaningful participation in water planning and management including beneficial ownership of water entitlements. (https://www.environment.gov.au/water/legislation/water-act-review)*

In the same review, the Northern Basin Aboriginal Nations (NBAN) explicitly refer to unjust patterns of access based on previous appropriations and the historic accumulation of water rights by non-Indigenous land-owners:

> *Aboriginal people were removed from their lands, including lands adjacent to rivers and streams and access denied ... Under the old English law used in Australia (riparian*

rights), unless you owned land adjacent to, or land which had access to water such as groundwater, or via an irrigation scheme, you were unable to hold a water licence.

When the COAG reforms were implemented in each state, unless a person held a prior water licence or entitlement, they were excluded from owning a new licence under the new water management regime unless they purchased it.

Accordingly, Aboriginal people were largely excluded from owning the new form of water entitlements, or shares and accessing their water dependent cultural places. Aboriginal people should have an equal right to participate in water management and use and should be allocated shares accordingly.

NBAN calls for the Basin Plan and Water Resource Plans to: 'Facilitate Aboriginal Peoples' ownership of a fair and equitable proportion of commercial and environmental water licences' and propose several measures that would, in their estimation, remedy the historical injustice.

Water governance and the long shadow of colonialism

Water policy is dominated by the view that questions of access and distribution are simply technical problems to be resolved by economic models and more efficient hydrological engineering. It fails to appreciate that 'water justice is embedded and specific to historical and socio-cultural contexts' (Zwarteveen and Boelens 2014: 143). This section concentrates more closely on historical events and provides a more detailed, albeit still very brief, account that, like NBAN's, connects our current system of water governance to the policies, laws and norms that resulted in the appropriation of Indigenous land and water rights and repression of Indigenous water cultures. For the sake of brevity, this review is organised into three phases: (i) the riparian phase; (ii) the state administration phase; and (iii) the neoliberal phase. Indigenous systems of water governance of course pre-date and co-exist with this periodisation, but space constraints prevent any explication.

The riparian phase

Australian colonies adhered to common law riparianism as the basis of their water law for at least a century (Harris 2007). The riparian doctrine 'dictated that only individuals owning land that came in contact with a water source could acquire the right to make use of its water' (Harris 2007: 121). Aboriginal occupation of Australia was not originally recognised by colonial law and so, as landless people, Indigenous communities were not entitled to exercise common law riparian rights. White riparian landowners enjoyed the security of 'reasonable use' of water, as well as substantial government subsidies to build farming and grazing economies (Connell 2007).

The state administration phase

Over time, riparianism came to be viewed as an inappropriate institutional basis of water management. In the mid-1880s, the Victorian Minister for Water Resources, Alfred Deakin, presided over a Royal Commission that recommended legislation to abolish riparianism, nationalise water by transferring rights in water to the Crown and establish a state system of licenced private diversions (Harris 2007). Indigenous property rights or interests in either land or waters were not referred to in the Royal Commission's background materials, findings, transcripts of hearings or in its recommendations for a restructure of water rights. Indigenous need for water was being articulated at this time, however. A need for water was implicit in Indigenous political campaigns waged during this era for land for family farms in Victoria and NSW (Goodall 1996). It was also recognised to some extent in early government actions to protect or reserve waterbodies for Indigenous use to mitigate vulnerability to the environmental consequences of agricultural expansion and intensification.

Deakin's reforms were to alter the use, management and socio- political organisation of Victoria's water cycle for the benefit for the white settler and similar changes were to have a corresponding impact in New South Wales and other states. Legislation and civil law interpretation placed the use of water in the public domain under a social logic of allocation (Hussey and Dovers 2006) so that the government would have jurisdiction over the use of water. Each state resolutely maintained control of its water resources beyond the era of Federation through systems that licenced private diversions. Licensees came to regard these entitlements 'almost as rights in perpetuity' (Tan 2002). By virtue of the fact that Aboriginal people in those states had been dispossessed of their land, they had no opportunity to obtain water licences (McAvoy 2006). In other states, too, water legislation did not provide rights to water for Indigenous groups (e.g. Western Australia's *Rights in Water and Irrigation Act 1914* (Bartlett 1997)).

The neoliberal phase

This phase took hold in the 1990s when the first tranche of Council of Australian Government (COAG) reforms created a tradeable water market founded upon the existence of property rights in water and placed limits on diversions in the MDB (the Cap; see Jackson 2011). Further changes were made when Australian governments agreed to the NWI in 2004. Lukasiewicz *et al.* (2013) state that 'The NWI has a dual purpose: to establish sustainable levels of water extraction and to protect the rights of existing users' (p. 1316). Changes to the allocation regime were subject to significant advocacy by the agricultural lobby in the formulation stages (Hussey and Dovers 2006), resulting in a 'windfall gain of perpetual water rights' (Howard 2010: 361). This latter point is confirmed by McAvoy's NSW analysis:

> *Upon the de-coupling of water rights from the land it was those people who had water licenses under the Water Act 1912 who were able to then convert such into water access licences under the Water Management Act 2000. The result was that Aboriginal people were also dispossessed of their water rights. (McAvoy 2006: 101)*

Looking at the initial period of this phase brings to light a crucial historical coincidence in the timing of several events: the COAG reforms, the Cap, the incomplete processes of land restitution under statutory land claims processes and Native Title law. The numerous implications of this coincidence were not considered by the COAG leaders:

(i) Land rights legislation had been instituted as a restoration measure in most states in the 1970s and 1980s. For example, the *Aboriginal Land Rights Act 1983 (NSW)*, which recognised that Aboriginal people had been dispossessed (McAvoy 2006), had come into effect during emerging awareness of limits to water use. Hundreds of claims were lodged in the 1980s and many were still unresolved in the 1990s (Behrendt 2011). Where would the Cap and the separation of land and water titles leave those groups with unresolved claims? Would future land transfers be accompanied with water entitlements?

(ii) The *Native Title Act 1993* defined native title to include rights to water for customary purposes (Tan and Jackson 2013), yet the COAG reforms 'made no reference to native title or any other form of Aboriginal water rights' (McAvoy 2006: 93). By this time, water had become a scarce and costly resource in many regions. How would the COAG reforms, especially the pricing of water and water trading, affect those groups who were to lodge claims and prove the existence of native title, particularly their capacity to pursue livelihood opportunities reliant on water? Buying water on the market is not a trivial matter, with the high cost of water licences and the fact that one-third of the Basin is subject to native title application (Arthur 2010).

(iii) The *Native Title Act 1993* established another avenue to land restitution, via purchases by the Indigenous Land Corporation. It has become a principal route for acquiring larger parcels of rural land in the country's south and east (Arthur 2010). Would those people acquiring land under this mechanism continue to gain access to water as a result of the reforms?

(iv) The potential for conflict and uncertainty as a result of interactions between western and customary laws (see Altman 2004; Jackson and Altman 2009). The need for well-defined property rights underpinned these reforms and security was provided for environmental water of a level at least the same as water access entitlements for consumptive use (Hussey and Dovers 2006), but no similar thought appears to have been given to security of Indigenous uses.

Indigenous representatives were prevented from influencing the rules that govern the distribution of access to water. Although they had a substantial stake in these decisions, as far as I know, there was no dialogue or negotiation with Indigenous representatives at the national level until well after the NWI when, in 2010, an Indigenous advisory group was established.

The states implemented the reforms by changing their water legislation. The NSW Government for example passed the *Water Management Act 2000,* which, *inter alia,* decoupled water from land ownership and created a semi-regulated water market (McAvoy 2006). Although at that time the government did consider some of the above issues, McAvoy (2006) argues that it did so unsatisfactorily, refusing, for example, to act on calls from Aboriginal organisations to establish an Aboriginal Water Trust funded by a levy on water licences to buy water for Aboriginal benefit. In the Northern Territory also, amendments to water legislation took no account of the issues listed above and did not include any special measures to guarantee Indigenous access to water, despite the fact that Indigenous people own half of the land base. In that jurisdiction, and others, there is a real risk that the pattern of injustice evident in south-east Australia may be perpetuated as the water governance regime is rolled out (Jackson and Altman 2009; Nikolakis and Grafton 2014).

It was not until 2004 that Australian water policy recognised Indigenous rights and interests. The NWI contains several clauses designed to improve Indigenous access, but these are discretionary and, as noted by Tan and Jackson (2013), rely on interpretations of native title that constrain the scope of this newly recognised property right. Moreover, they argue that implementation of the NWI gives a low priority to Indigenous needs in over-allocated catchments, that its goals are prejudiced by delay and difficulties in native title determinations, and that outcomes generally preclude economic development (Tan and Jackson 2013).

Theories of justice

I have argued above that there is a poor appreciation within the water sector and among wider society of the importance of history, of the conditions underlying poor distributions, and for the need to accept responsibility for these and the other legacies from our colonial past. These legacies shape today's pattern of access to water (such as loss of land, the impact of unemployment and low education rates on social and financial capital). This lacuna is not restricted to the field of natural resource management. Indeed, it is a feature of justice studies as well. Ivison *et al.* (2000) state that 'The relevance of the history of relations between indigenous and non-indigenous peoples is something contemporary theories of justice have been slow to recognise' (p. 9). The passage of time between the original act of wrongdoing and the moment in which a remedy is sought raises particular challenges for doing justice in these contexts (Gibbs 2009).

How should we think about historical responsibility for colonial injustices and possible remedies and what do justice theories have to offer? Having thought about Maori rights in New Zealand, Jeremy Waldron developed a framework for tackling the question: 'What are we to do now about what has happened in the past' (Waldron 1992: 140)? Waldron's dynamic model of historical injustice:

> *assumes that the importance of the injustice has to do not only with the immediate impact of the events themselves (someone suffered; someone died; something was taken), but also with its connection to other events because she suffered she lost this opportunity to educate her children ... this land was not only taken from us, but never given back). Individuals live whole lives, not just series of momentary events; and acts of injustice not only hurt but blight those lives. (Waldron 1992: 146)*

In order to bring the connections between linked events to the fore, we need to attend to the structural basis of society. Structure is a concept that political theorists have analysed in studies of justice. Such studies focus on the structures and processes embodied in 'institutions, discourses, and practices' (Lu 2011: 267) that produce distributive patterns as well as the patterns themselves (Young 2006). According to Young, 'structural injustice exists when social processes put large categories of persons under a systematic threat of domination or deprivation of the means to develop and exercise their capacities, at the same time as these processes enable others to dominate or have a wide range of opportunities for developing and exercising their capacities' (Young 2006: 114). They are based on morally unacceptable values or belief systems (Lu 2011), such as colonialism, which in Australia and elsewhere relied on a premise of entitlement to conquer and subjugate and was buttressed by ideas of racial superiority.

In Australia, the consequences of colonisation – dispossession, subjugation and assimilation – are clearly evident in the differences between Indigenous and non-Indigenous communities in many measures of socio-economic standing, health and wellbeing (Balint *et al.* 2014). As the above shows, they have had and continue to have a discernible effect on the ability of Indigenous people to govern water, including rates of water ownership and capacities to benefit from its economic use and to participate in democratic processes of management.

A structural account of injustice acknowledges the role of institutional rules and accepted norms, focusing attention on those who 'participate in rules, institutions and social practices at various levels that enable, encourage and produce 'widespread and repeated' violations' (Lu 2011: 115). The restructuring that institutionalised new water access rules in the late 18th century and again 100 years later represent arenas in which injustice or justice could be enacted.

Waldron (1992) argues that a just course of action is one in which we: (i) act to alter the consequences that flow from historical injustice; and (ii) prevent the repetition or perpetuation of such. By this reasoning, the COAG leaders took an unjust course of action when they failed to consider and tackle the effect on water access of the alienation of Indigenous property rights in land and made no attempt to pursue strategies to prevent a continuation of historic injustices sustained through previous restructuring of Australian water law. South Africa's water allocation reform program, implemented after the demise of the apartheid regime, presents a model to which they might have turned for insight into restorative or reparative measures. Reformers in that case created a category of user referred to as Historically Disadvantaged Individuals and set targets for reallocation in order to remedy historical inequalities stemming from racial exclusion (Movik 2014).

An additional consideration is the failure to negotiate with Indigenous people as a compounding act that further strained relations between Indigenous and non-Indigenous Australians and re-enacted the exclusion of Indigenous voices that had occurred during previous acts of colonial resource appropriation (Altman 2004). This relational quality of justice is a dimension that has commanded the attention of some theorists (e.g. Gibbs 2009), but is beyond the scope of this chapter.

Conclusion

Colonial injustice expropriated land and water and rendered Indigenous people incapable of recognition as members of the public or as citizens with needs for water, or any entitlement to benefit from water use. The pattern of exclusion from water allocations imprinted by riparianism was entrenched in subsequent water allocations as rights to water were nationalised and then handed on from landholder to landholder under a state-administered allocation system (Jackson 2011). Injustice endured in water policy beyond the initial events of the colonial era, shaping the conditions of life experienced during the 20th century by both the dominant non-Indigenous and Indigenous societies to the detriment of the latter (Balint *et al.* 2014).

During the land rights and Native Title era state and Australian governments instituted statutory processes to restore some lands to their traditional owners and to recognise customary systems of law, which include rights to water. When water rights were detached from land titles, the intent of these acts of restitution was undermined. The effect of narrow definitions of customary rights to water in the latest water law and policy serve as a barrier to Indigenous access, alongside others (e.g. 1998 amendments to the *Native Title Act*) (Tan and Jackson 2013).

Indigenous Australians' rights and interests in water have been systematically and persistently marginalised because reparative mechanisms have not been established in any of the processes that have made and re-made water law and policy over the course of Australian history. In so far as the recent water reform agenda, ongoing governance processes, such as the review of the *Water Act*, and water policy discourse ignore these historical and structural issues, they will continue to conceal the violence enacted throughout the expropriation of Indigenous lands and creation of Australia's agricultural systems, as well as the injustice evident in today's pattern of water access distribution.

References

Altman JC (2004) Indigenous interests and water property rights. *Dialogue* **23**, 29–33.

Altman J, Jackson S (2014) Indigenous land and sea management: recognition, redistribution, representation. In *Ten Commitments Revisited.* (Eds D Lindenmayer, S Dovers and S Morton) pp 207–216. CSIRO Publishing, Melbourne.

Arthur B (2010) *The Murray-Darling Basin Regional and Basin Plans: Indigenous Water and Land Data.* Murray Darling Basin Authority, Canberra.

Balint J, Evansy J, McMillan N (2014) Rethinking transitional justice, redressing indigenous harm: a new conceptual approach. *The International Journal of Transitional Justice*, 1–23.

Barber M, Jackson S (2011) Indigenous people, water values and resource development pressures in the Pilbara region of north-west Australia. *Australian Aboriginal Studies* **2**, 32–49.

Bartlett R (1997) Native Title to water. In *Water Law in Western Australia: Comparative Studies and Options for Reform.* (Eds R Bartlett, A Gardner and S Mascher) pp. 125–160. University of Western Australia, Perth, WA.

Behrendt J (2011) Some emerging issues in relation to claims to land under the *Aboriginal Land Rights Act 1983 (NSW). The University of New South Wales Law Journal* **34**, 811–834.

Connell D (2007) *Water Politics in the Murray-Darling Basin.* The Federation Press, Sydney.

Dempsey J, Gould K, Sundberg J (2011) Changing land tenure defining subjects: neoliberalism and property regimes on Indigenous reserves. In *Rethinking the Great White North: Race, Nature and the Historical Geographies of Whiteness in Canada.* (Eds A Baldwin, L Cameron and A Kobayashi) pp. 233–235. University of British Columbia Press, Vancouver, Canada.

Gibbs M (2009) Using restorative justice to resolve historical injustices of Indigenous peoples. *Contemporary Justice Review* **12**, 45–57. doi:10.1080/10282580802681725

Goodall H (1996) *Invasion to Embassy: Land in Aboriginal Politics in New South Wales, 1770–1970.* Allen & Unwin, Sydney.

Harris E (2007) Institutional change and economic growth: the evolution of water rights in Victoria, Australia 1850–1886. *Economic Papers* **26**, 118–127. doi:10.1111/j.1759-3441.2007.tb01011.x

Howard J (2010) Managing for justice in community-based water planning: a conceptual framework. *Environmental Conservation* **37**, 356–363. doi:10.1017/S0376892910000627

Hussey K, Dovers S (2006) Trajectories in Australian water policy. *Journal of Contemporary Water Research & Education* **135**, 36–50. doi:10.1111/j.1936-704X.2006.mp135001005.x

Ivison D, Patton P, Sanders W (2000) *Political Theory and the Rights of Indigenous Peoples.* Cambridge University Press, Cambridge, UK.

Jackson S (2011) Aboriginal access to water in Australia: opportunities and constraints. In *Water Resources, Planning and Management.* (Eds Q Grafton and K Hussey) pp. 601–628. Cambridge University Press, Cambridge, UK.

Jackson S, Altman JC (2009) Indigenous rights and water policy: perspectives from tropical northern Australia. *Australian Indigenous Law Review* **13**, 27–48.

Jackson S, Barber M (2013) Indigenous water values and resource governance in Australia's Northern Territory: current progress and ongoing challenges for social justice in water planning. *Planning Theory & Practice* **14**, 435–454. doi:10.1080/14649357.2013.845684

Jackson S, Langton M (2012) Trends in the recognition of indigenous water needs in Australian water reform: the limitations of 'cultural' entitlements in achieving water equity. *Journal of Water Law* **22**, 109–123.

Jackson S, Morrison J (2007) Indigenous perspectives in water management, reforms and implementation. In *Managing Water for Australia: The Social and Institutional Challenges.* (Eds K Hussey and S Dovers) pp. 23–41. CSIRO Publishing, Melbourne.

Jackson S, Tan P, Mooney C, Hoverman S, White I (2012) Principles and guidelines for good practice in Indigenous engagement in water planning. *Journal of Hydrology* **474**, 57–65. doi:10.1016/j.jhydrol.2011.12.015

Jung C (2009) 'Transitional justice for indigenous people in a non-transitional society'. Research brief. International Centre for Transitional Justice, New York, USA.

Lu C (2011) Colonialism as structural injustice: Historical responsibility and contemporary redress. *Journal of Political Philosophy* **19**, 261–281. doi:10.1111/j.1467-9760.2011.00403.x

Lukasiewicz A, Bowmer K, Syme GS, Davidson P (2013) Assessing government intentions for Australian water reform using a social justice framework. *Society & Natural Resources* **26**, 1314–1329. doi:10.1080/08941920.2013.791903

McAvoy T (2006) Water: fluid perceptions. *Transforming Cultures E-Journal* **1**, 97–103, <http://www.nwc.gov.au/organisation/partners/fpwec>.

Meyer L, Roser D (2010) Climate justice and historical emissions. *Critical Review of International Social and Political Philosophy* **13**, 229–253. doi:10.1080/13698230903326349

Movik S (2014) A fair share? Perceptions of justice in South Africa's water allocation reform policy. *Geoforum* **54**, 187–195. doi:10.1016/j.geoforum.2013.03.003

Nikolakis W, Grafton RQ (2014) Fairness and justice in Indigenous water allocations: insights from Northern Australia. *Water Policy* **16**, 19–35. doi:10.2166/wp.2014.206

Nikolakis W, Grafton RQ, To H (2013) Indigenous values and water markets: survey insights from north Australia. *Journal of Hydrology* **500**, 12–20. doi:10.1016/j.jhydrol.2013.07.016

NWC (2011) *The National Water Initiative – Securing Australia's Water Future: 2011 Assessment.* National Water Commission, Canberra.

Ribot J, Peluso JC (2003) A theory of access. *Rural Sociology* **60**, 153–181.

Sheehan J, Small G (2007) *Aqua Nullius.* Presentation given at the *13th Pacific-Rim Real Estate Society Conference.* 21–24 January, Fremantle, WA, <http://www.prres.net/Conference/Conference2007/PRRES%2007%20Conference%20Program.pdf>.

Smith D (1998) *Water in Australia: Resources and Management.* Oxford University Press, Oxford, UK.

Tan PL (2002) Legal issues relating to water use. In *Property: Rights and Responsibilities Current Australian Thinking.* (Eds C Mobbs and K Moore) pp. 13–42. Land and Water Australia, Canberra.

Tan PL, Jackson S (2013) Impossible Dreaming – does Australia's water law and policy fulfil Indigenous aspirations? *Environment and Planning Law Journal* **30**, 132–149.

Waldron J (1992) Historic injustice: its remembrance and supersession. In *Justice, Ethics and New Zealand Society.* (Eds G Oddie and R Perrett) pp. 139–170. Oxford University Press, Oxford, UK.

Weir J (2009) *Murray River Country: An Ecological Dialogue with Traditional Owners.* Aboriginal Studies Press, Canberra.

Weir J (2011) Water planning and dispossession. In *Basin Futures: Water Reform in the Murray-Darling Basin.* (Eds Q Grafton and D Connell) pp. 179–191. ANU E-Press, Canberra.

Young IM (2006) Taking the basic structure seriously. *Perspectives on Politics* **4**, 91–97. doi:10.1017/S1537592706060099

Zwarteveen MZ, Boelens R (2014) Defining, researching and struggling for water justice: some conceptual building blocks for research and action. *Water International* **39**, 143–158. doi:10.1080/02508060.2014.891168

10

Animal property rights: justice or conservation?

John Hadley

Summary

In this chapter, I will explain two contrasting theories of animal property rights (Hadley 2015). Whereas each of the two theories is consistent with the idea of giving wild animals property rights in their habitats, only the first theory – what I will refer to as the 'justice theory' – is consistent with the idea that habitat rights are owed to animals as a matter of basic justice. What do I mean by justice? When rights are held as a matter of basic justice, it means that certain persons must act in prescribed ways and states must pass laws designed to ensure that the intended recipients enjoy the relevant rights. The ethical presupposition behind the concept of justice is that people are obliged to act (or not act) irrespective of whether acting (or not acting) in such a way suits them. In other words, justice is a normative or aspirational concept: it speaks to what people *should* do and how states *should* be governed.[1] As far as animal property rights are concerned, the kernel of the justice theory is that people have an obligation to refrain from violating the habitat rights of wild animals. Below, I will say more about how such an obligation might translate in the real world, but suffice to say now it means that human landholders must act in good faith and take into account the interests of animals during land-management decision making.

The basic structure of animal property rights system

Before explaining the two theories of animal property rights, I will first explain the basic structure of an animal property rights system. Whether animals are given property rights as a matter of justice, or simply because the rights could be a useful conservation tool, the

1 A number of theoretical presuppositions constitute the philosophical background of such a concept of justice. Although constraints of space prevent me from presenting a thorough explication and defence of the conceptual and methodological foundations of 'analytic' philosophical ethics, three presuppositions are worth mentioning. The first is that the long-standing feature of human life commonly known as morality or ethics is real and a viable enterprise. In line with moral realism (Miller 2003), moral terms such as 'good', 'bad', 'right' and 'wrong' pick out bona fide values existing in the world. The second presupposition is that under the right circumstances human beings are capable of acting altruistically. Depending upon the theory of moral psychology that prevails, people may be motivated to act altruistically by either reason alone or feelings of sympathy. The third presupposition is the logical separation of morality from legality and the ceding of lexical priority to the former. The inappropriateness of equating morality with legality is meant to be obvious after a moment's reflection upon the injustices sanctioned by laws in apartheid South Africa, antebellum USA and Nazi Germany.

basic structure of an animal property rights system is the same. There are two key elements: guardianship and territorial behaviour (Hadley 2005; 2015).

Guardianship

If a landholder clears vegetation for agricultural or urban development, their activities damage the habitat of resident animals. The increasing rates of biodiversity decline illustrate that the ways in which animal interests are presently accounted for in habitat conservation systems are inadequate. An animal property rights system is designed to give animals a voice during the land-management decision-making process that puts their lives at risk. Obviously, animals cannot speak for themselves and some mechanism is required to facilitate the concept of an animal voice. Instead of simply requiring the landholder to report their land-management objectives to a regulatory agency, an animal property rights system impairs the landholder's rights by requiring them to hear a submission made by a human guardian on the animal's behalf. The basic idea is that the landholder would have to respond in good faith at an out of court mediation session. The requirement to have the landholder respond in person is necessary in order to do justice to the idea of giving animals a voice. If a landholder was not obliged to respond to a submission made on the animal's behalf, there is no guarantee that they have acted in good faith and taken the time to 'hear' the animals.

As radical as such a proposal may seem, it is not very different to the kind of vegetation planning and licensing systems that presently dominate habitat conservation systems for land outside of protected areas. Instead of a regulatory agency representative scrutinising the development application against a range of considerations in a politically charged context, the person charged with making the case for animals during the development approval process in an animal property rights system is independent. The animal property rights guardian is removed from the political fray that is so conducive to politically expedient administrative decisions.

An obvious question is who will be the guardians and how will the system be administered (Hadley 2015: 16–24)? In the absence of an animal property rights guardianship accreditation system, a person's eligibility to serve as an animal property rights guardian would require sufficient knowledge and skills in relevant fields such as ecology, animal welfare, law and land management. Most importantly, the guardian must be prepared to act in the best interests of the animals. Given it is foreseeable that mediation sessions between landholders and animal property rights guardians could be adversarial and emotionally charged, superior interpersonal and communication skills would also be an imperative. It is important not to overstate the potential for conflict, however; indeed, just as some landholders take out conservation covenants over their property (Office of Environment and Heritage 2016) and others willingly enter into land use agreements with Indigenous titleholders that have co-existing title to the same parcels of land (National Native Title Tribunal 2016), some landholders may be enthusiastically willing to embrace animal property rights. It is worth stressing that it is envisaged animal property rights will only operate over private land, and public land located outside of existing protected area systems. One of the logical implications of the basic needs argument is that animal guardianship mediation need only occur at times when land-management decision making puts animals at risk. Assuming that protected area systems are indeed protecting habitat, then there is no pressing moral reason to extend animal property rights to those areas. The effect of extending the scope of an animal property rights to protected areas would be purely symbolic.

Territorial or species-typical behaviour

The second key conceptual element of an animal property rights system is territorial behaviour (Hadley 2015: 24–28). Instead of using human property as a guide and drawing the boundaries of animal property in ecologically arbitrary or administratively expedient ways, the natural behaviour patterns of animals are used. In theory, the boundary of an animal's property will be co-extensive to their territorial boundary. Although the identification of territory is a contested enterprise, evidence from ecologists is regularly used in a wide range of land-use determinations. Given that territorial determinations are widely used in existing land-use planning systems, territory is not so nebulous a property identification measure as to be unsuitable as the mechanism for establishing workable animal property boundaries. Bear in mind that the impetus for the implementation of an animal property rights system is either considerations of basic justice or the failure of existing systems to achieve conservation goals. Some mechanism is needed, therefore, to ensure that the system works in an environmentally credible way, and territorial behaviour is well suited to the task. The vagaries of territorial determination ought not to preclude implementation of an animal property rights system when the costs associated with doing nothing are likely to be high.

An obvious implication of using territorial behaviour to determine the boundary of animal property is that the same guardian will be participating in mediation with several stakeholders. Indeed, in cases when the territorial range of a species stretches for hundreds of square kilometres, the idea of face-to-face mediation between each landholder and guardian would be impractical. Perhaps in such cases it is best to use core area or some other unit smaller than territory as the basis for the property boundary. Alternatively, mediation could take place in a series of town hall style meetings between guardians and landholders. Again, it is worth pointing out that the system is put in place because either animals are owed rights as a matter of justice, or existing conservation policy instruments are not working. While there are foreseeable problems associated with the design of an animal property rights system, the relevant question is whether such a system would make matters worse.

The justice theory of animal property rights

The basic needs argument

The strongest justice argument in favour of giving habitat rights to wild animals is based upon an animal's interest in using natural 'goods' (e.g. trees, shrubs, grasses, soils, rocks and water) to meet their basic needs (Hadley 2015; Cooke, in press). The argument is, therefore, what philosophers call an interest-based argument (Feinberg 1976; Cochrane 2012). An interest-based argument is an argument that appeals to the wellbeing of the individuals concerned. The argument I will be defending appeals to the importance that access to natural goods, and the impact of losing access to those goods, has upon the lives of resident animals. Given that animals use natural goods to meet their basic needs, and given also that denying them access to natural goods has an adverse impact upon their welfare, it is meaningful to say that animals 'have an interest' in using natural goods. This interest, in turn, is a *property* interest because property, by definition, is an institution concerned with regulating access to and usage of natural resources (Waldron 2004). Therefore, if animals have an interest in using natural goods, then they have a property interest.

The first premise of the basic needs argument is an uncontroversial descriptive claim: animals have an interest in using natural goods to meet their basic needs. I will assume that I do not need to defend this claim because it is a truism of ecology and biology that animals hunt, forage, roam, mark territories, build nests and shelters, and play and mate with other animals. In other words, the basic needs argument presupposes that animals are the living creatures that ecologists and conservation biologists tell us they are, and their characteristics are likewise the products of evolutionary forces. Given that animals use natural goods to meet their basic needs, it is also unproblematic to suggest that access to the goods concerned is beneficial to them and foreclosing access to the goods makes them worse off to some extent. That access and the denial of access to natural goods impacts upon animals in good and bad ways is recognition that habitat modification is a wellbeing issue and, therefore, animals have an interest in accessing and using the goods concerned.

The second premise of the basic needs argument is the controversial premise: the animal interest in using natural goods to meet their basic needs is at least as important as a *non-critical* human interest that, taking the contemporary institution of property as our guide, would be important enough to give rise to a property right in natural resources. The strategy behind the second premise is to make a cross-species comparison and then argue that animals cross the threshold level of moral significance requisite to qualify for a property right. This strategy is a little artful because it trades upon the relative ease by which human interests cross the requisite threshold level of significance to warrant property rights. In other words, only because it is relatively easy for human beings to acquire property rights in land do animals acquire them at all. The relative ease with which humans can acquire property opens the logical door for animals. The important normative claim implicit in the second premise is that the vital interests of animals matter at least as much as non-critical human interests.

There are two ways to demonstrate that the vital interests of animals are of comparable importance to the non-critical interests of human beings (Hadley 2015: 53–58). The first appeals to the significance of basic needs in the lives of the animals themselves. Having their basic needs met is a precondition for leading any kind of life at all. The widespread inclusion of the satisfaction of basic needs in animal welfare statutes and codes of practice around the world bolsters this argument. The satisfaction of basic needs is a mainstream ethical consideration, and even people who use and kill animals for a living can acknowledge the satisfaction of basic needs as a requisite condition for life. The second way to demonstrate that the vital interests of animals are of comparable importance to non-critical human interests is to focus on the desire to have basic needs satisfied and argue that the satisfaction of the desire alone is important enough to cross the threshold level of importance. Each of these two ways of arguing for the moral importance of the vital interests of animals is grounded in orthodox theories of wellbeing (Crisp 2015): the former is in line with the so-called objective list theory of wellbeing; the latter is in line with the desire theory of wellbeing.[2]

2 If an individual's life contains certain 'goods' deemed to be intrinsically valuable, such as friendship, love, artistic appreciation, the satisfaction of basic needs (Griffin 1986: 42), knowledge and respect, then, according to proponents of the objective list theory, the individual enjoys high levels of wellbeing irrespective of whether they desire to have the goods concerned. In contrast, the desire theory of wellbeing makes desire satisfaction the measure of wellbeing and an individual will enjoy high levels of wellbeing in so far as they are having their desires satisfied.

The key question is whether the satisfaction of basic needs is a consideration important enough to cross the requisite threshold of importance necessary for property rights status. I think that there are good grounds for answering in the affirmative because: (i) basic needs are an intuitively compelling consideration; and (ii) in theory and practice, the threshold is set quite low to begin with. Consider what it takes to have property rights in land. Imagine a parcel of land that is presently unowned by any human being. By the lights of contemporary property rights theory, so long as any would-be owner sincerely intends to use the land in time-honoured ways, then there is a presumption in favour of allowing them to own it. Whether land is the kind of good that people ought to be allowed to own in the first place is not seriously questioned; instead, the key question in property rights theory concerns the original allocation of property (Becker 1977: 3; Alexander and Peńalver 2012: 6): who, among any would-be owners, is to be the bearer of the property rights? Once that question has been answered, it is relatively uncontroversial that ownership comes with a bundle of rights, including rights to use, attached to it.

The basic needs argument asks us to accept that the animal's interest in satisfying their basic needs matters at least as much as a human being's interest in owning land, when the person doesn't need to own land as a requisite condition for leading a good life. Relevant cases would be owners who already own property or non-owners who otherwise enjoy a good standard of living. Property rights theory allows these people to acquire land even though the acquisition of land is not a life-and-death situation for them. In order to establish whether animals meet the threshold of importance requisite for property rights, therefore, we need only to set the bar at the lowest point requisite for successful acquisition of land by human beings. And, rightly or wrongly, as far as contemporary property rights theory is concerned, the bar is set quite low. Of course, in reality, nation states may place all sorts of requirements upon the availability of land for ownership, but these do not enter into any assessment of the theoretical eligibility of animals to own parcels of land that *are* deemed as appropriate objects of acquisition. If a state effectively makes it harder for people to acquire property by legislating to place conditions on the eligibility of human beings to own land and, in the process, raises the bar for animal eligibility, then the state shows how far away from contemporary property rights theory it has moved.

Someone might object that the interests of animals do not cross the threshold level of importance needed to qualify property rights because an animal's basic needs are not as important as the non-critical interests of a human being. Such a view, however, cannot make sense of the presence of animals in welfare and anti-cruelty statutes and codes of practice, let alone the lives of many people. Animal welfare and anti-cruelty statutes are designed to protect animals from the actions of human beings (Garner 2010: 130). Legal statutes function to check human behaviour and protect the animals themselves. Implicit in the rationale of such instruments is recognition that animals are not pieces of furniture but morally considerable beings. The basic needs argument takes the status of animals in liberal democracies as support for the normative stance reflected in the second premise. Notice that it is not an implication of the basic needs argument that the basic needs of humans and animals are equally important; rather, the second class status of animals in liberal democracies is reflected by the fact that the basic needs of animals matters as much as the non-critical interest of human beings (Hadley 2015: 58–59).

A further objection to the second premise is to reject making the comparison altogether. The objector is, in effect, saying that you cannot compare the interest of an animal with the interest of a human being because they are different kinds of considerations: one is a consideration pertaining to a non-person; the other is a consideration pertaining to a person. The objector is suggesting that the comparison is akin to trying to compare apples with oranges. But, consider the comparison in more detail. Imagine a hypothetical case involving a man called Ron (Hadley 2015: 7). Ron has a family that cares about him. He owns a house and has a secure job. He enjoys good health, has friends and one or two hobbies that he is passionate about. By any measure, Ron has a good life. Ron's interest in owning land, considered either simply as a desire meriting satisfaction or as a facilitating condition for an even better life than he now enjoys, must be weighed against the satisfaction of an animal's basic needs. I will not describe a particular animal because any number of specific descriptions is implicit in the uncontroversial first premise of the basic needs argument. In the process of describing, side by side, Ron's interest with the animal interest, aren't we, in effect, making the relevant comparison and putting ourselves in the position to make the judgment about their relative significance? Only someone who thought the process of judging was ring-fenced from the process of description, or someone who rejected the rationale of welfare statutes and norms, could deny that the comparison is legitimate. It is just incorrect as a psychological fact that we cannot make the comparison and, as the presence of animals in animal welfare and anti-cruelty statutes and codes reflects, it is wrong to say that it shouldn't be made.

Problems with the justice theory

There are two major problems with the justice theory of animal property rights: the multiple owner problem (Hadley 2015: 59–60) and the 'problematic' species problem (Hadley 2015: 76–80). Both of these problems are traceable to the justice theory's basic needs justification: if all it takes to be eligible for a property right in natural goods is an interest in using those goods to meet basic needs, then potentially all animals will be owed property rights. This entails, first, that in any given parcel of land there will likely be many dozens of species populations with many (possibly hundreds) of individuals, each one the legitimate bearer of property rights. Unless some kind of group ownership mechanism can be found to represent all the relevant individuals, then the animal property rights system will be theoretically complex and very difficult to administer in the real world. Second, the attribution of property rights will be species-blind and no consideration will be given to whether the animals belong to an exotic, native, invasive or over-abundant native species. Such an implication will be unwelcome to people that prioritise native animal species over problematic species, and may well turn off many people otherwise sympathetic to the idea of giving habitat rights to animals. It is worth emphasising this implication of the justice theory of animal property rights: individuals from problematic species will be as eligible for property rights as individuals from endangered or threatened species.

The conservation theory of animal property rights

The conservation theory of animal property rights is not vulnerable to the same theoretical and practical problems as the justice theory. In line with the conservation theory, animals are given habitat rights, not because they are owed them as a matter of basic justice, but simply because giving them property rights serves specific human purposes (Hadley 2015: 104–109). Animal property rights, in line with the conservation theory, are tools for the achievement of human objectives. Objectives for which animal property rights could be instrumentally used include the protection of an endangered species, the mitigation of human–wildlife conflict, and the promotion of more sustainable land-use decision making. Whereas the justice theory is direct because it extends rights to animals directly in virtue of what they are like; the conservation theory is 'indirect' because animals are merely the indirect beneficiaries of humans having particular goals that they want to achieve.

As a conservation tool, and not as a matter of basic justice, animal property rights are a versatile instrument. They can be rolled out only to favoured species or only in biodiversity hotspots. In areas in which habitat destruction is recognised as a problem, an animal property rights system can be implemented as an emergency measure. They can be legislated to override, or co-exist with, existing property rights, or designed to complement, or supplant, existing conservation initiatives. There is a degree of flexibility associated with a conservation-focused animal property system that is lacking in the justice-focused approach. In line with the conservation theory, there is no requirement to give all animals property rights because the rights are not the product of abstract reasoning about animals and their interests; instead, the reason for giving animals rights in the first place is an avowedly human-centred concern to achieve some objective.

A conservation theory of animal property rights is prey neither to the multiple owner problem nor the problem of problematic species. The conservation theory allows for property rights to be given only to a keystone species, or an individual or population that has a very large territory. Giving property rights to animals with large territories protects the territories of all the other species and populations who reside within the area protected by the property right. Practically, the conservation theory makes organising mediation between animal guardians and landholders a much simpler exercise. In cases when a single territory encompasses the property of many human landholders, then a representative of the stakeholders can be chosen to participate in the face-to-face mediation session or prepare the submission in response to the guardian's concerns about development.

Objection to the conservation theory of animal property rights

Not surprisingly, the main objection to the conservation theory of animal property rights concern its indirectness (Hadley 2015: 107). Under a conservation-based animal property rights system, giving animals property rights is conditional upon human beings having conservation goals that they want to achieve and for which animal property rights can be instrumentally useful in helping to achieve; without such goals, however, animals will be denied entitlements altogether, and this would be unjust.

The response to this objection is to draw attention to the theoretical gap between ideal theory and non-ideal theory (Garner 2013). The justice approach is in line with ideal theory because it is based upon a consideration – basic needs – that, upon reflection in the abstract, is regarded as having moral significance. In contrast, the conservation approach is in line with non-ideal theory because it is based upon a consideration – conservation goals – that is promoted in legislation and codes of practice in the real world. Notice that in each case, however, there is an intermediate step between the animal and the property rights. The intermediate step is a crucial intervention by a human being. In the justice approach, the intervention takes the form of a judgment about the moral significance of basic needs after abstract reflection; in the conservation approach, the intervention takes the form of a judgment about the instrumental value of giving animals property rights as a means of bringing about real-world objectives. Given that human beings play a decisive role in the derivation of the property rights in both theories, if indirectness is a problem for the conservation theory then it will also be a problem for the justice theory.

Case study: eastern water dragons in suburban Sydney

For as long as I can remember, a family of eastern water dragons have lived under a rock ledge adjacent to the pool in the backyard of my old family home. Even though they spend time in the water and sun themselves on the paving, the animals are ostensibly wild; no body feeds them or offers them assistance of any kind. Generation after generation of water dragons have been raised by the pool, co-existing successfully alongside human residents and visitors. The block of land backs on to an extensive reserve that traces the path of a creek leading into Middle Harbour. Presumably, the dragons have survived by availing themselves of the natural 'goods' that the backyard, and the reserve, has to offer.

At this point, some readers might be thinking: he can't possibly be about to suggest that the eastern water dragons are owed property rights as a matter of basic justice. But, think about this case in more depth. If the population of dragons were on a threatened or endangered species list (say, if they were like the population of long-nosed bandicoots purportedly resident in remnant strips of land along the Cooks River in the inner-Sydney suburb of Marrickville (Marrickville Council 2012), or the small number of koalas purportedly living in the beachside suburb of Avalon (Office of Environment and Heritage 2011)), then they would already be covered by protective instruments that are supposedly designed to protect their interests. Going the extra step and giving a property right to such animals would simply represent a tightening up of existing protections and be in line with any normative sentiment already implicit in such measures. Therefore, if giving property rights to the dragons is controversial, it cannot be because of the goods on offer, but rather because the population of dragons is not on a threatened species list or, in this particular case, they happen to be residing on private land. A justice approach to animal property rights, however, logically cannot play favourites – the fact the dragons are not on a threatened species list or that they reside on private land does not matter; all that matters is that they have an interest in using natural goods to satisfy their basic needs. Any landholder whose development activities placed the animals at risk would be obliged to take the interests of the animals into account and this, if we take the theory outlined above as a guide, would require them to meet with a designated guardian. The animal's guardian would have an opportunity to persuade the landholder to alter, delay, or scrap their development plans; indeed, in the case of transferring ownership of a property from one human being to another, the guardian would have an opportunity to spell out to any prospective buyer the consequences for the animals. This may have the effect of sensitising the owners to the plight of animals or it may not: the point is not necessarily to stop development, but to give animals a chance at survival by giving them a voice during the decision-making process that places them at risk.

Further objections to animal property rights

Some readers might be thinking that the animal property rights system described above does not go far enough and we should have an animal property rights system that allows for the dragons' property rights to trump those of the human owner. Others might argue the system goes too far and that we should allow the human owner's rights to extinguish the dragons' property rights every time. Both of these outcomes are potential options for any animal property rights system; it all depends on the weight given to the interests of the stakeholders concerned and the values the system aims to promote. The system outlined above aspires to be pragmatic: that is, to work with the existing institution of property and the prevailing ethical and policy priorities of liberal democracies. It would be as unrealistic to suggest that the system should be read straight off animal rights ideals as it would to maintain the unreflectively 'pro-human' status quo.

A further objection is that an animal property rights system of the kind envisaged above duplicates existing protective mechanisms. But this charge fails on two counts. First, the animal property rights system facilitates personal contact between stakeholders, whereas under existing provisions contact is simply administrative: that is, 'on paper'. Under the existing mechanisms, a representative of a regulatory agency runs an eye over a document; the regulator does not get to meet, let alone argue with or implore, a landholder face to face. Second, the guardian is independent or animal-focused in contrast to a regulatory agency representative that is meant to be balanced. The independence of the guardian, coupled with their duty to act in the animal's best interest, means that they can be a genuine advocate for the animals. For the guardian, animals are not merely one stakeholder group among many but the interest group of sole concern. Under the animal property rights system, the guardian assumes a functional role analogous to a partisan legal advocate, not an impartial judge.

Conclusion

In this chapter, I outlined two contrasting theories of animal property rights: the justice theory and the conservation theory. The justice theory affords a direct argument for animal property rights. By 'direct' I mean that the reason for giving property rights to animals is a consideration pertaining directly to the animals themselves. In other words, in line with the justice theory, animals qualify for property rights because of what they are like, and not because we choose to give property rights to them simply because it suits us. The justice theory is in contrast to the conservation theory of property rights. The conservation theory gives animals property rights because it is instrumentally useful for us to do so. In line with the conservation theory, animals are the recipients of property rights, but only because we think it is expedient for us to give them rights and not because they qualify in their own right. The conservation theory of animal property rights is an 'indirect' theory because animals are simply indirect beneficiaries of human projects and concerns. I suggested that giving animals property rights indirectly has several distinct advantages: most notably, it allows for the selective roll-out of property rights to only certain species and populations. If animal property rights are simply tools for the achievement of human purposes and not a matter of basic justice, then there is no obligation to give them to all animals and problematic species (e.g. exotic, invasive or over-abundant native) can be denied property rights in a theoretically consistent way.

References

Alexander S, Peñalver EM (2012) *An Introduction of the Theory of Property*. Cambridge University Press, Cambridge, UK.

Becker L (1977) *Property Rights: Philosophic Foundations*. Routledge and Keegan Paul, Boston MA, USA.

Cochrane A (2012) *Animal Rights Without Liberation*. Columbia University Press, New York, USA.

Cooke S (in press) Animal kingdoms: On habitat rights for nonhuman animals. *Environmental Values*.

Crisp R (2015) Well-Being. In *The Stanford Encyclopedia of Philosophy* (Ed EN Zalta) Stanford University, Stanford CA, USA, <http://plato.stanford.edu/archives/sum2015/entries/well-being/>.

Feinberg J (1976) Can animals have rights? In *Animal Rights and Human Obligations.* (Eds T Regan and P Singer) pp. 190–196. Prentice Hall, Englewood Cliffs, NJ, USA.

Garner R (2010) A defense of a broad animal protectionism. In *The Animal Rights Debate: Abolition or Regulation?* (Eds GL Francione and R Garner) pp. 103–173. Columbia University Press, New York, USA.

Garner R (2013) *A Theory of Justice for Animals.* Oxford University Press, Oxford, UK.

Griffin J (1986) *Well-being.* Oxford University Press, Oxford, UK.

Hadley J (2005) Nonhuman animal property: reconciling environmentalism and animal rights. *Journal of Social Philosophy* **36**(3), 305–315. doi:10.1111/j.1467-9833.2005.00277.x

Hadley J (2015) *Animal Property Rights: a Theory of Habitat Rights for Wild Animals.* Lexington Books, Lanham MD, USA.

Marrickville Council (2012) *Our Nature, Our Vision: Biodiversity Strategy 2011–2021.* Marrickville Council, Sydney, <http://www.marrickville.nsw.gov.au/Global/marr/marrwr/_assets/main/lib65149/our%20nature,%20our%20place.pdf>.

Miller A (2003) *An Introduction to Contemporary Metaethics.* Polity, Cambridge, UK.

National Native Title Tribunal (2016) *About Indigenous Land Use Agreements (ILUAs).* Native Title Lands Tribunal, Sydney, <http://www.nntt.gov.au/ILUAs/Pages/default.aspx>.

Office of Environment and Heritage (2011) *Koala Population, Pittwater Local Government Area – Endangered Population Listing.* Office of Environment and Heritage, Sydney, <http://www.environment.nsw.gov.au/determinations/KoalaPopulationPittwaterEndPopListing.htm>.

Office of Environment and Heritage (2016) *Conservation Agreements.* Office of Environment and Heritage, Sydney, <http://www.environment.nsw.gov.au/cpp/ConservationAgreements.htm>.

Waldron J (2004) Property and Ownership. In *The Stanford Encyclopedia of Philosophy* (Ed. EN Zalta) Stanford University, Stanford CA, USA, <http://plato.stanford.edu/archives/spr2012/entries/property/>.

11

Justice in water resource management

Claudia Baldwin

Summary

Water and its management are essential to quality of life and productive capacity and evoke deeply held values and emotions. Further, as a normative concept, what is considered fair or just is contestable. This chapter draws on several of my research projects in Australia, through six case studies, which used innovative visual research methods – participatory GIS, photovoice and 3D visualisation, in addition to interviews – to dig deep into concepts of justice surrounding water management. In each of the cases, innate locational characteristics and extreme events combined to create vulnerable situations that were influenced by the institutional context and either hindered or supported through appropriate application of elements of the Social Justice Framework (SJF). Implications of SJF principles for reducing conflict and delivering community wellbeing in adapting to a water-variable future are discussed.

Introduction

Water, as an increasingly scarce and contested resource, evokes deeply held values and emotions that touch powerful human interests and basic needs for security, economic wellbeing, a sense of belonging, recognition and control over one's life (Fisher *et al.* 1991). Meeting these needs contributes to individual and community wellbeing. Threats to these – for example through inequitable processes and outcomes – can generate conflict.

Studies of international water disputes have found that the likelihood and intensity of dispute increases as the rate of change within a basin exceeds the institutional capacity to absorb that change (Wolf *et al.* 2002). Tisdell (2003) argues that if the water doctrine (i.e. legal framework) of a state or country conforms to social justice principles then the water policies derived from it should produce just outcomes. However, justice is a normative concept. Community attitudes and norms of what is equitable vary, aptly illustrated in case study 3 described in this chapter. What is in the public interest is contestable: the greatest benefit for the largest number of people; employment growth in regional areas; or long-term environmental health? How differential impacts, costs and benefits are shared, and trade-offs made, has been core to just decision making (Pincetl and Hogue 2015).

The international water governance agenda has given primacy to a neoliberal approach to allocating water to where marginal economic returns are highest, justified in terms of efficiency, effectiveness or productivity (Joy *et al.* 2014). Of less importance is social justice or consideration of benefits that are less easily quantifiable or produce a lower return in market terms, yet play an important role to a segment of society. In developing countries such as India (Joy *et al.* 2014) or Indonesia (Sjah and Baldwin 2014), a trend is to see water being re-allocated from producing food crops or domestic use in rural areas to urban water reticulation, bottled water companies or tourist resorts, where access to and control of water is increasingly concentrated in the hands of a privileged few who have technological means to access it, and to the detriment of many others (Joy *et al.* 2014). So providing equitable access to water for urban dwellers to meet Millennium Development Goals might be at the cost of small landholders in rural areas, which could be mitigated with appropriate cognisance and processes put in place.

Water justice is thus scale- and place-sensitive, with justice at one scale sometimes implying injustice at another. Because of its biophysical characteristics and the spatial boundaries that shape how water is harvested, stored, used and managed, what is considered fair upstream might not be so downstream. Locational disadvantage leading to increased vulnerability to risk is a characteristic of all of the case studies referred to here. The histories and practices of water resource exploitation or development also have a role. The crux of the issue is that, although allocation outcomes are important, the process determines whether those with the least power, rights and voice are further marginalised (Joy *et al.* 2014). The popular neoliberal focus often means that certain voices have been taken into account more than others.

As a normative concept though, it is important that the variety of local views are taken into account. Joy *et al.* (2014: 16) suggests that 'justice is always deeply relational; it is about reaching agreement about how the costs and benefits of accessing water is or should be shared between different actors, or between people and the environment'. With plurality as a starting point, political participation and recognition provide the context against which to search for common ground at higher levels, consistent with more universal concepts of justice (Joy *et al.* 2014). Reconciling recognition of the intrinsic localness of justice with the aim for more universal and overarching frameworks is thus a challenge (Joy *et al.* 2014). It is not clear-cut and is frequently contested. Justice studies benefit from interdisciplinary approaches. Water rights and the multiple ways in which rights are claimed, negotiated, defined and contested are central to an understanding of water justice (Joy *et al.* 2014).

Inherent to any discussion of water justice is understanding and managing risk. Climate change means that effects on variability and availability of water resources are unpredictable and may exacerbate current fluctuation between drought and floods in Australia. It will test design standards of water infrastructure (as in case study 4). Institutions may be stretched to cope, and many adaptations may be through spontaneous actions to perceived and actual risks in the environment (Adger 2001) rather than considered and planned approaches enhanced by consensus. I argue here that adaptive capacity and resilience of institutions and communities in times of change would be enhanced through adoption of the fairness principles of the SJF, which include distributional, procedural and interactional justice (Lukasiewicz and Baldwin 2014). The three components of SJF are described below (see also Chapter 17). Other authors support the three elements in substance using slightly different terminology for different purposes (see e.g. Boulle 2001 regarding causes of conflict).

Distributive justice refers to how resources are distributed, taking into account need, equity, self-interest and efficiency. It may consider substantive organisational, structural or concrete outcomes such as differential rewards, income and conditions of employment.

Procedural justice is about the decision-making process and rules that ensure fairness, such as transparency and good communication, neutrality of decision makers, accuracy, accountability, inclusiveness and timeliness. It reveals influence of power and power-brokers on decisions. It may include the way complaints or a dispute has been dealt with.

Interactional justice refers to interpersonal relations among various stakeholders and with decision makers, including trust, respect, recognition of social standing and dignified treatment, truthfulness and propriety. Poorly dealt with, it can result in hurt emotions, vindictiveness and conflict.

This paper examines six case studies in which the author has been involved in research that raises questions about fairness. Each of the cases are characterised by innate locational attributes that make people more vulnerable to extreme events, such as droughts and floods. These water resource and/or management cases have sorely tested existing institutional arrangements. The case studies are examined according to the SJF to identify fairness in relation to distribution of costs and benefits and implications for future adaptation to a water-variable future. Although the environment is a 'loser' in all of the cases to some extent, my focus is on analysing inequity among water users or residents using the concepts of distributional, procedural and interactional justice.

Characteristics of the six research case studies

Six case study areas have been chosen to illustrate aspects of justice in relation to water resources, and to stimulate a conversation about how justice could be better served in such

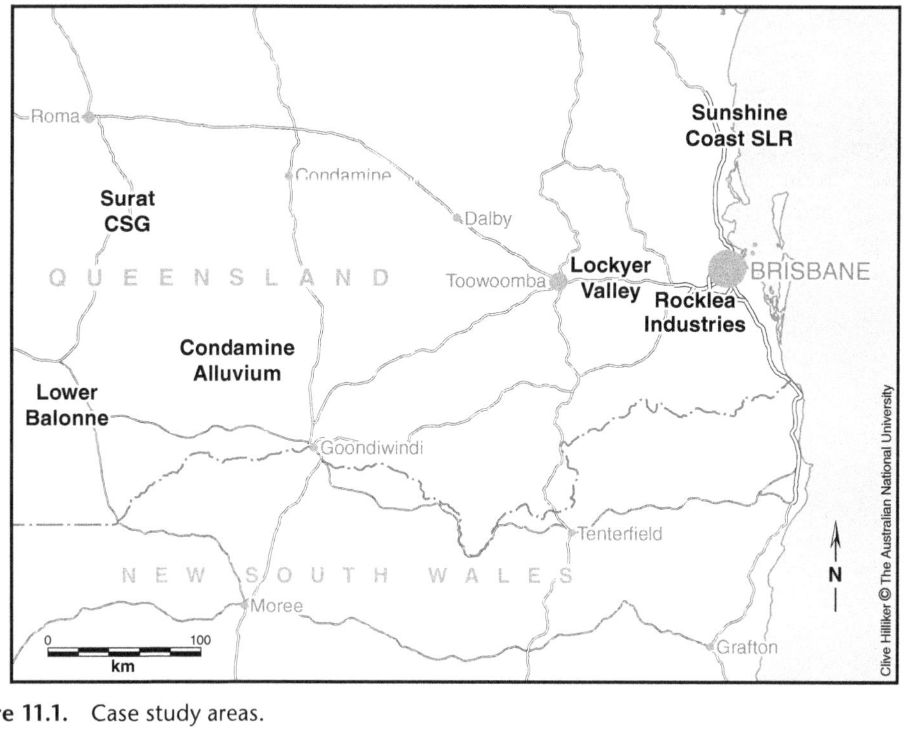

Figure 11.1. Case study areas.

Table 11.1. Case studies, methods used and locational characteristics

Case study (and references)	Methods	Locational characteristics
1. Lockyer Valley irrigators (Baldwin 2008a; Sarker *et al.* 2009; Baldwin 2011; Baldwin and Ross 2012)	Photovoice interviews and focus groups of 33 key stakeholders	Upstream *v.* downstream; over-extraction of surface and groundwater; uneven regulation; drought
2. Lower Balonne WAP stakeholders (Baldwin 2008b; Lukasiewicz and Baldwin 2014)	Photovoice interviews of 34 key stakeholders	Upstream *v.* downstream; floodplain; seasonal variation resulting in capture of overland flow
3. Condamine Alluvium (Tan *et al.* 2012)	Irrigator questionnaire survey (43 responses) and MCA with advisory panel, mainly irrigators	Over-allocated groundwater; some parts of Alluvium more affected
4. Rocklea industries (Baldwin 2013)	Semi-structured interviews with eight business owners; photos	Floodplain flooded in extreme event; Wivenhoe Dam management queried
5. Surat CSG development (Marshall *et al.* unpublished; Tan *et al.* 2012)	Document analysis; Condamine Alluvium survey	Lack of participation of mining companies in water allocation planning process; no limits on water extraction (compared with irrigators)
6. Potential sea-level rise in a Sunshine Coast canal estate (Grant *et al.* 2015; Lieske *et al.* 2015)	Workshops using photovoice; flood maps; 3D visualisation fly-through; and interactive mapping	Adjacent to an estuary; vulnerable to flooding from increased storm events and sea-level rise due to climate change

situations (Fig. 11.1). Table 11.1 summarises the methods used in each research case study (CS) and the locational attributes relevant to their vulnerability to extreme events and justice issues. Following the table, the relevant context of each case study is described according to geographical location and an extreme situation that has tested, or will test, resilience (e.g. drought, flood or sea level rise). The method of study is briefly described. Standard qualitative methods of semi-structured interviews, surveys and focus groups were combined with visual techniques in four of the case studies (CS1, 2, 4, 6). One other (CS3) employed a multi-criteria analysis with an advisory panel. For a deeper understanding of each of the case studies, refer to the relevant published articles.

Case study 1. Lockyer Valley

(Sources: Baldwin and Ross 2012; Baldwin 2011; Sarker *et al.* 2009; Baldwin 2008a)

The Lockyer catchment in South-East Queensland is a major horticultural area, located an hour by road west of Brisbane. Irrigation has intensified over the years to meet market demands for a year-round supply of vegetables and fruit, resulting in reduced streamflow and increasing reliance on groundwater. Groundwater used for agriculture was regulated (licenced and metered) in about one-third of the catchment, primarily the central Lockyer Valley. During the 'Millennium Drought' (2001–2007), most upstream farmers, who had captured water in large farm dams, had sufficient water for crops, whereas lower in the catchment up to half of the productive land was estimated to be temporarily withdrawn from cultivation due to insufficient surface and groundwater. On some properties, groundwater supply had become too saline to use.

The study took place at the commencement of the Queensland Government's water-planning process for this region, which is used to determine volumetric allocations of water to the environment and among various consumptive users. I used the photovoice method whereby participants were asked to take photos according to themes: in this case around sustainable use of water. Interviews of 33 stakeholders based around the photos identify underlying interests and values about water.

Case study 2. Lower Balonne WAP stakeholders

(Sources: Lukasiewicz and Baldwin 2014; Baldwin 2008b)

The Lower Balonne River is part of the Condamine-Balonne river catchment at the headwaters of the Murray–Darling Basin. The vast floodplain used to be regularly nourished (approximately every 18 months) by moderate flooding, and less frequently by large floods that slowly make their way into the Murray–Darling system. Downstream graziers, primarily in New South Wales (NSW), relied on floodwaters to regularly nourish sheep and cattle grazing land and for opportunistic cropping. The upper part of the Lower Balonne around St George-Dirribandi in Queensland was transformed by development of large private dams to capture floodwaters and bunding (i.e. low levees) to retain the overland flow water to provide for intensive irrigation of cotton. The combination of this development and the drought meant that, by 2007, there had been no floodwaters into northern NSW for the past 7 years, creating an economic and personal crisis for the downstream grazing industry.

This research took place in 2005 at the start of the second phase of the Queensland Government's water-planning process (i.e. the Resource Operation Plan) in which invited downstream graziers refused to participate due to perceived bias of government towards cotton farmers in the process. Similar to the Lockyer Valley study, I used photovoice interviews of 33 key stakeholders to identify underlying values about sustainable water management.

Case study 3. Central Condamine Alluvium

(Source: Tan *et al.* 2012)

This fertile agricultural area at the head of the Murray–Darling system in Queensland relies on groundwater from the Central Condamine Alluvium. The best scientific data estimates that the extraction needs to be reduced by 27 GL/annum to achieve a sustainable yield (CSIRO 2008). While there was little debate among landholders about the need for reduction, how that would be achieved was an issue.

The team undertaking the research used a 3D groundwater visualisation tool as a basis for discussion with stakeholders about functioning of regional aquifers, along with a survey of groundwater users. A multi-criteria analysis (MCA) was also conducted with a Community Reference Panel to seek views about setting and apportioning cuts to achieve sustainable yield. The results fed into the pre-planning for the Queensland Government's water planning process in 2011.

Case study 4. Flooding of Rocklea industrial area during Brisbane floods

(Source: Baldwin 2013)

As part of severe flooding from the Brisbane River from 12 to 14 January 2012, the Brisbane industrial suburb of Rocklea on the Oxley Creek floodplain was severely impacted. Although flooding in general was not as high as in the previous 1974 flood, in Rocklea some areas were inundated higher than levels predicted by Brisbane City Council, possibly due to stormwater back up. The completion of Wivenhoe dam in 1985 had been widely expected to mitigate future downstream flooding.

Eight in-depth interviews with business owners 6 months post-flood, supported by photos provided by them, found that warnings were inaccurate, not timely and inappropriate.

Case study 5. Surat Coal Seam Gas (CSG) development

(Sources: Marshall *et al.* unpublished; Tan *et al.* 2012)
The CSG industry has expanded rapidly since 2010 in the Western Downs and Maranoa regions in southern Queensland, resulting in 7000 wells by 2014 and another 40 000 wells estimated over the next 20 to 25 years (CSIRO 2011). Among the many concerns expressed about CSG development are potential impacts of extraction of groundwater, estimated at 75 GL per year to 480 GL per year (Biggs *et al.* 2012), and possible contamination of agricultural aquifers (NWC 2010; QWC 2012). Criticism has been lodged at the Queensland Government for inadequate regulation of the industry and consequent impact on water resources. Queensland CSG operators are exempt from needing a Water Extraction Licence (*Water Act 2000 (Qld)* s 3.5; 8.4) and an Underground Water Impact Report is not required until 14 months after production has commenced. Not only are there potential local impacts on individual wells and farmers, Tan *et al.* (2015) point out that there had been limited recognition of cumulative risk management of CSG at a regional scale and inadequate monitoring and ability to reverse impacts if found.

This case study is informed by a desk-top study of literature and legislation, and also draws on results of case study three.

Case study 6. Investigating effects of sea level rise in a Sunshine Coast canal estate

(Sources: Grant *et al.* 2015; Lieske *et al.* 2015)
Since the 1970s, low lying coastal areas in South-East Queensland have been developed into high-value waterfront canal estates from the Sunshine Coast south through Redland city, to the Gold Coast. During much of the period, little consideration was given to potential effects of sea-level rise on these low-lying coastal properties. Some, but not all, canal estates, are vulnerable to an increase in sea level rise, estimated to be 1.2 m by 2100. A Sunshine Coast canal estate was chosen as a subject of research due to recent community opposition to a proposal for adjacent development that would aggravate likelihood of flooding. Responsibility for tackling potential impacts of flooding from sea level rise is contested: the Queensland Government, which ultimately approved the original canal estate development; the Local Council, which has taken over ownership and maintenance of vacant land/parks, canals and some infrastructure; the developer who benefitted from creating developable land; and the landholder who is responsible for their property and revetment wall.

The study investigated the perceptions of risk due to sea-level rise in a Sunshine Coast canal estate through use of four visual methods in two focus groups. It is unique among the case studies in that it looks into the possible means of resolving a future problem using the SJF.

Common characteristics that contributed to inequity in these case study locations

Each of these case studies were characterised by a physical constraint due to geographical location that affected their vulnerability. In the Lockyer Valley, aquifer flow limited groundwater for downstream horticulture; the Lower Balonne floodplain relied on regular flooding of the flat land; those most affected in the Condamine Alluvium were located in an aquifer depression; Surat farmers were located above CSG deposits; Rocklea industries were located on a floodplain with history of flooding; and sea-level rise had not been considered when the Sunshine Coast canal estate was developed. Although people have choices about where to settle, the choices are often affected by historical factors and/or financial reasons, and risks are often unknown, underestimated or downplayed.

Moreover each of the cases experienced an extreme situation (or forecast a potential extreme event in the case of sea-level rise) that revealed inequities and tested resilience of the environment, institutional arrangements, individuals and the community. Drought drove high competition for scarce water resources, floods impacted some more than others, rapid growth of CSG affected agricultural landowners, and sea-level rise will have a potential, yet imprecisely known, effect on canal estates.

Analysis of case studies according to SJF criteria

Each case has been analysed according to the SJF criteria to identify common issues and supportive or unhelpful actions.

Distributive justice

As noted above, inherent characteristics of an area or a population can influence distributive justice, but it is also often affected by process, or a lack of good process, and therefore interacts with procedural justice. I first examine fairness in resource allocation or distribution, taking into account need, equity, self-interest and efficiency.

All of the case studies demonstrated the maximising of private gain or benefit, to the detriment of public interest or smaller companies and individuals. This affected deep values of economic wellbeing and loss of control over one's life. Inequitable distribution of costs and benefits was influenced by large corporations, whether cotton growers (CS2), SEQWater (CS1, 4), CSG companies (CS5) or insurance companies (CS4, 6). In a couple of cases, government was not perceived as being impartial (CS2, CS5), affecting individuals' sense of security. Some proponents were perceived to have greater access to power (such as CSG industry) and entitlements (cotton, CSG industries).

Institutional arrangements in terms of regulations enabled inequities to develop. Building of private dam infrastructure permitted capture of benefits upstream in both the Lockyer Valley (CS1) and Lower Balonne (CS2). Similarly, in the Surat Basin (CS5), institutional arrangements enabled mining companies to access resources with variable compensation to, and consideration of, landholders, their property and health. The CSG companies needed no licence to access water in this area, whereas agricultural landholders did. In the case of flooding in Rocklea (CS4), few of my study participants were able to claim on insurance. While participants in the canal estate research (CS6) appreciated the benefits of waterfront land, they soon became aware of potential risks to their investment and were already experiencing an increase in insurance premiums or had been refused insurance.

Of interest is community (normative) interpretation of justice. In the Condamine Alluvium (CS3), survey respondents and participants in MCA workshops were asked to choose among ways of reducing groundwater allocation: sharing cuts equally across all users (including urban); or variable reductions to equalise ML/ha take across sub-areas by considering recharge and deepening depressions in parts of the aquifer; or demonstration of efficient use of water (no wastage). The response from both survey respondents and Panel members in the MCA workshops recommended equal reduction in allocations across all users including urban, illustrating their local interpretation of fairness and equity. Survey respondents accepted the reduction to sustainable yield but did not support additional precautionary limitations based on restoring the aquifer, or to take into account uncertainties such as lack of data or potential climate change effects. The Panel identified fairness and equity, loss in income, availability of water for trade and other factors as important criteria for determining allocation cuts, but did not agree to taking into account uncertainty.

Procedural justice

In examining procedural justice, I examined aspects of the decision-making process such as inclusiveness, power and influence on decisions, rules that ensure fairness such as transparency, neutrality of decision maker, accuracy, accountability and timeliness. Each of the case studies demonstrated aspects of procedural justice that affected individuals' sense of belonging and feeling of being recognised, affecting the trust and relationships among parties, and hence interactional justice.

The major procedural issue in the Lockyer Valley (CS1) was that regulation did not extend upstream, and this disadvantaged downstream users. In negotiating with government during the water-planning process, however, the Lockyer irrigators presented a proposition on behalf of all irrigators for co-management of the water resource to ensure regulation of all.

In the Lower Balonne (CS2), issues regarding the water resource operation planning process disadvantaged downstream graziers who protested against the appointment of the chair of the water advisory committee who, as an irrigator, was not seen as neutral. The proposed terms of reference for the committee also made it difficult for single operator graziers to easily participate, touching on issues of inclusiveness and power.

All participants (landholders) in the Condamine Alluvium research (CS3) protested about the lack of involvement of the CSG industry in the water-planning process, the lack of transparency and information about CSG development and effect on the water resource, as well as inadequate regulation of the industry. Although landholders participate in a 2-year water planning process, which will affect their allocations, there were no such constraints on the CSG industry. This inequity continues to affect trust in government and with the industry.

In the Surat Basin (CS5), for example, extensive water planning to tackle over-allocation of water resources has led to reduced water allocations for irrigation, monitoring of water extraction and moratoriums on further development of agricultural bores. Yet, Tan *et al.* (2015: 1) refers to the legislation providing 'an enabling environment for [the CSG] industry, allowing unlimited volumes of groundwater to be extracted as a by-product'. Queensland is the only state that allows CSG companies unlimited rights to extract groundwater and not require a licence (Hamman 2013). The lack of transparency and accountability in relation to this industry means that landholders feel their needs are not adequately recognised and their future livelihood is at risk.

Among the Rocklea participants (CS4), transparency and accuracy of information played a role in trust of government. Predictions of flood impact were too late and imprecise for appropriate response. Many thought they were safe since Wivenhoe Dam had been built. Several business owners had communication difficulties (such as deafness or dyslexia) that made it difficult to access information and understand risk.

In regards to the canal estates (CS6), although flood risk maps are available on the local Council website, none of the participants had accessed them and few were aware that even under current sea level conditions with a severe storm event, transportation corridors throughout the estate would be impassable. Participants were concerned at a lack of awareness among residents and lack of an evacuation plan.

Interactional justice

An examination of interactional justice focuses on interpersonal relations among participants (community, industry and decision-making stakeholders) in the process. This includes issues of trust, respect, recognition of social standing and dignified treatment, truthfulness and propriety.

Lack of interactive justice was so strong in two of the case studies that parties withdrew from the process. Downstream graziers in the Lower Balonne (CS2) were so wary of government bias towards upstream irrigators that they refused to participate in the water-planning process even though this would guarantee that their voices would not be heard. In the Surat (CS5), opposition movements, such as Lock the Gate, formed to provide support for those aiming to prevent access of CSG industry to their land.

On the other hand, demonstration of good examples of building trust and respect took place in other case studies. In the Condamine Alluvium (CS3), Panel members spoke with trust and respect in terms of dealing with the issues and trying to find an equitable process for reducing allocations. In the Lockyer Valley (CS1), 17 sub-groups of irrigators came together based on an understanding of common ground to develop a mission around triple-bottom-line sustainability and speak with one voice about the rules they wanted for water allocation in the Lockyer Valley. In Rocklea (CS4), the lack of trust in the dam operator (quasi-government) being able to control the flood event was apparent, and ambivalence towards Brisbane City Council was voiced through outrage about inadequate warnings. But these feelings were mitigated somewhat by Council's efforts to support clean-up after the event and a desire that processes be put in place so that such flooding would not happen again.

The canal estate case study (CS6) is an exception to the others in that the participants have not yet been affected by a major event. Putting in place a plan to mitigate potential sea level rise effects would reduce potential interactional justice issues.

Discussion

Each of the case studies represents groups in vulnerable locations detrimentally affected by an extreme event. As mentioned, water justice is scale- and place-sensitive. These situations are expected to increase under climate change. Inadequate or inappropriate responses by institutions (public, quasi-public and private) will only serve to further alienate or disadvantage those affected. In some of these cases, the process and outcomes can be attributed to unintentional oversight; in others (CS2 and 5), deliberate decisions were made that knowingly disadvantaged certain segments of the community to the benefit of those with highest returns as measured primarily from an economic perspective.

Procedural fairness had implications for both distributive and interactional justice. If the process does not allow for access to information, participation, dialogue and input to decision making, then there is little opportunity to consider the normative aspects of water justice, and for actors to come closer to finding common ground. Soundly facilitated interaction makes it possible for actors to better understand each other; unfairness has difficulty in maintaining itself in the face of connection, thus giving meaning to Joy *et al.*'s (2014) comment that justice is relational, reinforcing the interactional component of the SJF.

The successful application of the SJF to these case studies post-completion reveals the robustness of the concept, and of the original qualitative methods that facilitated application of the SJF. The role of the researcher should not be underestimated in supporting procedural justice and reporting distributional fairness.

The inclusion of the canal estate case study allowed examination of the utility of the SJF to frame a possible future way forward. Including all stakeholders (landholders, government, developer and insurance companies) in a transparent dialogue, well informed about risk could assist in planning an adaptive response to flooding caused by extreme events combined with sea-level rise. A good process needs to be used to facilitate understanding and valuing others' perspectives, building respect and trust, advocating transparency and information sharing. Importantly, credible third parties have a role.

University researchers using sound and impactful methods can function as a boundary organisation, with the capacity to work outside conventional divisions of social knowledge and labour. Being an 'honest broker', creating a safe space for dialogue, and co-production of knowledge enhances credibility and effectiveness (Cash *et al.* 2006; McNie *et al.* 2008). Delivering or mediating technical information and different viewpoints between knowledge makers, decision makers and the community (Clarke *et al.* 2013) can create 'salience, credibility and legitimacy across boundaries' (Cash *et al.* 2006: 468), resulting in relationships built on mutual respect and trust.

Conclusion

Application of the SJF to the six case studies in southern Queensland has demonstrated the robustness of the concept to transparently illustrate components of justice in water resource management. It is particularly salient for populations in locations that are vulnerable to extreme events. It is incumbent on institutions to act fairly in these situations. The SJF provides guiding principles for reducing conflict and delivering community wellbeing in adapting to a water-variable future.

References

Adger N (2001) Scale of governance and environmental justice for adaptation and mitigation of climate change. *Journal of International Development* **13**, 921–931. doi:10.1002/jid.833

Baldwin C (2008a) Rules for the magic pudding: managing Lockyer groundwater. *Social Alternatives* **27**(3), 26–31.

Baldwin C (2008b) Integrating values and interests in water planning using a consensus-building approach. PhD thesis. University of Queensland, Brisbane.

Baldwin C (2011) Understanding the social obligations of farmers. In *Defending the Social Licence of Farming: Issues, Challenges and New Directions for Agriculture*. (Eds J Williams and P Martin) pp. 13 -21. CSIRO Publishing, Melbourne.

Baldwin C (2013) Impact of the 2011 flood event on a Brisbane industrial area. In *National Climate Change Adaptation Research Conference*, 25–27 June, Sydney.

Baldwin C, Ross H (2012) Bridging Troubled Waters: applying consensus-building techniques to water planning. *Society and Natural Resources: An International Journal* **25**(3), 217–234. doi:10.1080/08941920.2011.578120

Biggs A, Witheyman S, Williams K, Cupples N, de Voil C, Power R, *et al.* (2012) 'Assessing the salinity impacts of coal seam gas water on landscapes and surface streams'. Final report of

Activity 3 of the Healthy HeadWaters coal seam gas water feasibility study. DNRM, Toowoomba.
Boulle A (2001) *Mediation – Skills and Techniques*. Butterworths, Melbourne.
Cash D, Borck J, Patt A (2006) Countering the loading-dock approach to linking science and decision making: comparative analysis of El Niño/Southern Oscillation (ENSO) forecasting systems. *Science, Technology & Human Values* **31**(4), 465–494. doi:10.1177/0162243906287547
Clarke B, Stocker L, Coffey B, Leith P, Harvey N, Baldwin C, *et al.* (2013) Enhancing the knowledge -governance interface: coasts, climate and collaboration. *Ocean and Coastal Management* **86**, 88–99. doi:10.1016/j.ocecoaman.2013.02.009
CSIRO (2008) 'Water availability in the Condamine-Balonne. A report to the Australian Government from the CSIRO Murray Darling Basin Sustainable Yields project'. CSIRO, Australia.
CSIRO (2011) *Coal Seam Gas Developments – Predicting Impacts.* Fact sheet. CSIRO, Australia. <http://www.clw.csiro.au/publications/waterforahealthycountry/MM110611_CLW_CSGFactsheet_GPredictingImpacts_WEB_111031.pdf>.
Hamman E (Ed.) (2013) *Mining and Coal Seam Gas Law in Queensland: A Guide for the Community.* Queensland Environmental Defenders Office, Brisbane.
Fisher R, Ury W, Patton B (1991) *Getting to Yes: Negotiating an Agreement Without Giving in.* 2nd edn, Random House, Sydney.
Grant B, Baldwin C, Lieske S, Martin K (2015) Using participatory visual methods for information exchange about climate risk in canal estate communities. *Australian Journal of Maritime and Ocean Affairs* **7**(1), 23–37. doi:10.1080/18366503.2015.1014012
Joy K, Kulkarni S, Roth D, Zwarteveen M (2014) Re-politicising water governance: exploring water re-allocations in terms of justice. *Local Environment: The International Journal of Justice and Sustainability* **19**(9), 954–973. doi:10.1080/13549839.2013.870542
Lieske SN, Martin K, Grant B, Baldwin C (2015) Visualization methods for linking scientific and local knowledge of climate change impacts. In *Planning Support Systems for Smart Cities.* (Eds S Geertman, J Stillwell, J Ferreira and R Goodspeed) pp. 373 –389. Springer, New York, USA.
Lukasiewicz A, Baldwin C (2014) Voice, power and history: ensuring social justice for all stakeholders in water decision-making. *Local Environment: The International Journal of Justice and Sustainability* [online] 1–22. doi:10.1080/13549839.2014.942261
McNie EC, van Noordwijk M, Clark W, Dickson N, Sakuntaladewi N, Suyanto J, *et al.* (2008) *Boundary Organisations, Objects and Agents: Linking Knowledge with Action in Agroforestry Watersheds* Working Paper no. 80, Harvard University, Boston USA.
NWC (2010) *Position Statement: The Coal Seam Gas and Water Challenge.* National Water Commission, Canberra.
Pincetl S, Hogue T (2015) California's new normal? Recurring drought: addressing winners and losers. *Local Environment: The International Journal of Justice and Sustainability* **20**(7), 850–854. doi:10.1080/13549839.2015.1042778
Queensland Water Commission (2012) *Underground Water Impact Report for the Surat Cumulative Management Area.* Queensland Water Commission, Brisbane.
Sarker A, Baldwin C, Ross H (2009) Managing groundwater as a common-pool resource: an Australian case study. *Water Policy* **11**, 598–614. doi:10.2166/wp.2009.076
Sjah T, Baldwin C (2014) Options for future effective water management in Lombok: a multi-level nested framework. *Journal of Hydrology* **519**, 2448–2455. doi:10.1016/j.jhydrol.2014.07.006

Tan P-L, Baldwin C, White I, Burry K (2012) Water planning in the Condamine Alluvium, Queensland: sharing information and eliciting views in a context of overallocation. *Journal of Hydrology* **474**, 38–46. doi:10.1016/j.jhydrol.2012.01.004

Tan P-L, George D, Comino M (2015) Cumulative risk management, coal seam gas, sustainable water, and agriculture in Australia. *International Journal of Water Resources Development* **31**(4), 682–700. doi:10.1080/07900627.2014.994593

Tisdell J (2003) Equity and social justice in water doctrines. *Social Justice Research* **16**(4), 401–416. doi:10.1023/A:1026365700255

Wolf A, Yoffe S, Giordano M (2002) *International Waters: Indicators for Identifying Basins at Risk*. Technical documents in Hydrology PC-CP series No. 20. UNESCO IHP WWAP, Paris, France.

12

Legal determinations, geography and justice in Australia's coal seam gas debate

David J. Turton

Summary

Social and environmental justice issues loom large in Australia's coal seam gas (CSG) debate. This chapter will draw on relevant Australian judicial decisions and related literature to show how aspects of procedural and distributive justice have arisen for a variety of stakeholders associated with CSG extraction. Noting that social justice questions have been considered by geographers in the past and that aspects of spatial justice figure prominently in the work of legal geographers, this chapter focuses on the spatial and temporal features of CSG in Australia from an environmental and social justice perspective – using court judgments as short, illustrative case studies. Because both law and landscapes are shaped by contests over what is deemed just and unjust in society, this chapter also responds to recent calls for greater attention to justice questions by legal geography and justice scholars.

Introduction

Coal seam gas (CSG) is a regular feature of Australian newspaper headlines and is a politically charged topic for many communities, particularly in Queensland and New South Wales (e.g. Manning 2012; Munro 2012; Sherval and Hardiman 2014). Calls to expand this industry, which involves the extraction of methane from underground coal seams, have been driven partly by arguments that the adoption of CSG is a necessary step towards reducing fossil fuel emissions in the wider context of combating anthropogenic climate change (Lacey and Lamont 2014). It is notable that research into the possible greenhouse benefits of CSG over more conventional fossil fuel sources (such as oil and coal) remain ongoing against a need for greater consistency in the environmental impact assessment of CSG projects in Australia (Vickas *et al.* 2015). The potential of unconventional gas (encompassing CSG, tight gas and shale gas) to serve as alternatives to coal and oil in a lower carbon emitting energy future has also been questioned (e.g. Stephenson *et al.* 2012). The anti-CSG social movement Lock the Gate has sought to challenge the political support that exists for this industry by drawing attention to private property rights, the potential of CSG to threaten the economic viability of productive farming land and the ability of farmers to determine whether or not the industry should proceed on an individual's land (Colvin *et al.* 2015). Support for Lock the Gate's position exists among some federal politicians (e.g. ABC News 2015c), but as approvals for CSG developments lie largely with state governments, calls by these individuals for a farmers' right to 'say no' to CSG have been described by one commentator as politically opportunistic 'flip-flopping' (Windsor 2015: 193).

Other criticisms of the CSG sector relate to potential risks to water aquifers, above-ground spills of so-called 'produced water' arising from the extractive process (including 'fracking'), fugitive methane emissions, and the possibility of cumulative risks to ecosystems caused by necessary infrastructure such as roads and pipelines (e.g. Tan *et al.* 2015). It is important to note that the CSG industry is not without its supporters in rural and regional Australia, with some commentators arguing that CSG represents an economic boon to communities with potentially positive socio-economic impacts (e.g. Office of the Chief Economist 2015; Gribbin 2015; Schwartz 2015). In view of these contested perspectives, it is not surprising that notions of justice and fairness loom large in this community debate (e.g. Christie 2012; Munro 2012; Maloney 2014). Although Australia's CSG debate is certainly played out in urban centres – for example, through protests at sites of CSG extraction (such as AGL's Camden operation near Sydney in New South Wales, see Smith 2016) – arguments for and against the development of CSG revolve primarily around differing visions for the future of rural and regional Australia. Justice in this context is revealed in the tension between those who view CSG as a resource to be developed in the national interest – with potentially positive spill-over benefits into parts of rural and regional Australia – and those who fear that expanding this industry risks harming the future viability of existing agricultural business operations and the communities that exist around them. The challenge of finding an equitable path between these viewpoints remains ongoing. It is worth noting that arguments for and against CSG have been made against a larger background of declining populations, reduced service provision, drought, economic restructuring and cultural divisions in rural and regional Australia (e.g. Gross 2015), with some people contending that CSG provides a means of reversing and ameliorating these losses. These divergent aspirations for the future are also present in the existing CSG litigation discussed below.

Focusing on judicial determinations relating to CSG necessarily means reducing the scope of the chapter, with an emphasis towards what Marcel Wissenburg has described as the *legal* aspects of justice – 'more or less synonymous with positive law' – rather than moral or normative elements of justice (quoted in Gross 2014: 24). Narrowing in on judicial determinations also runs the risk of categorising the judiciary as a coherent, unified body, which is not necessarily the case (Hammerslev 2005). Judges are not disinterested observers in the CSG debate either, but rather key actors who nonetheless claim to have a neutral stance on the contentious issues brought before them (Delaney 2010). In focusing on questions of justice in the realm of CSG-related litigation, it is noted from the outset that law and justice do not always intersect, yet have the capacity to do so – as observed recently by Ronald Sackville QC, a former Judge of the Federal Court of Australia (Sackville 2014: 1162):

> *Law can be a powerful instrument of injustice, particularly in its impact on poor and vulnerable groups in the community. Yet law can also be a powerful force for the elimination or amelioration of injustice, especially through carefully designed legislative reforms supported by adequate mechanisms for their implementation.*

Australian litigation relating to CSG has involved not only government and anti-CSG community groups (e.g. *Fullerton Cove Residents Action Group Inc v Dart Energy Ltd (No 2) [2013] NSWLEC 38*; *Police v Rankin; Police v Roberts [2013] NSWLC 25*), but also CSG companies and their contractors (*AGL Sales (Qld) P/L v Dawson Sales P/L & Ors [2009] QCA 262*; *Saipem Australia Pty Ltd v GLNG Operations Pty Ltd [2014] QSC 310*). Such trends are replicated internationally and seem likely to continue (Preston 2014; Mulcahy 2015). Other cases have considered: freedom of information requests about the police response to anti-CSG protests (*Shoebridge v The Office of the Minister for Police and Emergency Services [2014] NSWCATAD 189*); the distribution of financial benefits of CSG to local government in the form of rates (*APT Petroleum Pipelines Pty Ltd v Western Downs Regional Council [2014] QLC 18*); and the use of fines as a deterrent following a failure to report environmental harm (*Connell v Santos NSW Pty Limited [2014] NSWLEC 1*). Not all of these cases can be considered here, nor are all of these judgments relevant to procedural and distributive justice – despite being indicative of a sometimes vitriolic community debate. An example of this can be seen in a recent defamation suit involving comments made by a talkback radio host about CSG and the then Deputy Premier of New South Wales, Andrew Stoner (*Stoner v Jones [2015] NSWSC 585*).

This chapter therefore draws upon relevant Australian judicial decisions to showcase how aspects of procedural and distributive justice have arisen for a variety of stakeholders associated with CSG extraction. By doing so, this chapter contributes to existing literature in the sub-discipline of legal geography and builds upon a growing interest in justice-related research in that field. Contending that these judicial cases are symptomatic of key themes in the wider CSG community debate (such as public participation), it is argued that the potential risks and benefits of CSG provide an opportunity to examine procedural and distributive justice issues in this public debate from various temporal and spatial perspectives, adding to existing geographical studies of environmental and social justice (Smith 1994; Cotton *et al.* 2014) – as well as emerging justice literature on unconventional gas more broadly (e.g. Fry *et al.* 2015; Ogneva-Himmelberger and Huang 2015; Clough and Bell 2016).

Going beyond a purely legalistic analysis of CSG court cases (e.g. Preston 2014), this chapter highlights the socio-political context that surrounds CSG litigation, because the court cases discussed are in many ways manifestations of larger themes in Australia's CSG discussion. Engaging with the existing domestic and international unconventional gas literature, it will be argued that Australian courts play an important role in driving improvements to stakeholder interaction and ultimately the quality of justice outcomes sought by various parties in the Australian CSG debate – despite their frequent reluctance to comment on the merits of CSG per se. A short clarification is necessary before canvasing the first judicial case study in this chapter. This concerns the concept of legal geography. For the purposes of this chapter, legal geography is approached as:

> *[A] stream of scholarship that makes the interconnections between law and spatiality, and especially their reciprocal construction, into core objects of inquiry. Legal geographers contend that in the world of lived social relations and experience, aspects of the law that are analytically identified as either legal or spatial are conjoined and co-constitutive. (Braverman* et al. *2014: 1)*

With a long tradition of commentators exploring the connections between law and justice (Kearns and Sarat 1996; Sackville 2014), it is not surprising that legal geographers have also delved into the realm of justice, often by examining spatial injustice and the law's role in constraining and enabling such power dynamics (Delaney 2016). Law and landscapes are shaped 'by contestations over what is considered just and unjust in different societies' (Jones 2006: 1). Indeed many legal geographers seek to engage with 'systemic *asymmetries* of power, such as domination, exploitation, and marginalization both in the world and with respect to access to law' (Delaney 2016: 268; see also Silbey 2001). The two key aspects of justice to be considered below are procedural and distributive justice. An early instance of procedural justice issues emerging in the context of CSG can be seen in the *Baker* decision, analysed below.

Procedural justice: *Baker v Minister for Employment, Skills and Mining & Another*

Legal meanings are projected on to 'every segment of the physical world' (Braverman *et al.* 2014: 1). These meanings are not fixed, but are instead shaped and reshaped by 'contested social practices and material realities' (Jepson 2012: 616). As an element of the legal process, litigation has a significant role to play in creating spaces, through judicial interpretations of time and geography (Benson 2014). Discussions of space are also relevant in negotiations between landholders and CSG companies, particularly where it serves as grounds for judicial review of government decision making. In 2011, QCLNG Pipeline Pty Ltd was engaged in negotiations with the Eidsvoid grazier Michael Baker to install a CSG pipeline across a portion of his property in the North Burnett region of Queensland. This was part of QCLNG's plan to construct a CSG pipeline from the Surat Basin to the industrial port of Gladstone in central Queensland. QCLNG's failure to provide Baker with an appropriately scaled map of the property that would be used for the pipeline was found to be a denial of procedural fairness. Procedural fairness is a fundamental concept of Australian administrative law and is premised on the idea that a person who might be adversely affected by an administrative decision must be given a fair hearing before a decision is made. In this case, procedural fairness required that QCLNG provide Baker with sufficient information to be able to participate meaningfully in the decision-making process with the company and also be permitted a reasonable opportunity to respond to QCLNG's pipeline application before a decision was made by the Queensland Department of Natural Resources and Mines. In her findings, Justice Jean Dalton observed that:

> *Remarkably enough, the application lodged on behalf of QCLNG did not contain a precise description of the land over which ... permission [for access] was sought, either in words, or in the two maps which were part of the application ... The realigned pipeline was nowhere shown on the material in the application and the map showing the original pipeline was at such a scale as to be meaningless to someone trying to ascertain where exactly the proposed pipeline land would run over Mr Baker's property. (Baker v Minister for Employment, Skills and Mining & Another [2012] QSC 160)*

Because Baker lacked a precise map of QCLNG's pipeline route and was not given accurate mapping information in sufficient time to respond to QCLNG's application to the Department (only being informed of the modified scale arrangements proposed by QCLNG a mere 2 days before the 28-day statutory notification period expired), Justice Dalton concluded Baker did not receive procedural fairness. As a consequence, she ruled that the Department's decision to grant QCLNG access in that particular location was void (*Baker v Minister for Employment, Skills and Mining & Another [2012] QSC 160*). In this instance, judicial understandings of time and space converged to create legal consequences for QCLNG and Baker (Turton 2015a). Scale is, of course, a politically constructed concept and its use in the legal arena is frequent, particularly in relation to matters of jurisdiction (Valverde 2009). Following Justice Dalton's decision, QCLNG claimed that their application failed on a mere 'technicality' (Gribbin 2012), but since this decision was handed down, no similar cause of action has yet been pursued by a landholder against a CSG company – suggesting that CSG operators have learned from the experience and become far more cautious in providing accurate geographical information to those with whom they are seeking to negotiate access for pipeline purposes and beyond.

Similar issues regarding the adequacy of geographical information arose before the *Baker* decision in *O'Connor & O'Connor v Arrow (Daandine) Pty Ltd [2009] QSC 432*, when an entry notice for the construction of a CSG water treatment pipeline was found to be void for failing to give sufficient particulars to a landholder. As Geritz *et al.* (2012) warned:

> *Given the conclusions in [the] Arrow Daandine and Baker [judgments], project proponents should be very careful in the descriptions that they provide to owners (and occupiers where relevant) in notices (including negotiation notices, Consultation Notices and notices of entry) under the various pieces of resources legislation. The lack of such precision could result in the notice being found to be void and the activities subsequently conducted being unlawful.*

A greater attention to land access information by CSG companies has perhaps also been encouraged by wide coverage of Justice Dalton's decision in *Baker*, both within legal circles and the press (e.g. Geritz *et al.* 2012; Hairsine and McCarthy 2012; Wilson and Jacques 2012). The above cases consider aspects of procedural justice in a geographical context, whether in terms of open and transparent information-sharing or allowing sufficient time for stakeholders to reach a considered decision. It is stressed that legal commentary on procedural justice as it relates to CSG in Australia has not been limited to the courts and professional journals, but can also be found in less formal settings, including CSG community forums that have featured lawyers as presenters on speaker panels (Turton 2015b).

Despite their importance in improving justice processes and outcomes, judges have emphasised their limited capacity to resolve the larger community debate surrounding CSG, as observed by Justice Pepper of the New South Wales Land and Environment Court:

> *Coal seam gas is contentious. This judgment will, however, do little to quell the current anxiety surrounding the coal seam gas mining debate. In this regard it must be understood that the merits, or otherwise, of the use of this resource are irrelevant to the issues raised for determination by these judicial review proceedings, concerning, as they do, only the lawfulness of the approval under challenge. (Barrington-Gloucester-Stroud Preservation Alliance Inc v Minister for Planning and Infrastructure [2012] 194 LGERA 113)*

This is not to suggest that the judiciary has been blind to the political undercurrent that surrounds CSG. Indeed, in one case concerning a vexatious prosecution launched by the New South Wales police against two anti-CSG protesters in 2013, the presiding magistrate strongly criticised the role of politics as a potential source of interference with proper criminal justice procedures and a waste of the court's time and resources (*Police v Rankin; Police v Roberts* [2013] NSWLC 25). This privileged arm of the legal profession has also been required to consider geographical questions in relation to particular spaces, which can indirectly influence the community debate. The law is bound up with space in surprising ways and this extends to the ways in which spatial tactics are infused with legal terms and offences. Criminal law, for instance, is filled with prohibitive measures designed to inhibit movement through space – indeed criminal offences such as homicide are laden with spatial and temporal features (e.g. Sylvestre *et al.* 2015), particularly in setting out the requisite elements needed to establish a criminal act (Farmer 2010).

Spatial and legal interplay can be seen in the types of relief sought between CSG operators and community group litigants: for example, the seeking of an injunction (a remedy long associated with the restraint of behaviour in particular places, Blandy and Sibley 2010) by an anti-CSG community group, so as to prevent a CSG operator from constructing infrastructure required for a pilot exploration program until substantive legal arguments about the environmental impacts of the proposal could be heard (*Fullerton Cove Residents Action Group Incorporated v Dart Energy Ltd [2012] NSWLEC 207*). In granting the injunction, Justice Sheahan ordered a temporary halt to the potential physical and economic geography of the exploration site (Platt 2014). This freeze on exploration activity was subsequently lifted on appeal (*Fullerton Cove Residents Action Group Inc v Dart Energy Ltd (No 2) [2013] NSWLEC 38*), but the proponent, Dart Energy, ceased exploration work in 2013. At the time of writing, the CSG licence for this region remains current, with some residents of Fullerton in the Lower Hunter Valley region suggesting that the New South Wales Government should cancel the licence following the expiration of a CSG licence buyback scheme instituted after community backlash against the CSG industry in that state (ABC News 2015a). Another important case concerning matters central to procedural justice in CSG decision making was the decision of Justice Button in *Metgasco Limited v Minister for Resources and Energy [2015] NSWSC 453*. This judgment (and its aftermath) poses particular questions about the nature of public participation: a core tenet of procedural justice.

'Effective' consultation: *Metgasco Limited v Minister for Resources and Energy*

Metgasco was a CSG company that held a licence for petroleum exploration at Bentley in the New South Wales North Coast region. This licence was initially issued under the *Petroleum (Onshore) Act 1991* (NSW) in November 1996. After presenting a CSG exploration program to the responsible Minister for Resources and Energy in March 2013, a delegate of the Minister approved the company's exploration plans on 6 February 2014. The CSG exploration licence related to a drilling site known as Rosella Well, located on freehold private land within a former gravel quarry. Approval for the licence was partly contingent upon Metgasco demonstrating adequate community consultation efforts, in addition to preparing the site sufficiently for exploration purposes. Despite Metgasco's efforts to engage the community, community opposition to CSG at Bentley remained strong – with public demonstrations at Metgasco's drilling site before the New South Wales state election. On 14 May 2014, the Minister's delegate, Rachel Connell, informed Metgasco of her decision to suspend operations at the Rosella site under section 22 of the *Petroleum (Onshore) Act*, due to an alleged failure by Metgasco to fulfil conditions to their CSG licence. Specifically, Rachel Connell determined that Metgasco had breached New South Wales Strategic Regional Land Use Policy delivery guidelines, entitled 'Guidelines for community consultation requirements for the extraction of coal and petroleum, including coal seam gas' ('the Guidelines'). Connell contended that Metgasco had failed in its community consultation obligations under the Guidelines, developing a 'defeatist attitude' to community consultation and claimed that Metgasco's representatives 'lacked the necessary skills to engage with an often hostile audience' (*Metgasco Limited v Minister for Resources and Energy [2015] NSWSC 453*). The court proceedings arose when Metgasco applied for judicial review of the Minister's decision to suspend their activities, after failing to secure an internal 'review' of the decision with the delegate concerned. Metgasco contended that the delegate's decision was unlawful and sought a declaration from the Court to this effect. The Guidelines stipulate that licence holders should seek to engage in 'genuine and effective participation', as 'an integral component' of their exploration activities (*Metgasco Limited v Minister for Resources and Energy [2015] NSWSC 453*).

Justice Button noted that Metgasco's consultation activities with the community included one-on-one meetings with landholders, letter box information drops, the signing of land access agreements, open community meetings in Casino and telephone communications to deal with concerns. Discussing the history of community opposition to CSG at the Rosella site, including the existence of a protest camp adjacent to drilling operations, Button observed that the Minister had not permitted Metgasco sufficient time to respond to the delegate's decision to suspend the Rosella licence, as required under the Act. Justice Button described the May 2014 suspension of Metgasco's licence as 'a bolt from the blue' as far as the company was concerned (*Metgasco Limited v Minister for Resources and Energy [2015] NSWSC 453*). The Minister contended that Metgasco had failed to engage in 'effective consultation' because, despite its community consultation program, widespread opposition to the Rosella project continued. On this point Justice Button took the view that:

> *[A] reading of the guidelines as a whole shows that they are speaking of what is required in terms of the* activities *of the person or body engaging in consultation, rather than focusing on the* results *of the consultation upon the minds of the persons being consulted...[T]he guidelines are not prescriptive and ... have the tone of constructive suggestions rather than firm commands. And construing them as a whole, their reference to* 'effective consultation' *to my mind focuses on the quality of the process of consultation, rather than on any outcome whereby the persons who are the focus of the consultation are persuaded by it. (Metgasco Limited v Minister for Resources and Energy [2015] NSWSC 453. Emphasis in original)*

Justice Button concluded that the delegate's decision to suspend Metgasco's licence had attached a substantial significance to anti-CSG community activism in the region, instead of focusing on Metgasco's efforts to consult with the community. In doing so, the Minister had given attention to an irrelevant consideration when reaching a decision about licence suspension. For these reasons, the suspension of Metgasco's licence was declared to be invalid. The *Metgasco* decision is significant for its examination of public participation as a key aspect of procedural justice, attending to the nature of proper decision making by government authorities who are faced with community protest over a controversial resource.

The case is also an illustration of how litigation can be used in large-scale community debates as an important testing ground for core concepts of justice, providing an opportunity for all parties to re-assess their approach and perhaps improve community consultation and planning post-judgment. Litigation can be used as a means of giving authoritative expression to procedural justice principles (such as equity and fairness), allowing room for improved processes to support future interactions and improved outcomes between stakeholders (Lacey and Lamont 2014). Stakeholder anxieties leading up to a court case can also be allayed following a judgment, as shown by the *Metgasco* decision. Before Justice Button's findings were delivered, the Australian Petroleum Production and Exploration Association was concerned that the presence of protesters at a CSG site might mean that genuine community consultation could be construed as ineffective by government decision makers (McCullough Robertson 2015). Although litigation is not necessarily desirable in all cases, or indeed a stress-free exercise for those involved (e.g. Christie 2012), judges do provide crucial feedback on significant public issues – which can lead to improved processes and ultimately justice outcomes.

Following Justice Button's decision, Metgasco entered into compensation negotiations with the New South Wales Government for the expense involved in litigation and economic loss resulting from its halted exploration activities. The New South Wales Government offered Metgasco a A$25 million settlement/buyback arrangement, in exchange for relinquishing its three Northern Rivers petroleum exploration licences – including its Rosella site (Metgasco 2015). This offer was presented to Metgasco's shareholders for a vote in December 2015, with Metgasco's board of directors unanimous in their support for the settlement offer (Metgasco 2015). The New South Wales Government took the view that the settlement was 'fair and reasonable', balancing 'the interests of NSW taxpayers and Metgasco's shareholders' (New South Wales Government 2015). Ultimately, the government's compensation offer was accepted by Metgasco's shareholders – who indicated the company would investigate future investment options outside New South Wales (ABC News 2015e). Beyond procedural justice questions and acting as a catalyst for further negotiation between different stakeholders, the courts have also delved into matters of distributive justice and it is to here that this chapter now turns.

Distributive justice: *APT Petroleum Pipelines Pty Ltd v Western Downs Regional Council*

As noted above in the introduction, support for CSG exists in communities that foresee economic and social benefits from the extraction of this resource. In the case of Australian local governments, opinions are divided as to the relative merits of CSG, with some local government authorities openly adopting an anti-CSG line (e.g. ABC News 2015b) – while others embrace the sector as advantageous to their community (Anonymous 2014). Presented with few legal powers with which to control CSG development, some councils have nonetheless found themselves embroiled in CSG disputes relating to town planning matters, such as a CSG wastewater storage pond (*Golder v Maranoa Regional Council & Ors [2014] QPEC 68*; *Westrex Services Pty Ltd & Anor v Maranoa Regional Council & Anor [2014] QPEC 30*). Yet even in places where there is strong support for the industry, the law can be used to dispute the value of landscapes associated with this resource, as was highlighted in the *APT Petroleum Pipelines* decision.

Ray Brown, Mayor for the Western Downs Regional Council in south-west Queensland until March 2016, has been prominent in the media as a strong supporter of the CSG industry – earning the nickname 'the Mayor for Mining' (Gribbin 2015). In September 2012, the Council resolved to modify their rate categories after adopting a 2012–13 Revenue

Statement (a document that established rate codes and categories for rateable land within the Council's boundaries) in line with section 104 of the Queensland *Local Government Act 2009*. Their decision impacted upon the company APT Petroleum Pipelines (APT), the operator of the Roma/Brisbane Gas Pipeline. Part of APT's operation included five gas compressor sites within the boundaries of the Council – infrastructure designed to facilitate the flow of gas to the pipeline. The Council categorised these five compressor sites in a rates notice issued to APT, defining these localities as 'Petroleum under 400 ha' – with land in this category intended for gas extraction or processing (or for purposes ancillary or associated with gas extraction or processing, for example, water storage and pipelines). In response, APT formally objected to the rates categorisations the Council had placed upon the five parcels of land relating to the gas compressors. An internal review of the rates decision by a delegate from the Council found no cause for re-categorisation in October 2012. This became the focus for an appeal to the Queensland Land Court. The question the Court faced was whether these compressor sites were appropriately categorised for rates purposes by the Council.

In court, APT argued that its compression stations were not 'CSG processing facilities' as the Council had zoned them. Instead, the company argued they should be zoned under the separate rates category of 'transport and storage' – which attracted a lesser fee. The Council countered by arguing that because APT was using the compressor sites for a pipeline that transported gas, it was a use associated with gas extraction and/or processing because the pipeline provided the means of transport of the gas from the extraction and processing locations to gas distributors and users. As an alternative argument, the Council countered that the use of the compressor sites was ancillary to gas extraction and/or processing, because the sites were incidental to and necessarily associated with gas extraction – the fact that APT did not itself extract and/or process the gas at the compressor sites was irrelevant in the Council's view.

Focusing on the question of whether or not the transportation of gas was ancillary or associated with gas extraction or processing, the Court concluded in March 2014 that APT's pipeline was a large structure and, as a matter of scale, not something readily thought of as ancillary or associated with gas extraction or processing. The Court accepted APT's argument that the activities performed at the compressor sites did not involve any refining or purifying of the gas that was supplied for transport by the pipeline. As a consequence, the Land Court found that the five gas compressor sites should be placed within the industrial transport/storage category for rate purposes (*APT Petroleum Pipelines Pty Ltd v Western Downs Regional Council [2014] QLC 18*).

From a distributive justice perspective, this case demonstrates that, even when local governments support CSG development, they may still turn to the law to attempt to secure what they believe to be a reasonable share of the economic benefits arising from the resource. Despite a costs order subsequently being imposed against the Council by the Land Court in August 2014 (*APT Petroleum Pipelines Pty Ltd v Western Downs Regional Council (No. 2) [2014] QLC 27*), there do not appear to be ongoing tensions between the Council and APT as a result of the legal challenge over rates. As Ray Brown later commented:

> *I don't think there is going to be any long term broader implications of the court's decision ... These [CSG] companies have an impact on our region and we simply attempted to rate it accordingly, to cover the cost of using it and the future sterilisation of that area. (Anonymous 2014)*

Crucially, the *APT* judgment considered the distributive justice aspect of CSG in Australia, underscoring the fact that particular rural communities in Queensland and elsewhere have needed to respond to the impacts of CSG, both positive and negative, as a result of a broader state and national energy preference for the industry's expansion. The wider Australian population stands to benefit from CSG royalties, due to the Crown's ownership of mineral resources, but this has not been without cost for some local governments in CSG impacted regions. The manner in which local government authorities such as the Western Downs seek to distribute the burdens and benefits of CSG is an ongoing exercise – provoking questions of equitable treatment by state governments and drawing attention to whether or not local governments are appropriately equipped to bear both the burdens and economic prospects of CSG-driven development. Legislative changes that reduce the ability of New South Wales local governments to participate in planning matters associated with CSG activities are examples of ongoing challenges to local control by state governments (ABC News 2015d). The desire of the Western Downs Regional Council to seek a larger share of the economic benefits arising from CSG, in this case through a levy on the land occupied by CSG-related infrastructure, is therefore symptomatic of larger governance challenges facing Australian local governments in mining regions (e.g. Cheshire *et al.* 2014).

Conclusion

There has yet to be a 'flood' of litigation surrounding CSG in Australia, but from the existing decisions it can be seen that procedural and distributive justice matters have arisen over this contested resource – whether the parties be landholders, social movements, CSG operators, contractors or governments. It can be expected that further judicial determinations will be made as this industry continues to expand, providing litigants with important opportunities to clarify their expectations of the CSG sector and attend to the larger social licence project confronting this industry in Australia. Future litigation may include elements currently missing from the corpus of cases covered above. An obvious example of this being landholder compensation. Although members of the Australian judiciary have been careful to avoid commenting on the merits of CSG in their findings, they are nonetheless powerful actors in this community discussion, with the capacity to demand more from community–company interactions through their rulings. Judicial determinations can draw upon geographical concepts to bring attention to justice processes and outcomes in the context of individual stakeholder disputes, thereby offering an important moderating voice in the ongoing community debate over CSG in Australia. Although uncertainties will remain in this national discussion – particularly given the differences between individual state laws surrounding CSG and their specific political proclivities – judgments from the courts can drive some positive changes to the negotiation dynamic between all parties in the longer term.

References

ABC News (2015a) PEL buyback ends, but questions remain over Lower Hunter CSG exploration. *ABC News*, 2 October 2015, <http://www.abc.net.au/news/2015-10-02/pel-buyback-ends2c-but-questions-remain-over-lower-hunter-csg-/6821954>.

ABC News (2015b) Lismore City Council votes to deny Metgasco access for seismic testing. *ABC News*, 14 October 2015, <http://www.abc.net.au/news/2015-10-14/lismmore-vs-metgasco/6852482>.

ABC News (2015c) Daughter of anti-CSG farmer George Bender, who took his own life, blasts lack of political will on issue. *ABC News*, 26 October 2015, <http://www.abc.net.au/news/2015-10-26/daughter-of-anti-csg-campaigner-blasts-politicians/6886902>.

ABC News (2015d) Lismore Council's plans to stop Metgasco from using their roads for seismic testing are in jeopardy. *ABC News*, 27 October 2015, <http://www.abc.net.au/news/2015-10-27/csg-changes/6889806>.

ABC News (2015e) Metgasco accepts $25m buyback offer, flags investment outside NSW. *ABC News*, 17 December 2015, <http: //www.abc.net.au/news/2015-12-16/metgasco-not-confident-to-invest-in-nsw/7034640>.

Anonymous (2014) Decision on CSG definition. *Shift Miner*, 19 August 2014, <http://shiftminer.com/news_item/decision-on-csg-definition-831>.

Benson MH (2014) Rules of engagement: the spatiality of judicial review. In *The Expanding Spaces of Law: a Timely Legal Geography*. (Eds I Braverman, N Blomley, D Delaney and A Kedar) pp. 215–238. Stanford University Press, Stanford CA, USA.

Blandy S, Sibley D (2010) Law, boundaries and the production of space. *Social & Legal Studies* **19**, 275–284. doi:10.1177/0964663910372178.

Braverman I, Blomley N, Delaney D, Kedar A (2014) Introduction: expanding the spaces of law. In *The Expanding Spaces of Law: a Timely Legal Geography*. (Eds I Braverman, N Blomley, D Delaney and A Kedar) pp. 1–29. Stanford University Press, Stanford CA, USA.

Cheshire L, Everingham JA, Lawrence G (2014) Governing the impacts of mining and the impacts of mining governance: challenges for rural and regional local governments in Australia. *Journal of Rural Studies* **36**, 330–339. doi:10.1016/j.jrurstud.2013.10.010

Christie E (2012) Coal seam gas development, environmental legislation and power: towards a pathway for conflict resolution and environmental justice. *Queensland Lawyer* **32**, 225–247.

Clough E, Bell D (2016) Just fracking: a distributive environmental justice analysis of unconventional gas development in Pennsylvania, USA. *Environmental Research Letters* **11**, 025001. doi:10.1088/1748-9326/11/2/025001

Colvin RM, Witt GB, Lacey J (2015) Strange bedfellows or an aligning of values? Exploration of stakeholder values in an alliance of concerned citizens against coal seam gas mining. *Land Use Policy* **42**, 392–399. doi:10.1016/j.landusepol.2014.08.014

Cotton M, Rattle I, Van Alstine J (2014) Shale gas policy in the United Kingdom: an argumentative discourse analysis. *Energy Policy* **73**, 427–438. doi:10.1016/j.enpol.2014.05.031

Delaney D (2010) *The Spatial, the Legal and the Pragmatics of World-Making*. Routledge, London, UK.

Delaney D (2016) Legal geography II: discerning injustice. *Progress in Human Geography* **40**, 267–274. doi:10.1177/0309132515571725

Farmer L (2010) Time and space in criminal law. *New Criminal Law Review: An International and Interdisciplinary Journal* **13**, 333–356.

Fry M, Briggle A, Kincaid J (2015) Fracking and environmental (in)justice in a Texas city. *Ecological Economics* **117**, 97–107. doi:10.1016/j.ecolecon.2015.06.012

Geritz M, Aro T, Nielsen T (2012) 'Near enough not good enough when consulting with land owners'. 19 July 2012, Clayton Utz, Sydney, <http://www.claytonutz.com/publications/edition/19_july_2012/20120719/near_enough_not_good_enough_when_consulting_with_land_owners.page>.

Gribbin C (2012) Court denies CSG company access to Queensland farm. *ABC Rural*, 26 June 2012, <http://www.abc.net.au/site-archive/rural/news/content/201206/s3533088.htm>.

Gribbin C (2015) After the gas rush. *ABC Landline*, 7 June 2015, <http://www.abc.net.au/landline/content/2015/s4263145.htm>.

Gross C (2014) *Fairness and Justice in Environmental Decision Making: Water under the bridge.* Routledge, London, UK.

Gross C (2015) Understanding the aspirations of people living in rural and regional Australia. In *Rural and Regional Futures.* (Eds A Hogan and M Young) pp. 299–318. Routledge, London, UK.

Hairsine K, McCarthy L (2012) Supreme Court declares part 5 permission void. Ashurst Australia Resources Law Update, 2 July 2012, <https://www.ashurst.com/doc.aspx?id_Content=7950>.

Hammerslev O (2005) How to study Danish judges. In *Theory and Method in Socio-Legal Research.* (Eds R Banakar and M Travers) pp. 203–214. Hart Publishing, Oxford, UK.

Jepson W (2012) Claiming space, claiming water: contested legal geographies of water in South Texas. *Annals of the Association of American Geographers* **102**, 614–631. doi:10.1080/00045608.2011.641897

Jones M (2006) Landscape, law and justice – concepts and issues. *Norsk Geografisk Tidsskrift – Norwegian. The Journal of Geography* **60**, 1–14. doi:10.1080/00291950600618726

Kearns TR, Sarat A (1996) Legal justice and injustice: towards a situated perspective. In *Justice and Injustice in Law and Legal Theory.* (Eds A Sarat and TR Kearns) pp. 1–17. University of Michigan Press, Ann Arbor MI, USA.

Lacey J, Lamont J (2014) Using social contract to inform social licence to operate: an application in the Australian coal seam gas industry. *Journal of Cleaner Production* **84**, 831–839. doi:10.1016/j.jclepro.2013.11.047

Maloney M (2014) Environmental justice network Australia – where to from here? *Asia Pacific Journal of Environmental Law* **17**, 155–161.

Manning P (2012) *What the Frack? Everything You Need to Know About Coal Seam Gas.* University of New South Wales Press, Sydney.

McCullough Robertson (2015) Metgasco Limited takes NSW government to court over suspension of activities under petroleum drilling licence, 7 May 2015. McCullough Robertson, Brisbane, <http://www.mccullough.com.au/icms_docs/216885_Metgasco_Limited_takes_NSW_government_to_court_over_suspension_of_activities_under_petroleum_drilling_licence.pdf>.

Metgasco (2015) Settlement/buy-back proposal, 2 November 2015, Metgasco, Sydney, <http://www.metgasco.com.au/asx-announcements/settlement-buy-back-proposal>.

Mulcahy C (2015) The changing face of disputes in the liquefied natural gas market. *Journal of Energy & Natural Resources Law* **33**, 271–287. doi:10.1080/02646811.2015.1043819

Munro S (2012) *Rich Land, Wasteland.* Macmillan, Sydney.

New South Wales Government (2015) *NSW Government Negotiations with Metgasco.* New South Wales Government, Sydney, <http://www.resourcesandenergy.nsw.gov.au/__data/assets/pdf_file/0004/583321/NSW-Government-negotiations-with-Metgasco.pdf>.

Office of the Chief Economist (2015) *Review of the Socioeconomic Impacts of Coal Seam Gas in Queensland.* Office of the Chief Economist, Australian Government Department of Industry, Innovation and Science, Canberra, <http://www.industry.gov.au/Office-of-the-Chief-Economist/Publications/Documents/coal-seam-gas/Socioeconomic-impacts-of-coal-seam-gas-in-Queensland.pdf>.

Ogneva-Himmelberger Y, Huang L (2015) Spatial distribution of unconventional gas wells and human populations in the Marcellus Shale in the United States: vulnerability analysis. *Applied Geography (Sevenoaks, England)* **60**, 165–174. doi:10.1016/j.apgeog.2015.03.011

Platt RH (2014) *Land Use and Society: Geography, Law, and Public Policy.* Island Press, Washington DC, USA.

Preston BJ (2014) Unconventional natural gas in the courts: an overview. *Journal of Energy & Natural Resources Law* **32**, 377–424. doi:10.1080/02646811.2014.11435369

Sackville R (2014) Law and justice: do they meet? Some personal reflections. *The University of New South Wales Law Journal* **37**, 1142–1163.

Schwartz D (2015) How one Queensland farm has made CSG work for them. *ABC Radio National AM*, 29 October 2015, <http://www.abc.net.au/am/content/2015/s4341166.htm>.

Sherval M, Hardiman K (2014) Competing perceptions of the rural idyll: responses to threats from coal seam gas development in Gloucester, NSW, Australia. *The Australian Geographer* **45**, 185–203. doi:10.1080/00049182.2014.899028

Silbey SS (2001) 'Let them eat cake': globalisation, postmodern colonialism, and the possibilities of justice. In *The Legal Geographies Reader.* (Eds N Blomley, D Delaney and RT Ford) pp. 256–275. Blackwell, Oxford, UK.

Smith DM (1994) *Geography and Social Justice*. Blackwell, Oxford, UK.

Smith K (2016) AGL accused of releasing wastewater into Sydney's sewerage system. *Green Left Weekly*, 5 March 2016, <https://www.greenleft.org.au/node/61242>.

Stephenson E, Doukas A, Shaw K (2012) Greenwashing gas: might a 'transition fuel' label legitimize carbon-intensive natural gas development? *Energy Policy* **46**, 452–459. doi:10.1016/j.enpol.2012.04.010

Sylvestre M, Damon W, Blomley N, Bellot C (2015) Spatial tactics in criminal courts and the politics of legal technicalities. *Antipode* **47**, 1346–1366. doi:10.1111/anti.12161

Tan PL, George D, Comino M (2015) Cumulative risk management, coal seam gas, sustainable water, and agriculture in Australia. *International Journal of Water Resources Development* **31**, 682–700. doi:10.1080/07900627.2014.994593

Turton DJ (2015*a*) Unconventional gas in Australia: towards a legal geography. *Geographical Research* **53**, 53–67. doi:10.1111/1745-5871.12101

Turton DJ (2015*b*) Lawyers in Australia's coal seam gas debate: a study of participation in recorded community forums. *The Extractive Industries and Society* **2**, 802–812. doi:10.1016/j.exis.2015.06.010

Valverde M (2009) Jurisdiction and scale: legal 'technicalities' as resources for theory. *Social & Legal Studies* **18**, 139–157. doi:10.1177/0964663909103622

Vickas M, McManus P, Dey C (2015) From the seam to the stove: greenhouse gas assessment and the coal seam gas industry in Australia. *The Australian Geographer* **46**, 73–90. doi:10.1080/00049182.2014.986791

Wilson R, Jacques O (2012) Case sets precedent for landholders. *The Queensland Times*, 25 June, <http://www.qt.com.au/news/case-sets-precedent-landholders/1430875/>.

Windsor T (2015) *Windsor's Way*. Melbourne University Press, Melbourne.

13

The justice implications of focusing on the economically efficient use of natural resources and their environmental impacts

Steven G.M. Schilizzi and Md. Sayed Iftekhar

Summary

This chapter endeavours to show how scarcity, in the form of limited budgets, as well as competing goals and assessment criteria, can affect the importance of social justice in environmental planning and policy. When distributing a resource or burden, efficiency and equity are often seen to come into conflict. Traditionally, economists have considered efficiency as the primary criterion for allocation decisions, with equity as a secondary concern, because doing otherwise would lead to welfare losses for society. The factors determining the existence of such trade-offs and their severity have not to date been thoroughly explored. This chapter uses two case studies to show that the relationships between social equity and a competing concern (such as economic efficiency) are generally not direct, but mediated by structural factors, such as stakeholders holding different valuation criteria or equity norms, the severity of budget constraints, or the proportion and types of stakeholders affected. Such factors strongly affect the role that equity concerns play in environmental decisions, in some cases resulting in a specific equity norm leading to outcomes that contradict the purpose of that norm; for instance, an egalitarian norm can lead to very unequal outcomes!

The nature of the problem

This chapter aims to show that, when there is more than one goal being pursued, there must be trade-offs, and each goal will typically have to yield some of its performance to that of the other(s). This is in particular true when concerns for equity or fairness coexist with concerns for economic efficiency (maximum value from available resources) or for on-the-ground environmental outcomes.

This chapter focuses explicitly on distributive justice, or equity, which is outcome-based, as opposed to process-based or otherwise. The fundamental philosophy of this chapter, as well as of Chapter 15, is not to define justice or equity directly, but to replace this 'dictator-like' approach by a 'democratic-like' approach, where (i) the existence of different views on, and definitions of, justice or equity is acknowledged as an empirical social fact, and (ii) this diversity among stakeholders' views of any given decision or policy is taken as the raw material from which an equity definition is constructed, such that it best reflects the constituency's diversity of views; Chapter 15 shows how this can be done, along with the caveats that go with it.

The study of distributive justice or equity has a long history in economics; in fact it goes back to the founding father of modern economic thinking, Adam Smith, who wrote two major books that in some way balance each other: *The Theory of Moral Sentiments* (Smith 1759) and *The Wealth of Nations* (Smith 1776). The latter focuses on economic efficiency and on how the market can steer the economy towards it; the first counter-balances *The Wealth of Nations* by focusing on the 'social preferences' of humans as citizens (rather than as producers and consumers), which dampen the forces of 'self-interest' guiding the 'invisible hand'. Now, 250 years later, the 'new institutional economics', linking with the theory of 'social capital' and game theory, is able to clarify just how that might be possible.

Until fairly recently, the study of justice and equity have remained at the fringe of economics – a bit like outcasts – but things are now changing, driven in particular by the rise of behavioural and experimental economics, where novel hypotheses about non-selfish and social behaviour can be tested and retested. This has led to a growing literature on the study of altruism, reciprocity, trust and giving, as evidenced by such choices as charitable donations, voluntarist involvement towards justice causes in the area of exploitation of women, children or ecosystems and the livelihoods that depend on them. The debate is ongoing among economists as to whether such behaviours are what they seem to be, or whether they express self-interest in disguised form (e.g. as a 'warm glow' effect, where one derives on utilitarian grounds psychological pleasure from 'doing good', or as 'moral satisfaction', where one derives on deontological grounds moral or religious benefits from 'doing good'). To an extent, such questions may just boil down to semantics; one way to dig deeper is to acknowledge their context-dependence and study how such behaviours change when context, suitably defined, is made to change in carefully designed experiments. For example, the authors have studied the extent to which self-interest itself, relative to genuine concerns for distributional equity, varies with changes in context.[1]

The work reported on here lies within the growing fields that study the economics of distributional equity, the nature of social preferences and the analysis of equity–efficiency trade-offs. Because of the inherent complexity of the subject, rather than a general treatment, this chapter will build on case studies, each providing an illustrative example of different aspects of the problem.

Summary of the relevant literature

In relation to distribution of a resource or burden, efficiency and equity are seen as often conflicting with each other. Traditionally, economists have considered economic efficiency as the primary criterion for allocation of resource or cost (Daniel 1975). For example, in his famous book *Equality and Efficiency: The Big Tradeoff*, Okun (1975) suggested that 'the conflict between equality and economic efficiency is inescapable' (p. 120). He suggested that any effort by the government to re-distribute and achieve equity will result in less efficient use of resources and a smaller level of total income, in particular because re-distribution induces administrative costs and changes in work effort, in savings and

1 Schilizzi and Iftekhar (2016), Self-serving Behavior in Distributional Choices: Context-Dependence in Multiple Dimensions, as yet an unpublished manuscript.

investment behaviour, and in people's attitudes (Blank 2002). However, later, through various examples, Blank (2002) showed that re-distribution of resources (to achieve equity) does not always lead to efficiency loss. In many cases, by appropriately designing the distribution programs, it is possible to reduce efficiency loss or even achieve efficiency gains.

There have been some attempts to include both efficiency and equity criteria in the same decision-making framework through the use of distributional weights in the standard benefit–cost analysis (BCA) (Drèze and Stern 1985; Adler and Posner 2009). Traditionally, BCA has been used to pursue efficiency goals (Loomis 2011). For example, traditional BCA would recommend a project if the total net benefit of the project is positive: it does not consider the distributional aspect of the benefit, so if poorer stakeholders were to be worse off, the project could still be selected, provided that the benefits to the richer section of the community are larger than the losses to the poorer section[2] (Adler 2013). The use of distributional weights, where costs and benefits are multiplied by the marginal utility of income and by the marginal social welfare of the affected individuals' utility, allows us to incorporate equity concerns in BCA (Weisbach 2015; Farrow 2011). However, there are arguments against using distributional weights, which stem from the fact that measuring marginal utility of income is very problematic, and in practice seldom feasible. As a result, distributional weights mostly reflect arbitrary choices and are so flexible that policy makers could justify any selection (Harberger 1978). Second, government agencies often use BCA to allocate projects within a limited budget, so that value for money is the primary concern. Maximisation of social welfare is often not the agency's objective: inclusion of distributional weights would deter from the main concern (Weisbach 2015).

In natural resource management, efficiency–equity trade-offs are prominent because environmental issues typically impact a particular section of society rather than the whole society. Also, government agencies are often able to select different combinations of environmental projects that would impact different sections of the community differently (Bento 2013). For example, Young *et al.* (1991) discussed two important dimensions of equity in the Conservation Reserve Program: cost distribution between government and farmers, as well as among farmers, and different project allocations resulting in different distributions of resources. Equity–efficiency trade-offs have been discussed in various other fields too, namely: water pricing (García-valiñas 2005); reducing fishing quotas to restore fish stocks (Bailey *et al.* 2013; Hanich *et al.* 2015); reducing water abstraction from rivers to meet environmental flows (Ward and Pulido-Velázquez 2008); and the allocation of global funds to reduce carbon emissions (Cattaneo *et al.* 2010). The main emphasis of these studies was to demonstrate how different allocation rules (based on equity and efficiency criteria) influence the wellbeing of the target groups, whether individuals, communities or whole countries.

Various techniques have been applied to study the trade-offs, ranging from theoretical analysis (Feldstein 1972) to field surveys (Akino and Hayami 1975). For example, Atkinson *et al.* (2000) reported a survey where they looked at people's preferences for different equity principles that defined the cost share of an environmental program. The equity principles differed as a function of 'responsibility for the environmental cost', 'benefits from

2 The rationale underlying this use of BCA, known as the Hicks-Kaldor or potential Pareto improvement (PPI) principle, rests with the idea that those who benefit could (be made to) compensate those who lose and still be better off than without the project. Economists have traditionally been of the view that actual compensation is the politician's job; the economist's is to provide the information that the project generates sufficient benefits for everyone to be better off (strict Pareto improvement), or at least no-one to be worse off (weak PI).

implementing the program' and 'the ability to pay'. They observed that it is often a mix of principles that best explain people's choices. For example, even though there was broad consensus on the 'polluter's pay' principle, consideration of 'ability to pay' somehow influenced this choice. In another survey, Haferkamp *et al.* (2009) studied the difference between lay people and economists in their preferences for certain labour market interventions: economists focused more on the efficiency aspects of the policies, whereas lay persons focused more on fairness aspects. On the more theoretical side, optimisation and simulation of the impacts of different policies are typically used to study trade-offs between efficiency and equity performance (e.g. Halpern *et al.* 2013).

A third methodology is the use of economic experiments. Examples of controlled laboratory experiments on equity–efficiency trade-offs abound (e.g. Engelmann and Strobel 2004; Bolton and Ockenfels 2006) and cover various domains: these include inter-generational equity (Fischer *et al.* 2004); preference for different equity norms and influence of self-serving bias in the choice of equity norms (Ubeda 2014); other-regarding preferences (Haisley and Weber 2010); differences between individuals and teams (Balafoutas *et al.* 2014); gender and socio-economic differences (Owens 2011), among many others. These studies typically focused on how different factors could influence people's choice of equity norms.

Case study 1: Allocating positively valued benefits – selecting conservation projects for funding

In the context of Australia's environmental conservation programs, large amounts of government money have been allocated across different states, regions and localities. For instance, the so-called NAPSWQ (National Action Plan for Salinity and Water Quality) allocated a total of A$1.4 billion over 7 years. One of the main relays for allocating funds for conservation projects has been through the Catchment Management Authorities (CMAs). Conservation projects are submitted by landholders or various conservation organisations and, at any given time, there typically are more submitted than can be funded. CMAs have had a varying degree of autonomy in selecting which projects would be funded and which would not. In the best of cases, CMA decision makers would have had a set of consistent selection criteria.[3]

Experience reveals that conservation projects generate several concerns, which typically do not sit well together. For example, project selection may be sensitive to: local power politics; the personal benefits that competing CMA decision makers may derive from the amount of funds they control; value-for-money considerations; and equity or fairness concerns. Let us assume the best of cases and that the decision makers would like to have a consistent set of selection criteria. Second, let us assume for simplicity that only two potentially conflicting concerns are present: value-for-money, which represents economic efficiency,[4] and a concern for justice, equity or fairness (JEF), restricted for the purpose of this example to distributional outcomes – hence 'equity' as the preferred term. Third, we assume that an equity metric has been worked out that, first, combines different notions of

3 In reality, direct experience suggests this is rarely the case and, when there are any criteria, these tend to be inconsistent, confounding and amalgamating different concerns in a haphazard manner.

4 Economic efficiency is defined for each project by the ratio of its (expected) economic benefits to its economic costs. Both costs and benefits are broader than just financial values and include so-called non-market values, which themselves include both use and non-use values. Environmental economics has developed methods for measuring such non-market (and non-financial) values.

Table 13.1. CMA projects with their efficiency (BCR) and equity metrics

Project	Total benefit (A$ million)	Total cost (A$ million)	BCR	Equity metric
Anna	3.9	2.8	1.4	1.56
Bill	9.9	2.6	3.8	1.00
Chelsea	4.4	4.0	1.1	2.10
Danny	16.1	4.6	3.5	2.12
Elsa	2.7	2.1	1.3	2.38
Fred	11.3	4.2	2.7	4.80
Georgia	35.5	7.4	4.8	4.61
Hugh	6.8	3.8	1.8	4.11
Ingrid	9.0	4.3	2.1	3.49
Jack	11.3	2.7	4.2	2.18

Note: How to compute an equity metric is the object of Chapter 15.

equity that different stakeholders might hold, and second, allows direct trade-offs with some other evaluation criterion – in this example (as further detailed in chapter 15), economic efficiency. What follows is based on work ongoing with a CMA in the state of Victoria.

A list of 10 projects (here named Anna to Jack) has been submitted to the CMA. Each project's own costs and expected benefits have been analysed using the INFFER package originally designed by Professor David Pannell of UWA,[5] and a benefit–cost ratio (BCR) has been calculated for each project, as shown in Table 13.1. If economic efficiency was the only concern, then the projects could easily be ranked accordingly, and the number selected, starting from the first-ranked, would depend on the available budget (e.g. if the budget allows only half of the 10 projects to be funded, the first 5 ranked would be selected). But if a second criterion such as equity is also present and the projects are ranked by their equity measure,[6] then the ranking would in general be different and a different set of projects would be selected. This is clearly the case with the projects shown in Table 13.1. If both criteria are used together, then the final ranking will depend on the relative weight given to efficiency and equity. But how important is this in practice, and what does the impact of different weightings depend on?

It turns out that the importance of weighting efficiency and equity differently depends on at least two factors. First, the rankings themselves matter only insofar as they determine which projects are selected and which are not: how a project is ranked within the selected set does not matter. Second, the size of the selection depends on the available budget: the greater it is, the larger the proportion of projects that can be selected. Thus, the tighter the budget, the greater the impacts on project selection from giving different weights to efficiency and equity.

Table 13.2, which builds on Table 13.1, illustrates how this happens. The horizontal lines represent the number of projects that can be selected given the budget constraint (above a given line). A very tight budget would allow for only 3 out of the 10 projects, a

5 See http://www.inffer.com.au/.

6 How to design an equity metric when different stakeholders hold different equity norms is the object of Chapter 15. We implicitly assume that different stakeholders are positioned to benefit more or less from different projects and that different stakeholders may not be deemed equally deserving.

Table 13.2. Effect of different efficiency–equity weightings on project selection

	Panel A				Panel B			
	Efficiency–equity weightings				Efficiency–equity weightings			
Rank	67–33%	33–67%	Same ranks	Rank distance	90–10%	10–90%	Same ranks	Rank distance
1	Georgia	Georgia	1	0	Georgia	Danny	0	3
2	Danny	Danny	1	0	Jack	Georgia	0	1
3	Jack	Ingrid	0	3	Bill	Ingrid	0	3
4	Bill	Fred	0	1	Danny	Fred	0	1
5	Fred	Jack	0	2	Fred	Jack	0	3
6	Ingrid	Bill	0	2	Ingrid	Bill	0	3
7	Elsa	Elsa	1	0	Hugh	Chelsea	0	3
8	Hugh	Chelsea	0	1	Anna	Elsa	0	1
9	Chelsea	Anna	0	1	Elsa	Anna	0	1
10	Anna	Hugh	0	2	Chelsea	Hugh	0	3
			3	12			0	22
			30%	24%			0%	44%

Note: The two percentage values under the two 'Rank distance' columns, in each panel, is the total relative to the maximum of 50, obtained by ranking the b%-a%-weighted list in reverse order to the a%-b%-weighted.

more generous one 5, and an even more generous one 6 (third line). The impact of the budget constraint is seen by comparing Panel B with Panel A: each is defined by different weightings of equity *v.* efficiency. Reversing each weighting and comparing each pair of opposites (67–33% *v.* 33–67% compared to 90–10% *v.* 10–90%) reveals the impact of the budget constraint. As the table shows, if the budget is generous enough to allow 6 out of the 10 projects to be funded, then it does not matter how the planners or decision makers weight equity *v.* economic efficiency. They could be divided into sub-groups with highly antagonistic philosophies or values, but it would make no difference in the end: the same 6 projects would be selected regardless. The tighter the budget, however, the greater the difference between the selections resulting from whether equity or efficiency is valued most.

Thus, depending on how constraining the budget is, the extent to which stakeholders or decision makers disagree on how they value social equity (or justice) against 'value-for-money' criteria may matter a lot, a little, or not at all. If it does not matter much, it would be unproductive to continue any argument as to which of the competing ethics mattered more for the selection of environmental conservation projects. It is unlikely that governments would tighten or relax budgets so as to influence justice-oriented or efficiency-oriented policies, but they would benefit from being aware of the implications when communities disagree about the importance of social justice relative to other concerns.

Case study 2: Prioritising support for house reconstruction after Pakistan's disastrous 2010 floods[7]

Following the disastrous floods that hit Pakistan in 2010 and destroyed much property, the Government of Pakistan (GOP) was faced with equity–efficiency trade-offs in prioritising which houses to help rebuild. Of the three types in rural areas and small towns, the '*kacha*', or poor man's house, costs ~100 000 rupees to rebuild; the mid-range house costs ~180 000 rupees; and the '*pakka*', or rich man's house (relatively speaking), costs about 415 000 rupees (column 1 in Table 13.3). As detailed in column 4 of Table 13.3, the more expensive types are expected to be more resistant to future floods: 3, 10 and 30 years of safety, respectively. But Table 13.4, column 8, also shows that poorer households are less likely to be able to rebuild their house. Houses are in principle to be rebuilt identically. This section simulates the equity–efficiency outcomes of several reconstruction policies that the kind of development that can, by inadequate land and soil management, amplify environmental catastrophes such as floods.

The GOP faces a budget constraint – a realistic figure for the Punjab's affected area is Rs. 100 million (about A$1.35 million), and considers four different subsidy rates: 100%, 75%, 50% and 25% (column 2 of Table 13.3). With Rs. 100 million, 1000 basic '*kachas*' can be 100% subsidised (column 3). The more high-quality '*pakkas*' subsidised, or the greater the subsidy per house, the fewer the number of homes the GOP can subsidise. There is a trade-off between the number of homes that can be built (i.e. the number of households who can benefit) and the quality of the homes in terms of resistance to future floods. One way to deal with this is to consider the total amount of flood protection measured in 'safe house-years', or SHYs (column 5 of Table 13.3), and decide to maximise that quantity. But

Table 13.3. Housing reconstruction options given a total budget of 100 million rupees

(1)	(2)	(3)	(4)	(5)	(6)
Reconstruction options	Financing = percentage subsidy	Cost to household (000 Rs.)	No. of units funded by district (base = 1000)	Level of risk = years safe (expected)	Quantity of flood prevention (safe house years) = (4) × (5)
Rebuild poor *kacha*	100%	0	1000	3	3000
Cost =	75%	25	1333	3	4000
Rs. 100 000/house	50%	50	2000	3	6000
	25%	75	4000	3	12 000
Build 'better'	100%	0	556	10	5556
house Cost =	75%	45	741	10	7407
Rs. 180 000/house	50%	90	1111	10	11 111
	25%	135	2222	10	22 222
Build safe *pakka*	100%	0	241	30	7229
Cost =	75%	104	321	30	9639
Rs. 415 000/house	50%	208	482	30	14 458
	25%	311	964	30	28 916

Note: The average cost of each type of house was estimated by the World Bank.

7 Md. Masood Azeem generated the data and results of this section as part of his work towards his PhD at UWA.

Table 13.4. Household's budget constraint and cost of rebuilding an equivalent new house

(1)	(2)	(3)	(4)	(5)	(6)	(7)	(8)*
Household categories	**Number of households (total 1000)**	**Annual expenses (Rs.)**	**Income spent on necessities**	**Annual income/ capita (Rs.)**	**Available budget (Rs.)**	**Type of house**	**Budget/ house cost**
Very poor	330	76 650	75%	102 200	25 550	*Kacha*	26%
Poor	240	82 125	63%	131 400	49 275	*Kacha*	49%
Middle	180	86 870	50%	173 740	86 870	Middle	48%
Mid rich	150	91 980	38%	245 280	153 300	Middle	85%
Rich	100	121 545	25%	486 180	364 635	*Pakka*	88%

Source: Authors' estimates based on statistical data for the Punjab in Pakistan.
* Column 8 is the ratio of household category's available budget (column 6) to the non-subsidised cost of reconstruction (Table 13.3 column 3).

pursuing this policy leads (of course) to the minimum 25% subsidy rate (to achieve the maximum 28 916 SHYs) but only of the highest quality 'pakkas'. With 25% subsidy, these would cost households Rs. 311 000, but only the 100 'rich' households, whose available income is over Rs. 364 000, can afford this: no-one else can. Out of the 964 *pakkas,* the GOP budget could afford (with a 25% subsidy), only 100 of them (10%). Any reconstruction program faces both its own budget constraint and each individual household's, which varies widely (Table 13.4): equity issues arise. The data is relevant to cost-effectiveness (or value-for-money) concerns (Table 13.3) and to equity concerns (Table 13.4). Equity concerns add a third dimension to the conflict between an aggregate economic goal and households' individual budget constraints.

Four equity norms were considered, given the costs of each house type, GOP's total budget and the different subsidisation options. An egalitarian norm (EG) allocated an equal proportion of homes across all household groups; inverse proportionality (IP) made the allocation inversely proportional to households' available income; vertical equity (VE) is a proportional rescaling of IP that allocates nothing to the richest category; and Pareto Compensation (PC) defines a maximum entitlement for each category: the cost of rebuilding its previous home (column 7 in Table 13.4). These equity norms result in the following allocations, in percentages of total houses affordable by the GOP's budget:

- Egalitarian: 20%; 20%; 20%; 20%; 20% (i.e. an equal share of the total budget)
- Inverse proportionality: 30%; 25%; 20%; 15%; 10% (as a function of column 6 in Table 13.4)
- Vertical equity: 50%; 30%; 15%; 5%; 0%
- Pareto (in 000s of Rs): 100; 100; 180; 180; 415 (these are maximum entitlements).

Once its total budget is fixed, the GOP's decision variables are the four subsidy rates. The policy deemed 'best' will depend on which evaluation criteria receive the greatest weight. There are four equity norms to choose from and several efficiency criteria either to

8 Another study, not reported here, where allocating negative 'takings' aimed at better conserving an over-exploited fishery, yielded additional insights. Not only do trade-offs between equity, efficiency and conservation outcomes depend on which type of equity norm is referred to, but those trade-offs also depend on what proportion of stakeholders are affected, on the degree of heterogeneity in resource quality, and on whether the economic and ecological values of the resource are positively or negatively correlated.

choose from (if one retained) or weigh up (if several are valued). These include the total number of houses built, the total amount of protection, the average cost per house, the average cost per unit of protection and proportion of budget spent. The first two are maximised while the last three are minimised.

From the many combinations of results, key features stand out. First, equity alone will not uniquely determine a policy: *many different policies are compatible with a given equity norm*. Both efficiency-related criteria and an equity norm need to be specified. This is because the two sets of criteria need to minimise incompatible outcomes. But compatibility is not always an issue. Thus, if we focus on maximising the proportion of fundable units that can meet households' budget constraints, it turns out that it does not matter whether we value the EG, the IP or the VE norms most: they all lead to the same outcome of 100% subsidisation of poor-quality *kacha* houses for everyone. Of course, this means that if middle class and rich people want to avoid replacing their previous house with a lower quality one, they must self-fund part of the reconstruction. Only if the PC norm is chosen do the results differ.

The interplay of this kind of problem's typical constraints can lead to unexpected results. For example:

- If GOP focuses on maximising the total amount of flood protection (measured in safe house-years) affordable with its Rs. 100 million budget, then the IP and VE norms, although of a similar kind, lead to drastically different outcomes.
- The IP (as well as the EG) norm results in all the reconstruction benefiting only the two richest categories (with a 75% subsidy level), while the VE norm spreads the reconstruction across all groups, except the richest, and heavily favours the two poorest groups, with a 100% subsidy.
- By contrast, when the policy goal is to minimise unspent budget money, the IP norm leads to a rather egalitarian allocation of houses, while the EG norm does not and indeed favours the richer households – a most counter-intuitive outcome!
- If the PC norm is chosen with maximum flood protection, the policy outcome suffers a whopping 43% loss in flood protection (from 7500 to 4300 SHYs). In this case, valuing both the amount of protection the GOP can provide and holding onto a PC notion of equity reveal some incompatibility, measurable by the percentage loss in criterion performance.

This study highlighted several unexpected outcomes when equity–efficiency trade-offs must be made in the context of binding constraints and heterogeneous stakeholders.

1. The constraints inherent to the situation can render preferences for different equity norms irrelevant: the outcome is *over-determined* by the chosen efficiency-related goals.
2. In other configurations, some equity norms can come into *conflict* with efficiency-related goals, and trying to achieve one comes at the cost of under-achieving the other.
3. In some configurations, *similar* equity norms can lead to drastically *different* outcomes, while radically *different* norms can lead to *similar* outcomes.
4. The choice of an equity norm can, surprisingly, lead to outcomes that appear to *contradict* the purpose of that norm, as when an egalitarian norm leads to very unequal outcomes!

Conclusions

The two studies presented here shed light on different aspects of the same problem, that of resolving the trade-offs between social equity, as a form of justice, and other competing concerns when these come into conflict; equity–efficiency trade-offs are a special case of particular importance for environmental, as well as social, policy. Although such conflict does not always happen, it most often does, and all the more so as the constraints on the allocation problem are greater. After all, if resources were unlimited, there would be no equity issue to worry about and, therefore, no equity–efficiency trade-offs.

The first study revealed that giving more weight to social equity than economic efficiency can be sometimes effective and sometimes pointless, depending on the proportion of projects that can be funded. The second study added depth and showed how the interplay of budget constraints, competing evaluation criteria and heterogeneous stakes can lead to either an over- or an under-determined role of social equity and, counter-intuitively, to equity norms that result in outcomes that contradict the sought-for equity goal.[8] This certainly throws a spanner in the works of any simplistic judgment about conservation justice! As shown by the first case study, concerns for equity or justice rarely, if ever, come in isolation, without other concerns also being present to some degree. How equity or justice are affected by these other concerns is not as straightforward as we are sometimes led to believe: structural factors such as the ones exemplified above can lead to unexpected conclusions. Evaluating the social equity of a conservation program should not prevent us from examining the determinants of the trade-offs involved.

The conclusions from this chapter can be put into a broader perspective: trade-offs between equity and efficiency goals are not the whole story. First, the mere existence of efficiency-related goals based on some 'value for money' principle can make choosing one equity norm over another redundant: efficiency concerns over-determine equity concerns when all equity norms lead irrespectively to the same outcome: the one determined by the efficiency-related goals. Second, the presence of goals other than social equity when distributions between stakeholders have several dimensions can lead to trade-offs between equity norms themselves, because one norm may lead to desirable outcomes in one dimension but not in another, with another norm doing the opposite. Third, in some configurations, similar equity norms can lead to drastically different outcomes while radically different norms can lead to similar outcomes. And fourthly, in the presence of severe constraints (such as limited budgets) and of competing goals such as efficiency, the choice of an equity norm can, surprisingly, lead to outcomes that contradict the purpose of that norm; for instance, an egalitarian norm can lead to very unequal outcomes!

Although this chapter sheds light on key factors underpinning trade-offs between equity and efficiency, it leaves unanswered several further questions, in particular:

- How does uncertainty affect equity–efficiency trade-offs?
- What is the nature of the relationships between the nature and structure of the constraints defining the situation and the severity of trade-offs to be negotiated? For example, what happens when the number of possible options increases from a handful to a continuum?[9]

9 In this study the number as well as levels of subsidy are given and fixed. If there were a hundred different levels in 1% increments, or if any subsidisation level was allowed, results in terms of equity would differ.

- In both case studies we assumed a suitable equity metric was available to evaluate and compare different allocations. Designing such a metric when different stakeholders hold different notions of equity is by no means a trivial matter. This is the subject of Chapter 15.

Centuries or millennia ago, budgets tended to take the form of available physical or food resources. If such resources were lacking, allocation in any other way than dictator-like, perhaps based on brute force, would have led to highly contentious decision making, and raises interesting questions, from an evolutionary perspective, on the emergence of social justice concepts in the context of scarcity. Economics is often said to differ from other social sciences by studying human behaviour under scarcity. Also, equity or justice are rarely the only concerns at stake: they come alongside, or in addition to, other concerns, such as budget constraints, competitiveness, public safety, or political expediency, often as corrective factors to a situation characterised by excesses likely to lead to social dissatisfaction and unrest, with potentially high political costs. Accordingly, this chapter, adopting an economic approach, has endeavoured to show how scarcity in the form of limited budgets, as well as competing goals and assessment criteria, can affect the importance of social justice in environmental planning and policy.

Acknowledgements

We gratefully acknowledge financial support from NERP (the National Environmental Research Program), for allowing us to generate data for the first case study. We also acknowledge valuable input by Geoff Park and David Pannell.

References

Adler MD (2013) *Cost-Benefit Analysis and Distributional Weights: an Overview.* Duke University, Durham NC, USA, <http://scholarship.law.duke.edu/cgi/viewcontent.cgi?article=5813&context=faculty_scholarship>.

Adler M, Posner EA (2009) New foundations of cost–benefit analysis. *Regulation & Governance* **3**, 72–83. doi:10.1111/j.1748-5991.2009.01045.x

Akino M, Hayami Y (1975) Efficiency and equity in public research: rice breeding in Japan's economic development. *American Journal of Agricultural Economics* **57**, 1–10. doi:10.2307/1238834

Atkinson G, Machado F, Mourato S (2000) Balancing competing principles of environmental equity. *Environment & Planning A* **32**, 1791–1806. doi:10.1068/a32106

Bailey M, Sumaila UR, Martell SJ (2013) Can cooperative management of tuna fisheries in the western Pacific solve the growth overfishing problem? *Strategic Behavior and the Environment* **3**, 31–66.

Balafoutas L, Kerschbamer R, Kocher M, Sutter M (2014) Revealed distributional preferences: individuals vs. teams. *Journal of Economic Behavior & Organization* **108**, 319–330. doi:10.1016/j.jebo.2013.12.012

Bento AM (2013) Equity impacts of environmental policy. *Annual Review of Resource Economics* **5**, 181–196. doi:10.1146/annurev-resource-091912-151925

Blank RM (2002) Can equity and efficiency complement each other? *Labour Economics* **9**, 451–468. doi:10.1016/S0927-5371(02)00011-8

Bolton GE, Ockenfels A (2006) Inequality aversion, efficiency, and maximin preferences in simple distribution experiments. *The American Economic Review* **96**, 1906–1911 [comment]. doi:10.1257/aer.96.5.1906

Cattaneo A, Lubowski R, Busch J, Creed A, Strassburg B, Boltz F, Ashton R (2010) On international equity in reducing emissions from deforestation. *Environmental Science & Policy* **13**, 742–753. doi:10.1016/j.envsci.2010.08.009

Daniel TE (1975) A revised concept of distributional equity. *Journal of Economic Theory* **11**, 94–109. doi:10.1016/0022-0531(75)90041-1

Drèze J, Stern N (1985) *The Theory of Cost-Benefit Analysis.* Department of Economics, University of Warwick, Warwick, UK.

Engelmann D, Strobel M (2004) Inequality aversion, efficiency, and maximin preferences in simple distribution experiments. *The American Economic Review* **94**, 857–869. doi:10.1257/0002828042002741

Farrow S (2011) Incorporating equity in regulatory and benefit-cost analysis using risk-based preferences. *Risk Analysis* **31**, 902–907. doi:10.1111/j.1539-6924.2010.01474.x

Feldstein MS (1972) Distributional equity and the optimal structure of public prices. *The American Economic Review* **62**, 32–36.

Fischer M-E, Irlenbusch B, Sadrieh A (2004) An intergenerational common pool resource experiment. *Journal of Environmental Economics and Management* **48**, 811–836. doi:10.1016/j.jeem.2003.12.002

García-valiñas MÁ (2005) Efficiency and equity in natural resources pricing: a proposal for urban water distribution service. *Environmental and Resource Economics* **32**, 183–204. doi:10.1007/s10640-005-3363-0

Haferkamp A, Fetchenhauer D, Belschak F, Enste D (2009) Efficiency versus fairness: the evaluation of labor market policies by economists and laypeople. *Journal of Economic Psychology* **30**, 527–539. doi:10.1016/j.joep.2009.03.010

Haisley EC, Weber RA (2010) Self-serving interpretations of ambiguity in other-regarding behavior. *Games and Economic Behavior* **68**, 614–625. doi:10.1016/j.geb.2009.08.002

Halpern BS, Klein CJ, Brown CJ, Beger M, Grantham HS, Mangubhai S, *et al.* (2013) Achieving the triple bottom line in the face of inherent trade-offs among social equity, economic return, and conservation. *Proceedings of the National Academy of Sciences of the United States of America* **110**, 6229–6234. doi:10.1073/pnas.1217689110

Hanich Q, Campbell B, Bailey M, Molenaar E (2015) Research into fisheries equity and fairness—addressing conservation burden concerns in transboundary fisheries. *Marine Policy* **51**, 302–304. doi:10.1016/j.marpol.2014.09.011

Harberger AC (1978) On the use of distributional weights in social cost-benefit analysis. *Journal of Political Economy* **86**, S87–S120. doi:10.1086/260696

Loomis JB (2011) Incorporating distributional issues into benefit cost analysis: why, how, and two empirical examples using non-market valuation. *Journal of Benefit-Cost Analysis* **2**, 1–24. doi:10.2202/2152-2812.1044

Okun AM (1975) *Equality and Efficiency: the Big Tradeoff.* Brookings Institution Press, Washington DC, USA.

Owens MF (2011) Do other-regarding preferences change with age? Evidence from a gift exchange experiment. *Journal of Socio-Economics* **40**, 868–878. doi:10.1016/j.socec.2011.08.022

Smith A (1759) *The Theory of Moral Sentiments.* Clarendon Press, Oxford, UK. Republished in 1976 by Oxford University Press, New York, USA.

Smith A (1776) *The Wealth of Nations*. Methuen, London, UK. Republished in 2016 by Xist Publishing, Irvine, CA, USA.

Ubeda P (2014) The consistency of fairness rules: an experimental study. *Journal of Economic Psychology* **41**, 88–100. doi:10.1016/j.joep.2012.12.007

Ward FA, Pulido-Velázquez M (2008) Efficiency, equity, and sustainability in a water quantity–quality optimization model in the Rio Grande basin. *Ecological Economics* **66**, 23–37. doi:10.1016/j.ecolecon.2007.08.018

Weisbach DA (2015) Distributionally weighted cost–benefit analysis: welfare economics meets organizational design. *The Journal of Legal Analysis* **7**, 151–182. doi:10.1093/jla/lau009

Young DL, Walker DJ, Kanjo PL (1991) Cost-effectiveness and equity aspects of soil conservation programs in a highly erodible region. *American Journal of Agricultural Economics* **73**, 1053–1062. doi:10.2307/1242433

PART 4

THE WAY FORWARD

14

Politics of innovation and the spirit of justice

Katherine A. Daniell, Ehsan Nabavi and Claudia F. Benham

Summary

Justice and injustice related to natural resource management (NRM) regimes are perceived differently by various actors who seek to effect, and focus their attention on different parts of a governance system. Each innovation that will change the NRM regime and governance system will therefore be subject to politics and struggles between who is likely to win or lose something from the change. This chapter seeks to uncover and investigate these often 'behind the scenes' struggles to expose some of the factors that facilitate or block the development and implementation of innovations aiming to enhance environmental and social justice, at least at one particular scale or for one sector or stakeholder type. It does this by bringing together a framework of analysis for the politics of innovation in multi-level governance systems related to three types of justice decision-making or policy-making processes (interactional, procedural and distributive), then using it for the analysis of five case studies from Australia, China, Bulgaria and Iran. The analyses show that the environment or marginalised and weak stakeholder groups require a strong coalition of sponsors if they are to achieve a fairer deal or equitability – the 'spirit of justice' – because economic development actors are often supported by existing governance regimes or strong financial interests.

Introduction

History of innovation in NRM: a quest for just and sustainable futures?

Effective management of natural resources – for example, water, land, air, biodiversity, minerals or energy – underlies the wellbeing and development of the societies who rely on them for maintenance of life and livelihoods. It is well recognised that any natural resources management (NRM) regime will have an associated justice regime[1] (e.g. Gross 2007; Lukasiewicz *et al.* 2013; Patrick 2014 and discussed in the other chapters of this book). Such justice, and often more importantly, injustice, is perceived differently by different

1 We use regime in this sense to refer to a set of rules, conditions and structures in a governance system. In terms of a 'justice regime', these rules conditions and structures are those that lead to certain actors perceiving different levels of justice or injustice accorded to them.

actors who seek to effect, and focus their attention on, different scales and levels within a governance system (e.g. Hillman and Howitt 2008; Walker 2009; Patrick 2014; Schneider *et al.* 2014).

Within this context, innovation is seen as key to overcoming NRM challenges, such as supplying clean water and energy, disposing of waste in an environmentally sensitive manner and developing strategies to conserve biodiversity and Earth systems processes in certain regimes (e.g. Ajami *et al.* 2014; OECD 2015). Yet any innovation, depending on what it is, how it is used and who has access to it, will also modify existing and potentially future justice regimes in complex ways, having positive and negative impacts on justice for certain actors (see also Buchanan *et al.* 2011). For example, there is a complex relationship between technological innovation uptake and inequality in society (Kuznets 1955; Cozzens 2008; Botta 2015). Water storage dams, windfarms, shale or coal-seam gas, and instating 'protected areas' are just some examples of innovation for which sustainability and justice for some (but not all) have resulted, and thus remain controversial (e.g. Gross 2014; Schneider *et al.* 2014; Patrick *et al.* 2014; Dignum *et al.* 2015). Whoever develops and chooses what innovations to implement and diffuse, when and why, thus has a distinctly political character, which can work for and against developing more just and sustainable outcomes for different actors, including the environment or elements within it.

Chapter overview

This chapter seeks to explore these politics of innovation and their associated justice ramifications. Specifically, it will provide some preliminary responses to the questions:

- What factors and actors facilitate or block the uptake of innovations aiming to enhance justice (from interactional, procedural and distributive points of view)?
- What are the cross-scale or cross-level ramifications of these innovation perturbations in the governance system?
- How does this affect the winners and losers (e.g. what is the change in distribution of justice or injustice and for whom)?

These questions will be investigated through comparative case study analyses, drawing on already published data on three cases (participatory process in Bulgaria; water-sensitive urban design in Australia; payments for ecosystem services in China), and new data from two further case studies (liquefied natural gas development in Australia and water conflict transformation in Iran). The next section outlines the research methodology in more detail and develops an analytical framework for the politics of innovation in multi-level governance systems related to three types of justice (interactional, procedural and distributive). The case study analyses using this framework are then presented in the following section. The chapter concludes with some insights on to what extent innovation for NRM and change towards improved justice regimes can be facilitated or blocked in different ways.

Analytical framework and research methodology

One of the challenges of studying the justice ramifications of the politics of innovation in NRM is that there is, to our knowledge, no easy to apply framework of analysis or easy to access data archives on this issue. Instead, components of theory, analytic frameworks and case studies partially considering these questions of politics and policy change, innovation and justice in multi-level or multi-scalar systems are scattered across an enormous variety of literature. Here, we thus draw together a new hybrid analytic framework and some case studies from our own and collaborative research, to test the framework and its capacity for developing useful knowledge and insights.

Development of an analytical framework for studying the justice ramifications of the politics of innovation

The framework proposed in this chapter is the coupling of the 'analytic grid for the study of politics of innovation uptake in multi-level governance systems' (Daniell *et al.* 2014) with existing frameworks for justice analysis (e.g. Lukasiewicz *et al.* 2013; Gross 2014; Patrick 2014), and simplifying and adapting them for the purposes of comparative case analysis in this chapter. The literature behind the main elements of the frameworks is first presented, then the hybrid analytical framework.

Theories of relevance to innovation uptake and associated politics in multi-level systems

Traditional theories of innovation focus on uptake through several key stages (e.g. knowledge, persuasion, decision, implementation and confirmation) (e.g. Rogers 1983; Nutley and Davies 2000; see also Godin 2015), whereas more recent theories attempt to understand the adaptation, translation or formalisation processes that occur in different innovation contexts (e.g. David 2000; Klerkx *et al.* 2010; Mukhtarov 2014). Innovation uptake for NRM, or sustainable development more broadly, is also studied under the related area of transition theories and management or 'co-evolution' (e.g. Senge 1990; Geels and Schot 2007; Rotmans *et al.* 2001; Rotmans and Loorbach 2009; Geels 2010; Audet 2012; Kilelu *et al.* 2013; de Haan *et al.* 2015), in particular looking at issues of path-dependency and lock-in and how niches can be found and environmental and political awareness exploited to drive systemic intervention (see OECD 2015). In comparison to the traditional 'innovation uptake' processes, which are still useful for understanding the state of an innovation (e.g. idea through to proven through repeated implementation; see also 'technology readiness levels' such as EARTO 2014), these more recent theories tend to acknowledge the importance of system structure, as well as innovation actor agency. This leads to them covering similar issues to some governance and policy change theories that are also relevant to this area.

Specifically, useful connections can be drawn to theories of multi-level governance: 'systems of continuous negotiation across tiers of government and sectors of governance (markets, science, civil society)' (Marks 1993; Hooghe and Marks 2001; Bache and Flinders 2004; see Fisher and Ury 1981 on negotiation) and multi-scalar systems and the dynamics induced within them by particular actors (e.g. Cash *et al.* 2006; Daniell and Barreteau 2014). Here, politics and actions associated with different levels across scales (e.g. spatial, temporal, administrative, institutional, management, network, knowledge, stakes or issues) and the externalities that these create on other parts of the system (e.g. physical or material flows, information and knowledge flows, political and social control, financial flows, human flows or an irreversibility effect) can be analysed. In particular, they allow explicit analysis of the different patterns of interaction and negotiation across systems and the impact of certain governance structures on constraining or supporting passage of particular implementations of innovations.

Likewise, policy change theories such as 'policy advocacy coalitions' (e.g. Sabatier and Jenkins-Smith 1993; see Biggs and Smith 1998 on ideas of coalition-building specifically linked to technology development), policy networks (e.g. Marsh and Rhodes 1992; Mintrom and Vergari 1998) or policy arrangements (e.g. Arts and Van Tatenhove 2004) can be useful for developing an explicit understanding of the underlying competing values and politics that are at play around proposed innovations and the associated changes to existing regimes, who will seek to gain or lose from successful passage of the innovation (see also Ingram and Fraser 2006) and who can be particularly valuable in supporting successful passage (e.g. Smith 2007).

Justice regime changes linked to innovations

Such innovations, whether predominantly of a governance and organisational nature, or a technical nature, will thus adjust existing justice regimes, either reinforcing existing inequities or righting injustices for particular groups. Researchers such as Gross (2014) and Lukasiewicz *et al.* (2013) provide useful overviews of current thinking and streams of thought within the 'core' justice and environmental justice research communities, highlighting three important areas – interactional justice, procedural justice and distributive justice – across which the dynamics of innovation implementation will have varied outcomes.

We take the definitions of these justice types from the literature as follows:

- *Interactional justice* is related to the way in which participants are treated in interpersonal interactions during the decision-making process (Gross 2014).
- *Procedural justice* relates to the perceived fairness and transparency of the process (Bowmer 2007; Gross 2014).
- *Distributive justice* is related to the outcomes of a decision-making process. It is achieved when participants feel that the right decision has been made, and that the outcome is fair (Bowmer 2007).

A further category of restorative or reparative justice could also be particularly important to consider in the case where there are obvious injustices (e.g. against Indigenous people or environmental systems through colonisation or other victims of significant harms); see, for example: Whiteman (2009) and Bryant (2011). We consider here that because different perceptions of what is 'just' or fair across these different justice domains can be linked to underlying core beliefs and policy values that different groups in the policy process hold, justice, including for the environment, is inextricably linked to politics and decision-making processes. The study of justice and its application in NRM can thus profit from close association with disciplines such as the political and management sciences, as well as those more traditionally aligned such as law, sociology, philosophy (e.g. ethics), geography and economics, because each discipline will allow complementary analyses and issues of justice to be examined and taken into account (see Nancarrow and Syme 2001; Syme and Nancarrow 2001). Indeed, there are many power asymmetries and competition between alternative framings of justice or 'equitability' (e.g. Deutsch 1975; Young 1994; Zeitoun *et al.* 2014), which we do not dig into in this chapter. However, for our purposes, we do find this idea of equitability, meaning 'spirit of justice', a valuable concept for developing crude understandings of alternative actor perceptions of justice.

Specifically on innovation and its justice regime aspects, research of relevance to our analyses has been blossoming in recent years. This includes justice-oriented innovation systems (Abolhasani *et al.* 2014), innovations specifically designed to support improvements in justice regimes for those currently marginalised, and the linked concepts of 'responsible innovation' (e.g. Taebi *et al.* 2014; Hellström 2003) and 'inclusive innovation' (e.g. Papaioannou 2011; OECD 2013). These authors highlight the importance of including actors susceptible of being impacted by innovation within the innovation process as it occurs. That is to say, more participatory and open forms of innovation lead themselves to innovation adaptation and translation into the local contexts and a more democratic distribution of outcomes (e.g. see Dignum *et al.* (2015) on the inclusion of public values in technology and institutional design around Shale Gas; and STEPS Centre (2010) and Leach (2012) on the generalities of this approach in the use of science, technology and innovation for development). This is in contrast to those who focus on the impacts of a more traditional 'present and defend' model of innovation uptake and diffusion, where justice researchers focus on issues of social acceptability and perceptions of fairness related to who will win and lose from the implementation innovation implementations (e.g. Gross 2007; Wüstenhagen *et al.* 2007; Sommerville *et al.* 2010).

Development of an analytic framework linking justice and the politics of innovation

We draw on this literature to build an analytic framework (see Fig. 14.1) comprised of four types of analysis parameters: macro-context; micro-context and political dynamics; innovation process dynamics; and justice regime dynamics. The specifics of the first three types of analysis parameters are explained in Daniell *et al.* (2014), then we draw on Gross (2007), Lukasiewicz *et al.* (2013), Patrick (2014) and Zeitoun *et al.* (2014) for the meanings behind our justice categories of 'equitability' perceptions for actors related to interactional, procedural and distributive justice.

Research methodology

To respond to our research questions (cf. Introduction), we have taken a transdisciplinary intervention and/or participatory research approach (Hatchuel and Molet 1986; Midgley 2000; Barreteau *et al.* 2010), coupled with comparative case study analysis (Yin 2013) employing the framework in Fig. 14.1.

As previously outlined, this work extends the three-case analysis in Daniell *et al.* (2014), which focused on the politics of innovation in the water sector, to look at issues of justice in these cases. It also adds two additional cases with prominent social and environmental justice issues based on the ongoing PhD studies of two of the authors. The cases, which are outlined in the next section, were chosen for diversity (e.g. technological, governance or

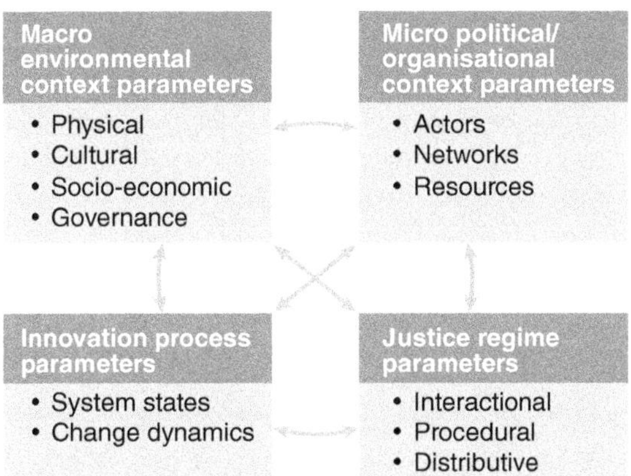

Figure 14.1. Justice and the politics of innovation analysis framework.

mixed innovation regimes) and researcher involvement that allowed the investigation of the cases from an 'engaged' and/or 'interventionist' perspective.

The data obtained and drawn on from across the cases was diverse, based on both researcher experiences (e.g. from experimental reports, interviews, surveys, participant observation and interactions, and experiments) and document analyses (e.g. of reports, email, media and meeting notes). This has allowed significant triangulation of data to occur for each case. The ethical considerations linked to this type of research approach are discussed in Daniell *et al.* (2009, 2014). All of the research stemming from the authors' PhD projects (including interviews and surveys) had also received ethical clearance through the ANU ethics committee procedures.

Comparative case study analysis

This chapter draws on five regional case studies that predominantly concern innovations linked to water, land use, energy and biodiversity or environmental quality. As previously outlined, it takes a multi-level governance (e.g. local to global) perspective on the context of specific recent research and policy interventions to promote or analyse a specific innovation and the impacts of its uptake (and/or lack thereof) on justice regimes and different stakeholders (e.g. local communities, industries, environment, media, NGOs, researchers/experts and different levels/ministries of governments). A fuller background to the first three case studies is outlined in Daniell *et al.* (2014) and background to the other two cases will be forthcoming in the PhD theses of the authors Nabavi and Benham so will only be briefly outlined here.

Case study 1. Integrated water cycle management in Australia

This case study in south-east Queensland, Australia, focuses on the proposed innovation of a new residential development that is designed with 'integrated water cycle management (IWCM)' systems throughout it, including rainwater capture for potable water and collection and treatment of wastewater for non-potable reuse. This innovative development was initially proposed and supported by a coalition of a designer, local councillor, some local communities and a scientific expert in IWCM in response to the need for more sustainable and 'water secure' development, considering water scarcity challenges during recent years of drought in the region. The coalition had also been able to secure a federal government grant to support the implementation of the project. However, the implementation of this innovation was successfully blocked by another coalition of actors who were supportive of the Queensland Government and water utilities' alternative innovation proposed for water security in the region: the 'south-east Queensland water grid', which includes large-scale infrastructure projects such as desalination plants, recycled water plants and interconnection of water storages. This coalition was mostly from the government and utilities 'water bureaucracy' and supported by engineering consultants with similar world views and economic interests, but finally was successful due to the local councillor initially in favour of IWCM changing allegiance to support the current state government policy agenda. The proposed development was thus implemented connected to the water grid, without the IWCM.

Case study 2. Liquefied natural gas (LNG) development in Australia

This case study in the Gladstone region of Queensland focuses on the impacts of a successfully implemented innovation: three large LNG plants aiming to support global transition to less-carbon-intensive fuels (as opposed to coal or petroleum) and a cleaner energy future, as well as driving Queensland's economy through gas exports. Despite significant and continuing opposition to the plants' environmental impacts on the Port of Gladstone and the Great Barrier Reef more generally, and socio-economic impacts on people living and working in the Gladstone area (see Benham 2016), the Queensland Government approved and continues to support the plants' owners. The plants' social and environmental impact statements clearly showed some of the potential risks to local society and environment, but did not comprehensively consider or manage impacts (Benham 2016). Since implementation of the innovation in the region, many of these impacts have been realised and efforts are now being made by various actor groups to minimise these through more effective ongoing management actions, primarily through financial support for community service providers and environmental monitoring and conservation measures.

Case study 3. Non-technical approaches to flood and drought risk management in Bulgaria

This case study in the Sofia region of Bulgaria focuses on a significant innovation in the Bulgarian governance context: the development of non-technical approaches to flood and drought risk management. Due to the highly technocratic and siloed government structures in Bulgaria, stemming in large part from the communist legacy before the fall of the Soviet Bloc, there has been a strong preference from almost all parts of Bulgarian society for technical solutions (e.g. dams, dykes, canals and hydro-climatic models) to manage flood and drought risks. However, as major flood and drought events over the last couple of decades have shown, the government and its technical solutions no longer have the capacity to protect and reduce communities' exposure and vulnerability to these risks. In the year of Bulgaria's entrance into the EU, a European Research project, 'AquaStress', developed a participatory process of engaging interested parties within and external to Bulgaria to improve their capacity to cope with water-related risks and eventually implementation of the EU Water Framework Directive. As the process filled a cultural and policy vacuum in the Bulgarian context, there was not a specific coalition of actors against the ideas of non-technical solutions (e.g. developing an insurance culture, education programmes, planning and development of community flood management groups) and, indeed, the process managed to build awareness and support for such alternatives. However, major barriers in the form of financing, lack of capacity and prevailing cultural norms meant that the eventual implementation of such innovations, put forward for implementation in the group's action planning, remains largely unrealised.

Case study 4. Payments for ecosystem services in China

This case study in the Beijing area of China focuses on the innovation of eco-compensation for drinking water catchment restoration. Specifically, the populated Miyun watershed upstream of Beijing that supplies much of its water has not been in a state to provide the ecosystem service of clean water, and hence eco-compensation payments have been implemented by a coalition of government levels, most prominently the National Government and Beijing Municipal Government, to restore the catchment and reduce water treatment costs. Although already implemented, there have been debates about how the particular form of the innovation's implementation could be improved to provide fairer outcomes to upstream inhabitants – many of whom are being expropriated – led by county officials and international NGOs, consultants and researchers. Although fruitful, these multi-level discussions and the individual champions involved in them faced significant opposition from the government structures and institutions (e.g. siloed ministries, hierarchical working relationships and difficulty with transmitting local information upwards through the administration). This led to a stalemate, with no improvements so far in the eco-compensation model to ensure more equal benefits between Beijing and upstream residents not yet receiving sufficient support.

Case study 5. Conflict transformation through participation in Iran

This case study in the Zayendeh-Rud basin of Iran focuses on the implementation of a governance innovation – a participatory process – for attempting to transform an intractable water conflict. Specifically, the downstream (Isfahan) and upstream (Charmahal) provinces have been in conflict over the use of water in the main river that flows through them and the cultural capital of Iran – Isfahan – for a couple of thousand years. This has typically been resolved by resorting to the construction of inter-basin transfers, but has ultimately been shown to be unsustainable as a solution for water management. The river – a symbol of Isfahan – is now dry and the conflict tense. Through a process of research and building trust with citizens, researchers, NGOs and government officials from the two provinces, a coalition of Iranian researchers (including from non-affected provinces) was able to implement a participatory 'forum theatre' process to bring together the two sides in the conflict to start to understand each other's issues, build trust and form the basis for working together. The process was able to fulfil these objectives, with the two key inter-provincial proponents of the conflict willing to try to work things out by themselves and to see if they could develop some alternative solutions to inter-basin transfers, without the interference of the national government (Nabavi 2015). These solutions are yet to be seen and the environment remains degraded and at risk.

Cross-case comparison

An overview of the theoretical framework is presented in Table 14.1 to enable the 5 case studies' comparative analyses. The main focus is on the justice components.

Discussion

The comparative case study analyses demonstrate that the attempted implementation of any particular innovation – whether it be of a technical, governance or hybrid nature – is surrounded by complex politics that affect the interactional, procedural and distributive justice regimes in place. Politics and competing needs and values related to NRM issues – as well as associated resourcing, cultural tendencies and prevailing governance structures – all work together to facilitate or block the passage of innovations and their resultant sought effects. In several of these cases (in Bulgaria, China and Iran), it was also apparent that external researchers to the governance systems where the innovations were to be contextualised were able to drive some improvements in procedural and interactional justice for some chosen actors, but in many cases lacked sufficient links to strong decision-maker champions and the cultural and governance system pre-requisites in key areas required for complete translation of the innovation and its uptake into desired distributional justice regimes. Other cases demonstrated that perceived positive environmental justice outcomes can vary substantially across scales, levels and actors, with local negative or positive environmental changes sometimes being seen as a necessary harm to enhance environmental justice across a border (e.g. between provinces in Iran), or between levels

Table 14.1. Cross-case comparison analysing politics of innovation working for and against justice

Case study	CS1. Australia	CS2. Australia	CS3. Bulgaria	CS4. China	CS5. Iran
Key macro-context parameters	Climate variability, increasing population; individualist/egalitarian cultural tendencies; differing beliefs on importance of centralised/decentralised control models; liberal democracy, fragmented governance across levels/sectors.	Fragile environment, resource exploitation potential for economic gains; individualist/egalitarian cultural tendencies; differing beliefs on precautionary/opportunistic approaches to environmental management and economic development; liberal democracy, fragmented governance across levels/sectors.	Climate variability; hierarchical culture dominant with individualist tendencies growing; strong belief in control of nature through technical means; parliamentary democracy formed in 1990 following single party socialist system; centralised governance (to national level, with provincial and municipal administrations).	High population density and pollution levels; strong collectivist and hierarchical cultural tendencies; strong belief in government capacity to control nature and social systems for collective benefit; single party socialist republic, strongly centralised governance across multiple administrative levels.	Harsh climate, environment under severe stress; strong hierarchical and competitive (individualist) cultural tendencies; strong belief in capacity to control nature for individual benefit; a political system based on a combination of parliamentary democracy and theocracy; centralised governance (to national level, with provincial and municipal administrations).
Innovation process parameters	Integrated water cycle management innovation formalised (detailed design and costing) but only partially contextualised through process; innovation driven by need to deal with drought and environmental movement; stalemate between competing pro centralised/decentralised innovations until federal government and a local councillor changed the strength of coalitions; no eventual uptake of IWCM.	Liquefied natural gas plant innovation fully formalised and contextualised; driven by need for cleaner energy and economic development; opponents of environmental/social impacts not in strong coalition to affect changes to innovation implementation; confirmation of innovation uptake and lock-in.	Non-technical innovative approaches to flood and drought risk management ranged from not at all formalised or well contextualised (e.g. community organisations) to well formalised and contextualised (e.g. insurance); driven by government incapacity to manage major flood drought prevention/impacts; insufficient coalition development/strength to drive innovation uptake; decision but little implementation of options.	Eco-compensation for catchment restoration innovation well formalised and contextualised through process, but not an equitable model; driven by need for drinking water-quality protection and more equitable proposition by upstream community disquiet; central and Beijing government coalition had power required to force uptake but government structures/institutions work against the equitable model; confirmation of traditional, inequitable model uptake.	Participatory process for water conflict transformation innovation formalised and contextualised through process (from a prior low level); driven by need to halt environmental degradation and return water to the river, not enflame the conflict further; researcher coalition built sufficient trust with both provinces' stakeholders to implement the innovation; confirmation of the usefulness of the innovation has not yet occurred.

(Continued)

Table 14.1. (Continued)

Case study	CS1. Australia	CS2. Australia	CS3. Bulgaria	CS4. China	CS5. Iran
Key micro-context parameters	Actors including: residents; environment (local); federal, state and local government; water monopoly planners/scientists split into pro-centralised/decentralised coalitions of almost equal strength with their own experts and financial resources.	Actors including: environment (multi-scale); rest of community; different levels of government and scientists not sufficiently organised or resources to make a significant difference to the LNG industry/state government coalition operations that have many more resources.	Actors including government officials, flooded citizens and EU researchers brought together but not in a sufficiently well resourced or strong coalition to support innovation uptake.	Actors including: environment (basin); upstream community residents; Beijing and central government officials in one strong coalition with resources to support the innovation uptake and a weak sub-coalition for the more equitable solution without resourcing to drive change in government structures.	Actors including: Isfahan (downstream) and Charmahal (upstream) officials and residents; Environment; NGOs; steel industry; typically in competing coalitions but were brought together under the radar by researchers with a key resource (trust) to implement the participatory process innovation.
Interactional equitability	Interactional equitability altered through the negotiations with the community and their coalitions' perceptions of it diminishing through the process and the state government and their coalition's perceptions improving (including being able to positively interact with a local councillor at the end).	Perceptions of interactional equitability differ between groups and contexts: community more likely to perceive interactions (e.g. in public meetings) as unjust compared with key stakeholders who interact with industry through other means (e.g. committees, formal processes); local background of consultation staff important in establishing credibility with public.	Generalised increase in interactional equitability, at least for the parties involved in the participatory process, although this was greater for actors with an interest in flood risk than those with an interest in drought risk; all actors were given opportunities to interact, voice their views and be listened to by others.	Varied perceptions of interactional equitability between stakeholders; upstream residents most likely to perceive injustice and environmental restoration campaigners (including Beijing Govt) to see an increase in justice; international research process promoted more interaction between groups than is typical (including having provincial officials representing upstream residents' interests).	General increase in interactional equitability for stakeholders due to the implementation of the participatory process innovation, except steel industry who were excluded from the process and unable to build new relationships and have their views considered.

Procedural equitability	Procedural equitability remained relatively low because the process followed typical coalition politics, including trial by media and few open stakeholder engagement processes between competing coalition groups.	Stakeholder/public engagement concentrated through written submissions; also public meetings and a key stakeholder committee; Perceived procedural equitability varies across these engagement types, because each type of engagement constrains participation in different ways.	Typically high levels of procedural equitability, except for excluded groups (e.g. the 'Roma'). The process itself was an example of increasing procedural equitability and giving citizens and stakeholders an opportunity to work with policy makers on action planning for flood and drought risk, although flood risk was given procedural preference.	Procedural equitability was enhanced by the international project partners' coordination efforts but remained moderate with different stakeholders, including the less powerful, being able to air their issues and solutions in a collective discussion forum, but not being able to have them seriously considered by key decision makers.	General increase in procedural equitability for stakeholders due to the design of the participatory process, except from the perspective of the steel industry, which was excluded from the process.
Distributive equitability	Win–lose negotiation. IWCM not implemented in development to the disappointment of the community involved with the designer. State government and partners were able to extend the water grid – which theoretically provides equality of water services to all in the region; environmental injustice outcome the largest.	Equitable distribution of the impacts and benefits of LNG development across spatial and temporal scales important in shaping local perceptions of the industry; stewardship of resources, both the gas resource and the marine environment, intergenerational equity, and long-term infrastructure investment in the local region were seen as important to local people.	This process led to the beginning of considerations to improve risk management for citizens (distributed more to flood than drought-affected citizens). Lack of access to EU flood risk funding led to many actions not being able to be realised; distributive equitability among the researchers from the EU project and achievement of their objectives was variable.	The implemented form of the eco-compensation mechanism led to improved distributive justice for the environment and Beijing residents but injustice for the upstream residents who did not believe the compensation scheme was adequate for their expropriation.	Innovation has not yet led to an adjustment in distributive equitability but has the potential to do so, since the two main protagonists in the conflict, leaders from Charmahal and Isfahan, decided to exchange phone numbers and see whether they could work together to transform the conflict and hence justice regime.

(Continued)

Table 14.1. (Continued)

Case study	CS1. Australia	CS2. Australia	CS3. Bulgaria	CS4. China	CS5. Iran
Key insights from individual cases					
What factors and actors facilitate or block the uptake of innovations aiming to enhance justice?	Coalition with the most resources supporting the innovation can support their innovations passage and block another one; lack of formal multi-stakeholder/level decision-aiding processes; competing scientific knowledge added to each of the coalitions and could not be used to form a 'consensus'; any actor may destabilise and alter the direction of justice perceptions for different actors: in this case one local government councillor switching coalitions.	Fragmented governance and lack of communication between governments (state, local) and industry/community through process; heavy reliance on written submissions and privileging of scientific over local and Indigenous knowledge marginalises key groups (see Voyer *et al.* 2012); justice is facilitated through informal external mechanisms (committees, funding to alleviate impacts).	Strong and well-financed external actors to the governance system required for innovation uptake and justice regime change were not deep enough for widespread empowerment; deep cultural values and preferences can block innovations requiring more capacity building and finance to support change; finding local and trusted exemplars of solutions where possible (e.g. Bulgarian insurance expert) led to support for 'developing an insurance culture' as a non-technical option.	Central government power to enable innovation and justice for the environment; inequitable distributive outcomes between poor/rich regions and residents; international researchers supported greater interactional and procedural justice, but impacts limited by deep structural blockages across scales/levels.	Intervention and governance innovation required to transform existing justice regime; significant improvement in interactional/procedural justice among chosen groups in typically non-conducive, entrenched governance systems that lack trust; more work needed for distributive justice improvements for both provinces.

What are the cross-scale/ level ramifications of these innovation perturbations in the governance system?	Federal government can step in to support local change and environmental justice but unless local political support is maintained for this intervention to support implementation, it may be futile; some loss of community and environmental justice; campaigners' hope that change in development patterns in this region is possible.	Loss of public confidence in decision making; public opposition to development; retroactive attempts to govern impacts through legislative change, policy and on-ground interventions; weak global environmental benefit – coal to LNG – but local environmental injustice so companies working to minimise social impact; inequitable distributive outcomes across local and global communities/actors.	The promotion of some of the proposed innovations sought to shift risk management responsibility from the national government to the local, community or individual level (and build the capacity of all levels to better prepare for, and respond to, floods and droughts). Procedurally and interactionally, all levels were brought together through this process to encourage multi-level and cross-sector relationship-building and working practices.	Inter-provincial justice distribution adjustments (between upstream/ downstream residents); Improved environmental management and impacts at all levels.	National level and steel industry by-passed by participatory innovation to provide a space for enhancing mutual upstream/down-stream understanding; further intervention required to understand environmental justice regime impacts across levels and water distribution and associated perceived justice between sectors.
How does this affect the winners and losers?	Little overall change. The incumbent continued its expansion using its preferred approach (water-grid) to the detriment of the more locally environmentally sensitive approach (IWCM) and community water prices.	Little change in justice/ injustice at the local scale; however, opposition and awareness-making has led to increased potential for justice in other, similar cases (e g. New South Wales, Northern Territory).	If implemented, these innovations could lead to sharing of risk and general reduction of flood and drought impacts on all stakeholders, yet due to a remaining lack of financing and capacity, there was little actual change in distributive justice for Bulgarian stakeholders.	Environment and Beijing residents the biggest winners and upstream expropriated residents suffered the greatest injustice. However, improvements in procedural justice outcomes for these upstream residents could be obtained if structural government challenges can be overcome.	Preliminary win in increasing interactional and procedural justice for both provinces and for enrolling researchers' issues on environmental sustainability and injustice into the participatory forum theatre process. Steel industry and National Government excluded but as yet have not suffered any distributional justice changes.

(e.g. LNG supporting a transition to a low carbon future at the global scale in Australia), or having negative impacts on social justice (e.g. upstream expropriated residents in China and on equality of water service to households in Australia). In all cases, the strength of coalitions and their connections to resources, including decision-making authority and finance, was vital in ensuring the passage of that coalition's desired innovation and associated justice regimes. More generally, across all of the case studies, it could be seen that innovation uptake in NRM is not just about one innovation, but that there are often competing innovations or system-type conflicts (e.g. centralised/de-centralised, technical/non-technical, participatory/technocratic management, and triple-bottom line/deep ecological sustainability) and associated justice regimes. Some of these may set a lock-in path for the future, requiring significant political or research action around the alternative innovation to right injustices of the marginalised. As underlined by increasing numbers of authors, such as Kilelu *et al.* (2013), innovation uptake cannot be represented as a simple linear process, but instead its formalisation and contextualisation in any targeted governance environment is a 'co-evolutionary process, resulting from alignment of technical, social, institutional and organizational dimensions' (Kilelu *et al.* 2013).

There is thus a need for carefully thought out meta-governance and co-engineering (organisation of collective processes – see Daniell 2012) that create an enabling environment and incentives for facilitating these innovations and their associated enhanced justice regimes. This signals that justice researchers hoping to implement innovations and change justice regimes in their specific areas of interest will need to learn how to better engage with key decision makers and those with the resources to support their passage. Due to the ramifications of changes in justice regimes across stakeholders, governance levels and scales, the ethics of this practice will require greater in-depth reflection (see e.g. Daniell *et al.* 2009, 2014; Cahill *et al.* 2007).

Conclusions

This chapter has sought to expose and discuss some of the factors that facilitate or block the development and implementation of innovations aiming to enhance environmental and social justice, at least at one particular scale or for one sector or stakeholder type. It has done this by bringing together a framework of analysis for the politics of innovation in multi-level governance systems related to three types of justice decision-making or policy-making processes (interactional, procedural and distributive), then using it for the analysis of five case studies from Australia and around the world.

These analyses show that there is a need to open up questioning about scales of justice and associated actors linked to often-competing innovations and how cross-scalar coalitions for the promotion of innovations seeking to ensure more environmentally and socially just regimes can be most effectively built in different contexts and governance regimes. Each of the cases demonstrate different interactional, distributive and procedural outcomes from the innovation formalisation and contextualisation into the prevailing governance structures and to what extent such change towards greater environmental and social justice can be facilitated or blocked in different ways.

We have demonstrated that there are typically direct relationships between policy change or innovation mechanisms and justice outcomes. Specifically, the introduction of something new – the technological, governance or hybrid innovation – is likely to benefit some actors, potentially to the detriment of others. Specifically, the environment or marginalised and weak stakeholder groups require a strong coalition of sponsors if they are to achieve a fairer deal (specifically related to economic development) because development actors are often supported by existing governance regimes or strong financial interests. Transcending key values and interests to find real win-win situations for distributive justice in many of these contexts is often a pipe-dream, so more attention needs to be given to politics, ethics and influencing the innovation game to develop specific environmental and social justice outcomes and fairer decisions in the spirit of justice for specific stakeholder communities.

Acknowledgements

We would like to thank Peter Coombes and Ian White who were the contributors of the first Australian, and Chinese, case studies as part of Daniell *et al.* (2014), the analyses of which have been reframed here. Thanks also to all of our other collaborators and the stakeholders of each of these case studies. Any errors or misinterpretations are the authors' alone.

References

Abolhasani Z, Hassanzadeh A, Ghazinoory SS, Pourezzat A (2014) A justice-oriented innovation system: a grounded theory approach. *Social Justice Research* **27**(3), 369–394. doi:10.1007/s11211-014-0218-2

Ajami NK, Thompson BH, Jr, Victor DG (2014) 'The path to water innovation'. Discussion Paper 2014–06. The Hamilton Project, Stanford Woods Institute for the Environment, Stanford CA, USA, <http://www.hamiltonproject.org/assets/legacy/files/downloads_and_links/path_to_water_innovation_thompson_paper_final.pdf>.

Arts B, Van Tatenhove J (2004) Policy and power: a conceptual framework between the 'Old' and 'New' policy idioms. *Policy Sciences* **37**(3–4), 339–356. doi:10.1007/s11077-005-0156-9

Audet R (2012) Transition as discourse and the issues of interventionism, justice and innovation. In *Proceedings of the ISEE2012 Conference: Ecological Economics and Rio+20: Contributions and Challenges for a Green Economy*. 10–13 May, Rio de Janeiro, Brazil. International Society of Ecological Economics, Rio de Janeiro, Brazil, <http://www.isecoeco.org/conferences/isee2012-versao3/pdf/716.pdf>.

Bache I, Flinders M (Eds) (2004) *Multi-level Governance*. Oxford University Press, Oxford, UK.

Barreteau O, Bots PWG, Daniell KA (2010) A framework for clarifying "participation" in participatory research to prevent its rejection for the wrong reasons. *Ecology and Society* **15**(2), 1. <http://www.ecologyandsociety.org/vol15/iss2/art1/>.

Benham CF (2016) Change, opportunity and grief: understanding the complex social-ecological impacts of gas development in the Australian coastal zone. *Energy Research and Social Science* **14**, 61–70.

Biggs S, Smith G (1998) Beyond methodologies: coalition-building for participatory technology development. *World Development* **26**, 239–248. doi:10.1016/S0305-750X(97)10041-9

Botta A (2015) Perverse and virtuous feedbacks between inequality and innovation: which role for public institutions and public investment? In *Proceedings of the 2015 INET Conference on Liberté, Equalité, and Fragilité*. 8–11 April, Paris, France. Institute for New Economic Thinking, Paris, France, <https://www.ineteconomics.org/uploads/papers/BOTTA.pdf >.

Bowmer K (2007) Water and conflict resolution: From smoke filled rooms to public participation. In *Proceedings of the 5th Australian Stream Management Conference: Australian Rivers: Making a Difference* 21–25 May. Charles Sturt University, Thurgoona,

NSW. (Eds AL Wilson, RL Dehaan, RJ Watts, KJ Page, KH Bowmer and A Curtis A). Institute for Land, Water & Society, Thurgoona, NSW, <https://www.csu.edu.au/__data/assets/pdf_file/0008/748277/Bowmer_Kathleen_25.pdf>.

Bryant E (2011) Innovation or degradation: an analysis of Hawai'i's cultural impact assessment process as a vehicle of environmental justice for Kanaka Maoli. *Asian Pacific Law and Policy Journal* **13**, 230–299.

Buchanan A, Cole T, Keohane, RO (2011) Justice in the diffusion of innovation. *Journal of Political Philosophy* **19**(3), 306–332. doi: 10.1111/j.1467-9760.2009.00348.x

Cahill C, Sultana F, Pain R (2007) Participatory ethics: politics, practices, institutions. *ACME: An International E-Journal for Critical Geographies* **6**(3), 304–318.

Cash DW, Adger W, Berkes F, Garden P, Lebel L, Olsson P, *et al.* (2006) Scale and cross-scale dynamics: governance and information in a multilevel world. *Ecology and Society* **11**(2), article 8. < http://www.ecologyandsociety.org/vol11/iss2/art8/>.

Cozzens SE (2008) 'Innovation and inequality'. Working Paper No. 30, School of Public Policy, Georgia Institute of Technology, Atlanta GA, USA, <https://smartech.gatech.edu/bitstream/handle/1853/23051/wp30.pdf?sequence=1>.

Daniell KA (2012) *Co-engineering and Participatory Water Management: Organisational Challenges for Water Governance.* Cambridge University Press, Cambridge, UK.

Daniell KA, Barreteau O (2014) Water governance across competing scales: coupling land and water management. *Journal of Hydrology* **519**(C), 2367–2380. doi:10.1016/j.jhydrol.2014.10.055

Daniell KA, White I, Rollin D (2009) Ethics and participatory water planning. In *Proceedings of the 32nd Hydrology and Water Resources Symposium: 'H2O09'.* 30 November–3 December 2009, Newcastle, NSW. pp. 1476–1487. Engineers Australia, Canberra.

Daniell KA, Coombes PJ, White I (2014) Politics of innovation in multi-level water governance systems. *Journal of Hydrology* **519**(C), 2415–2435. doi:10.1016/j.jhydrol.2014.08.058

David A (2000) La recherche-intervention, cadre général pour la recherche en sciences de gestion? In *Les Nouvelles Fondations des Sciences en Gestion.* (Eds David A, Hatchuel A, Laufer R) pp. 193–213. Vuibert, Paris, France.

de Haan F, Rogers B, Frantzeskaki N, Brown R (2015) Transitions through a lens of urban water. *Environmental Innovation and Societal Transitions* **15**. doi:10.1016/j.eist.2014.11.005.

Deutsch M (1975) Equity, equality, and need: what determines which value will be used as the basis of distributive justice? *The Journal of Social Issues* **31**(3), 137–149. doi:10.1111/j.1540-4560.1975.tb01000.x

Dignum M, Correljé A, Cuppen E, Pesch U, Taebi B (2015) Contested technologies and design for values: the case of shale gas. *Science and Engineering Ethics.* doi:10.1007/s11948-015-9685-6

EARTO (2014) *The TRL Scale as a Research & Innovation Policy Tool.* EARTO Recommendations. <http://www.earto.eu/fileadmin/content/03_Publications/The_TRL_Scale_as_a_R_I_Policy_Tool_-_EARTO_Recommendations_-_Final.pdf>.

Fisher R, Ury W (1981) *Getting to Yes: Negotiating Agreement Without Giving In.* Houghton Mifflin, Boston MA, USA.

Geels FW (2010) Ontologies, socio-technical transitions (to sustainability), and the multi-level perspective. *Research Policy* **39**, 495–510. doi:10.1016/j.respol.2010.01.022

Geels FW, Schot J (2007) Typology of sociotechnical transition pathways. *Research Policy* **36**, 399–417. doi:10.1016/j.respol.2007.01.003

Godin B (2015) Models of innovation: why models of innovation are models, or what work is being done in calling them models? *Social Studies of Science* **45**(4), 570–596. doi:10.1177/0306312715596852

Gross C (2007) Community perspectives of wind energy in Australia: the application of a justice and community fairness framework to increase social acceptance. *Energy Policy* **35**(5), 2727–2736. doi:10.1016/j.enpol.2006.12.013

Gross C (2014) *Fairness and Justice in Environmental Decision-Making: Water Under the Bridge.* Routledge, London, UK.

Hatchuel A, Molet H (1986) Rational modelling in understanding and aiding human decision-making: about two case studies. *European Journal of Operational Research* **24**, 178–186. doi:10.1016/j.ejor.2016.01.022

Hellström T (2003) Systemic innovation and risk: technology assessment and the challenge of responsible innovation. *Technology in Society* **25**(3), 369–384. doi:10.1016/S0160-791X(03)00041-1

Hillman M, Howitt R (2008) Institutional change in natural resource management in New South Wales, Australia: Sustaining capacity and justice. *Local Environment* **13**(1), 55–66. doi:10.1080/13549830701581929

Hooghe L, Marks G (2001) Types of Multi-Level Governance, European Integration online Papers, 5(11) <http://eiop.or.at/eiop/pdf/2001-011.pdf>.

Ingram H, Fraser L (2006) Path dependency and adroit innovation: the case of California water. In *Punctured Equilibrium and the Dynamics of US Environmental Policy.* (Eds R Repetto and JG Speth) pp. 78–109. Yale University Press, New Haven CT, USA.

Kilelu CW, Klerkx L, Leeuwis C (2013) Unravelling the role of innovation platforms in supporting co-evolution of innovation: contributions and tensions in a smallholder dairy development programme. *Agricultural Systems* **118**, 65–77. doi:10.1016/j.agsy.2013.03.003

Klerkx L, Aarts N, Leeuwis C (2010) Adaptive management in agricultural innovation systems: the interactions between innovation networks and their environment. *Agricultural Systems* **103**(6), 390–400. doi:10.1016/j.agsy.2010.03.012

Kuznets S (1955) Economic growth and income inequality. *The American Economic Review* **45**(1), 1–28.

Leach M (2012) Sustainability, development, social justice: towards a new politics of innovation technologies and innovations for development. In *Technologies and Innovations for Development.* (Eds J-C Bolay, M Schmid, G Tejada and E Hazboun) pp. 19–29. Springer, Paris, France.

Lukasiewicz A, Bowmer K, Syme GJ, Davidson P (2013) Assessing government intentions for Australian water reform using a social justice framework. *Society & Natural Resources* **26**(11), 1314–1329. doi:10.1080/08941920.2013.791903

Marks G (1993) Structural policy and multilevel governance in the EC. In *The State of the European Community. Vol. 2: The Maastricht Debates and Beyond.* (Eds AW Cafruny and GG Rosenthal), pp. 391– 410, Harlow Longman, Boulder CO, USA.

Marsh D, Rhodes RAW (Eds) (1992) *Policy Networks in British Government.* Clarendon Press, Oxford, UK.

Midgley G (2000) *Systemic Intervention: Philosophy, Methodology, and Practice.* Kluwer Academic/Plenum Publishers, New York, USA.

Mintrom M, Vergari S (1998) Policy networks and innovation diffusion: the case of state education reforms. *The Journal of Politics* **60**(1), 126–148. doi:10.2307/2648004

Mukhtarov F (2014) Rethinking the travel of ideas: policy translation in the water sector. *Policy and Politics* **42**(1), 71–88. doi:10.1332/030557312X655459

Nabavi E (2015) *Loading! Please Wait...: Evolution of Transboundary and Interprovincial Water Conflict in the Middle East.* Centre for Water and Development, SOAS, London, UK, <https://www.soas.ac.uk/water/event/24jun2015-loading-please-wait-evolution-of-transboundary--and-interprovincial-water-conflict-in-the-.html>.

Nancarrow B, Syme G (2001) Challenges in implementing justice research in the allocation of natural resources. *Social Justice Research* **14**(4), 441–452. doi:10.1023/A:1014606929949

Nutley S, Davies HTO (2000) Getting research into practice: making a reality of evidence-based practice: some lessons from the diffusion of innovations. *Public Money & Management* **20**(4), 35–42. doi:10.1111/1467-9302.00234

OECD (2013) 'Innovation and inclusive development'. Conference Discussion Report revisited February 2013. OECD, Paris, France, <http://www.oecd.org/sti/inno/oecd-inclusive-innovation.pdf>.

OECD (2015) 'System innovation: synthesis report'. OECD, Paris, France, <https://www.innovationpolicyplatform.org/sites/default/files/general/SYSTEMINNOVATION_FINALREPORT.pdf>.

Papaioannou T (2011) Technological innovation, global justice and politics of development. *Progress in Development Studies* **11**(4), 321–338. doi:10.1177/146499341001100404

Patrick MJ (2014) The cycles and spirals of justice in water-allocation decision making. *Water International* **39**(1), 63–80. doi:10.1080/02508060.2013.863646

Patrick MJ, Syme GJ, Horwitz P (2014) How reframing a water management issue across scales and levels impacts on perceptions of justice and injustice. *Journal of Hydrology* **519**(C), 2475–2482. doi:10.1016/j.jhydrol.2014.09.002

Rogers EM (1983) *Diffusion of Innovations*. Free Press, New York, USA.

Rotmans J, Loorbach D (2009) Complexity and transition management. *Journal of Industrial Ecology* **13**(2), 184–196. doi:10.1111/j.1530-9290.2009.00116.x

Rotmans J, Kemp R, van Asselt M (2001) More evolution than revolution: transition management in public policy. *Foresight* **3**, 15–31. doi:10.1108/14636680110803003

Sabatier PA, Jenkins-Smith H (1993) *Policy Change and Learning: An Advocacy Coalition Approach*. Westview Press, Boulder CO, USA.

Schneider F, Buser T, Graefe O (2014) Scales of justice in water governance: hydropower controversies in Switzerland. *Water Policy* **16**(2), 137–154. doi:10.2166/wp.2014.405

Senge P (1990) *The Fifth Discipline*. Doubleday/Currency, New York, USA.

Smith DJ (2007) The politics of innovation: why innovations need a godfather. *Technovation* **27**, 95–104. doi:10.1016/j.technovation.2006.05.001

Sommerville M, Jones JP, Rahajaharison M, Milner-Gulland E (2010) The role of fairness and benefit distribution in community-based Payment for Environmental Services interventions: a case study from Menabe, Madagascar. *Ecological Economics* **69**(6), 1262–1271. doi:10.1016/j.ecolecon.2009.11.005

STEPS Centre (2010) *Innovation, Sustainability, Development: A New Manifesto*. STEPS Centre, Brighton, UK.

Syme G, Nancarrow B (2001) Justice, sustainability, and integrated management: concluding thoughts. *Social Justice Research* **14**(4), 453–457. doi:10.1023/A:1014658914019

Taebi B, Correljé A, Cuppen E, Dignum M, Pesch U (2014) Responsible innovation as an endorsement of public values: the need for interdisciplinary research. *Journal of Responsible Innovation* **1**(1), 118–124. doi:10.1080/23299460.2014.882072

Voyer M, Gladstone W, Goodall H (2012) Methods of social assessment in Marine Protected Area planning: Is public participation enough? *Marine Policy* **36**(2), 432–439. doi:10.1016/j.marpol.2011.08.002

Walker G (2009) Beyond distribution and proximity: exploring the multiple spatialities of environmental justice. *Antipode* **41**(4), 614–636. doi:10.1111/j.1467-8330.2009.00691.x

Whiteman G (2009) All my relations: understanding perceptions of justice and conflict between companies and indigenous peoples. *Organization Studies* **30**(1), 101–120. doi:10.1177/0170840608100518

Wüstenhagen R, Wolsink M, Bürer MJ (2007) Social acceptance of renewable energy innovation: an introduction to the concept. *Energy Policy* **35**(5), 2683–2691. doi:10.1016/j.enpol.2006.12.001

Yin RK (2013) *Case Study Research: Design and Methods.* 5th edn. Sage publications, Los Angeles CA, USA.

Young IM (1994) *Justice and the Politics of Difference.* Princeton University Press, Princeton NJ, USA.

Zeitoun M, Warner J, Mirumachi N, Matthews N, McLaughlin K, Woodhouse M, *et al.* (2014) Transboundary water justice: a combined reading of literature on critical transboundary water interaction and 'justice', for analysis and diplomacy. *Water Policy* **16**(S2), 174–193. doi:10.2166/wp.2014.111

15

How can environmental justice be assessed when different stakeholders disagree on what is just? A practical solution

Steven G.M. Schilizzi

Summary

It has often been said that 'fairness lies in the eye of the beholder'. Contrary to decision criteria such as economic efficiency, the notion of fairness or equity cannot be given a definition independent from context. Economists have thus long shied away from including concerns about justice in their policy recommendations. Here we take context-dependence as a starting point and turn it into an advantage. We first establish that it is not random but highly structured by both social position-dependent and non-positional dimensions of context, and these can also interact in specific ways. We use laboratory experiments to explore the role of wealth and need (in the form of 'mouths to feed') and that of real financial incentives, veil-of-ignorance and equity-efficiency trade-offs. We also find that the very definition of equity norms is context-dependent in terms of implicit parameters and sub-rules. Insights from these results led to the construction of an 'equity metric' that is able to represent a collective's heterogeneous equity preferences that minimise total 'perceived inequity'. Although it has some inherent limitations, our metric increases the likelihood for equity and a competing policy goal to correlate positively, thereby reducing tensions and allowing higher achievements of both.

The nature of the problem

This chapter tackles the problem posed by the existence of several different, and sometimes incompatible, notions and definitions of justice, with a specific focus on outcome-based distributional equity.[1] A possible solution to this problem is the dictator's choice made by the decision maker (or decision-making group) on a specific definition, backed up with whatever arguments seem most appropriate. These could be based on logic, on local custom or tradition, on religion, on political expediency, or simply on brute force or

1 This chapter will focus specifically on distributional equity, but whenever a statement is likely to hold more generally for the broader notions of 'justice, equity or fairness', we shall refer to them by their initials 'JEF', *without*, however, implying any equivalence between the three terms: we focus on differences and diversity.

threats. Another way is to acknowledge this diversity, collect data from stakeholders[2] and construct an 'equity metric' that best reflects this diversity of views. This is the approach chosen here. The aim of such a metric is to allow a decision maker to decide which of two or more projects or policies is to be considered fairer or more equitable when stakeholders do not agree. In mathematical parlance, the problem is equivalent to transforming a vector quantity into a scalar one to make comparisons easier.

In order to be able to implement this approach, we must first be able to describe the diversity of equity views and understand the factors that determine them. The next section will thus explore their context-dependence in order to provide some insight into what information needs to be collected from the constituency so that a democratic equity metric may be constructed and used in a given context. Because people may hold an equity principle more or less strongly, the metric will also require a measure of (relative) preference intensity to be defined. In the subsequent section, it will appear that the foregoing observations also apply to the very definition of many equity norms[3] themselves, which can therefore also be context dependent. Finally, when this is clarified, we shall be in a position to construct an operational metric that can be applied in most standard contexts. To illustrate how the metric works in practice, we shall use data generated by controlled laboratory experiments. In a concluding section, the limitations of our approach and of the metric as currently constructed will be identified and formulated for future improvements.

This chapter builds on and extends recent work by Faravelli (2007), Cappelen *et al.* (2007), Durante *et al.* (2014) and Ubeda (2014) and is also an offshoot from Frohlich and Oppenheimer (1992) and Konow (2001). We also build on the set of wide-ranging equity norms used by Cazorla and Toman (2001), Ringius *et al.* (1998, 2002) and Rose and colleagues (Rose and Stevens 1998; Rose *et al.* 1998; Rose and Zhang 2002) in the context of international negotiations on greenhouse gas abatement around the Kyoto Protocol. One author (Le Grand 1990) has questioned the reality of an equity–efficiency trade-off, considering instead that (economic) efficiency is a specific equity principle. Although this can in some cases be true, it need not be so, as the existence and use of 'Pareto compensation' as an equity norm suggests. Indeed, we have used it ourselves in our work, but economic efficiency cannot be reduced to this specific norm.

Diversity and context-dependence – experimental disentanglement

From the literature cited above, as well as in Chapter 13, it is clear that several notions of what is considered 'JEF' (just, equitable or fair) are often in competition. We would expect this to be the case when decision makers (DMs) are personally involved and thus, as stakeholders, self-serving behaviour enters the fray[4]; but this also happens for 'impartial spectators' (Konow 2009), suggesting along the way that impartiality does not entail agreement on what is 'JEF'. Following Konow's early empirical work, in particular his 2001 paper 'Fair and square', it appears that differences and disagreements in equity judgments

2 We define here 'stakeholder' as any person or group (or sometimes animal or ecosystem), seen as entitled to certain rights by a specific community, and who is likely to benefit or suffer from the implementation of a policy.

3 In this chapter, we use the term equity 'norm', but other terms could also be used, such as 'rule', 'principle' or 'criterion'. At this stage of our thinking, they do not clearly differ for our purposes.

4 Self-serving behaviour (or SSB) differs from plain selfishness. The latter simply focuses on one's own benefits; the former may choose an equity principle that appears to reflect some form of justice but is chosen as a function of its expected benefits to oneself.

vary along two orthogonal dimensions, which map a space spanning situations and individuals. On a collective, 'horizontal' dimension, different individuals in the same situation may hold competing views about what is 'JEF'; on an individual 'vertical' dimension, the same individual may hold different and potentially inconsistent views about what is 'JEF' in different situations. Economists have traditionally shied away from including equity in their decision-making models specifically because of this double indeterminateness. Contrary to the concepts of economic efficiency or Pareto optimality, a rigorous definition independent of context (or situation) cannot be given to notions of equity. In our work, rather than considering this context-dependence as a dead-end and a full stop, we have chosen to view it as a starting point.

We start with Konow's (2001) initial observation that equity judgments are not 'context-specific' but 'context-dependent': in the first case, the nature of the dependence is, or is deemed to be, random, whereas in the latter case it is not, or is not deemed to be, random but at least to some degree systematic. One is thereby led to ask: 'if the dependence is not random, then what are the systematic elements that should be in principle predictable, at least in probabilistic terms?'

Initial experimental trials led us to conceive of context-dependence as being itself two-dimensional (Schilizzi and Black 2009): we later labelled one 'positional' and the other 'non-positional'. The first refers to the positions of individuals in society or in a given group, parameterised by such indicators as income or wealth, level of health, level of education and family size. Note that in principle this description refers to those *about whom* there are equity concerns, but in practice, if the DM is also a stakeholder, the description will include him or her too. The non-positional dimension of context-dependence refers to all other situational aspects that are not linked to stakeholders' position in society. These might include: the presence of other decision criteria that compete with equity concerns (economic efficiency, ecological effectiveness, urgency, safety, etc.); the magnitude of the values at stake; the knowledge or uncertainty of future consequences; the legitimacy of, or trust in, the DM; and so on.

We chose to investigate the double dimension of context-dependence using laboratory experiments because in the non-controlled conditions of the field, both dimensions are usually entangled and complex to disentangle, whereas in the laboratory they can be rigorously controlled for. Accordingly, in our initial runs, where the 'equity problem' was to allocate 'equitably' a sum of money in a group of 10 people, we designed and combined five contextual parameters: two positional and three non-positional. The first two defined individual wealth endowments and 'number of mouths to feed' (or family size) and included the DM as a stakeholder (i.e. as part of the group: every individual in the group differed in terms of these two positional parameters). The last three were defined by: the existence or not of real monetary stakes, as opposed to a hypothetical, survey-type situation[5]; the presence or not of competing decision criteria (economic efficiency); and whether the DM knew in advance or not what position he or she held within 'society' (i.e. the experimental group). The last treatment reflected Rawls' (1971) '*veil of ignorance*' situation; the trade-off situation was created by having a dollar given to a more richly endowed group member 'producing' a greater total amount available for distribution (the total amount then being a function of players' decisions). The combinations of the three non-positional treatments defined eight different scenarios. Later experiments added other (non-positional) dimensions: in one case, the DM was either part or not part of the

5 First-level judgments only were asked for, not second-level. The former represent 'what do you think is fair' types of questions, whereas the latter represent 'what do you think people should think is fair' questions.

group for which an equitable allocation was to be made; in another study, group members had either been randomly endowed with their initial wealth positions or they had had to earn them by performing a task requiring some effort.

To study the extent and nature of context-dependence, we defined nine different (distributional) equity norms as revealed from the economics literature reviewed above.[6] In each scenario, participants were asked to rate each norm separately on a Likert scale ranging from –10 to +10, where 0 represented indifference or hesitation and the minus and plus values represented, respectively, opposition to, and support of, a given equity norm. Such ratings made possible a measure of preference intensity that, although it poses some theoretical issues, was nonetheless acceptable for our purpose. The analysis then focused on how individual ratings differed as a function of all contextual dimensions. The experiment was carried out with many groups in order to produce sufficient data for statistical analysis, and we are still carrying out more work with new design variants.

The results have confirmed that, in our experiments at least, equity preferences as measured by individual ratings are indeed strongly context-dependent, with an overwhelming non-random component.[7] Regarding positional context, in particular the DM's initial wealth endowment relative to that of others in the group, some effects were to be expected. For example, greater wealth translated into stronger opposition against equity norms, which led to progressive allocation rates and stronger support for those leading to regressive allocation rates or giving weight to some function of the status quo. Others, however, were more surprising: the relationship between initial (relative) wealth and support for equality as a normative equity principle was not linear but U-shaped, and was opposed as much by 'rich' as by 'poor' participants. In addition to direct effects, interaction effects were also observed. Thus the impact of wealth on equity preferences differed depending on the non-positional context: it was stronger when equity-efficiency trade-offs were present, when Rawls' veil of ignorance was absent and when real monetary stakes were present. The last two effects are to be expected, but the role of the equity-efficiency trade-off is more subtle and rests on the relationship underpinning the existence of the trade-off: while the self-serving behaviour incentives for the richer players amplify each other (a larger personal share, as well as a greater total amount), they work in opposite directions for the poorer players (greater personal share but smaller total amount to be shared out). Thus the role of wealth on distributional equity reflects the net effect of both factors – something that can be expected to be quite general for this type of trade-off.

Regarding non-positional context, the absence of 'incentives' (real monetary stakes) had a somewhat unexpected effect: an analysis of the results based on measuring the extent and variability of self-serving behaviour showed that although such bias was smaller when no money was involved but only 'in principle' ratings were asked for, it was not much smaller (Schilizzi and Iftekhar 2015). In addition, the role of initial wealth was, as expected, smaller when no real money was present, but again not much smaller. Thus, even in the form of an opinion survey, participants showed substantial self-serving behaviour: a result that still warrants a clear explanation. Some authors have suggested referring to 'habit' or 'as-if play' (Becchetti *et al.* 2014). The existence of an equity-efficiency trade-off was found to amplify

6 These included equality (or egalitarian equity), Rawls' maxmin, inverse proportionality to initial endowments, vertical equity, horizontal equity (to reflect family size), an exceptions principle, Pareto compensation, sovereignty (which reflects the correlation between social position and decision-making power), and a proxy for intergenerational equity. We also included four process-based equity norms (consensus, sovereign bargaining, acquired rights and market justice, defined by the fact that a (sufficiently competitive) market yields 'just' outcomes), but these are not part of the present discussion.

7 This specification is based on our statistical analyses.

differences in preferences between equity norms, while the veil of ignorance condition greatly reduced them – to such an extent, in fact, that the role of other contextual factors became statistically insignificant in most cases. This now well-established result was first explained in terms of risk avoidance, but this explanation does not exhaust the possibilities and has been challenged, for example, by Schildberg-Hörisch (2010) and Krawczyk (2010).

Context-dependence and the very definitions of equity, fairness or justice

In designing our experiments and implementing our chosen equity norms, we realised that equity norms usually remain under-determined. Laboratory experiments do not allow for sloppiness! Although some norms are perfectly determined once one or more defining parameters have been operationally specified, others also require one or more sub-rules to be completely specified. Norms that do not require sub-rules to be perfectly defined include equality, inverse proportionality, vertical equity and sovereignty: 'equality of what?', as Amartya Sen liked to repeat; proportionality to what? The maxmin equity norm on the other hand also requires a sub-rule to be completely specified. Maxmin is defined in general terms as 'allocate the available benefits only to the least favoured category of people'. But this definition begs the question: who are those to be included in such a category and where does one draw the line? In addition, one needs to specify how the allocation should be made within the least favoured category.[8] In our experiments, who in our groups of 10 were to be considered as 'least favoured'? In theory, any choice between 'the one most poorly endowed' and 'the nine most poorly endowed' could be possible. The latter might make sense if there was a big jump in initial endowments between the ninth and the tenth group member, where, for example, the latter held as much starting wealth as all the other nine put together (a rather grim reflection of the 'top 1%' in the world's current wealth distribution). Once this choice made, say 'the poorest five', a rule for distributing the available benefits to these five is needed: for example, 'allocate it equally among them' if equality is chosen.

Horizontal equity is relevant when stakeholders are groups (such as families or nations) and makes the allocation a function of some parameter describing the group. For families, it could be family size; for nations the size of their population. Thus, in international negotiations on greenhouse gas abatement efforts, horizontal equity has sometimes been defined in terms of equal abatement expenditures per capita, but also per dollar of GDP (Cazorla and Toman 2001). Underlying this norm is equality on an individual basis. But, as this example illustrates, implementing this norm requires us to specify which parameter, such as population or the nation's GDP, must be taken as the basis for equalising across individuals. Until this is specified, horizontal equity remains an underspecified norm.

Other norms, such as 'interests of the future' (or intergenerational equity), require at least three parameters and a sub-rule for complete specification. *There is a marked difference between a conceptual definition and an operational implementation of an equity norm.* In our experiments, we implemented it as follows. If this norm was chosen, 20% of the available benefits were saved 'for the future', meaning they would be allocated after the session was ended to a pre-specified subgroup, in this case the five least favoured by the outcomes of the experiments. The saved 20% were to be allocated evenly across these five. As this example shows, one parameter is needed to specify what proportion of current

8 The sub-rule vanishes in this case if the least favoured category is reduced to a single person.

benefits should be saved for the future. A first sub-rule is needed to specify who should be the beneficiaries 'in the future' (i.e. at the end of the session). And a second sub-rule is needed to specify how the saved benefits should be allocated. In our experiments, the use of the maxmin norm conflated the two sub-rules. But in an experimental setting, other dimensions lie hidden and should be controlled for. As an example, the 20% left for the future were saved rather than invested: the 20% saved could have accrued interest and grown by the end of the game. Introducing such growth is the dynamic version of an equity-efficiency trade-off, even when future beneficiaries are not the same as current ones, because any weight put on future benefits gets amplified by their growth. By contrast, if the amount saved for the future risks eroding in value, or even risks destruction, the opposite effect is to be expected. Such considerations can easily be implemented in the laboratory while reflecting real challenges in sustainable resource use.

Because these were our first forays into this research area, and also to save time, we (as experimenters) made these choices as a 'dictator'. But the philosophy of the whole approach is to determine each equity norm's parameters and sub-rules based on participants' own preferences. Thus, the information-gathering procedure described above for the rating of each equity norm can be extended for the selection of a norm's specific parameters and sub-rules whenever required. Then the context-dependence question can be asked of the *definition* of those equity norms themselves. To our knowledge, no-one has yet carried out this research, but it is part of the follow-up to our current work.

But there is more to the definition and selection of equity norms. To the extent that equity only has meaning for entities endowed with certain rights or entitlements, understanding the factors that determine people's choice of equity norm in a given context suggests we relate such choices to perceptions of rights or entitlements. Thus, some preliminary surveys suggest that equity norms for access to, and cost of, higher education differ substantially from equity norms for access to, and cost of, health care, which themselves differ from 'rights to a clean environment'. Presumably, respondents do not endow a randomly chosen person with the same rights or entitlements to higher education as they do to health or environmental amenities. This link is as yet largely unexplored.

A new solution to an old problem: what is fair when people disagree on what is fair?

Given then that different people may uphold different equity norms in a given situation, how should a DM, whose decisions impact the community, decide on the most equitable policy? This question is often acute in the field of natural resource management and environmental policy, where stakeholders with different characteristics tend to be affected in different ways. This is largely due to the fact that the environment is mostly a public good. So between two environmental projects A and B with different distributional impacts in terms of benefits and in terms of costs, how does the policy maker decide which is 'fairer' if different stakeholders hold different, and potentially conflicting, views about what is 'fair' or equitable? A straightforward solution, though perhaps politically dangerous, would be for the DM (or decision-making group), to decide as a dictator: to decide unilaterally what is meant by fair or equitable and to compare projects A and B accordingly. But what if a majority of stakeholders think otherwise?

The starting premise of our approach is therefore that, by choice or by constraint, the DM decides not to 'go dictatorial' and to reflect, at least in part, the constituency's view of equity; that is, opts for a democratic solution. But by doing so the DM is faced with the 'horizontal' aspect of the disagreement problem: different notions of equity may be held in a given situation. The work reported on in the above sections provides a way to tackle this challenge. The detailed procedure of how to do this, akin to an algorithm, is rather complex and can be viewed in Schilizzi (2014); here we shall provide only the gist of it.

The two starting points are the allocations defined by each equity norm (e.g. equality allocates $100 as $10 to each of 10 group members), and the relative importance of each equity norm as expressed by individual ratings. In a given group, and for a given scenario, we first determine the importance to the group of an equity norm by summing the whole group's ratings for this norm, and we do this for all the norms.[9] The relative importance of each norm for this group in this scenario is then obtained by dividing the total ratings for this norm by the sum total ratings for all (nine) norms. In this manner, each norm receives a collective (group-specific) percentage weight that reflects all group members' equity preferences. Then, a given individual, instead of receiving a share of the total benefits (say $100) as determined by a single equity norm ($10 if this norm is equality), receives instead a share determined by all the equity norms, each weighted by its group's 'relative importance'. In other words, the individual receives a weighted average of norm-specific shares that reflects the group's equity preferences; and this is done for every individual. Because, for any given norm, an individual's share will differ as a function of his position in the group (e.g. richer or poorer), each group member will still receive different shares of the total. Thus, for group 1 in scenario 1, position 3 receives almost twice as much as position 8. The resulting 'weighted equity allocation' (WEA) across group members represents the 'fairest' or most equitable allocation the DM needed.[10] The procedure (or algorithm) just described that generates the WEA is the weighted equity metric, or WEM, and answers the question: 'What is fair when people do not agree what is fair?'

Conclusions and unresolved questions

How can environmental justice be assessed when different stakeholders disagree on what is just? This chapter has provided one possible solution to this problem, which requires a democratic consensus on process. Besides the implied political pre-requisites, this solution also relies on a technical innovation, which is represented by the WEM procedure. First, an appropriate method for obtaining the relevant information from stakeholders is needed, capable of recording differences in equity preferences both qualitatively and quantitatively (preference intensity). Here we used a Likert scale with a sufficiently fine-grained gradient (21 points). Second, a consistent method for processing this information is needed so as to derive the group's collective equity preferences. In Schilizzi (2014), it is shown that WEM procedure minimises the group's total degree of dissatisfaction in terms of 'perception of inequity'.[11] From the DM's point of view, if minimal community dissatisfaction translates into minimal monitoring and enforcement costs, greater perceived equity becomes positively correlated with greater economic efficiency. This in turn reduces the costs of trade-offs between equity and efficiency related goals, thereby increasing the chances of

9 It is important here to avoid any negative numbers: purely positive numbers are obtained by adding a fixed constant to all ratings: in this case +10. The ratings range is thus translated from [–10, +10] to [0, 20] without modifying their meaning.

10 With nine groups and eight scenarios, our experiments yielded 72 such WEAs.

11 Although important for the use of the WEM metric, it would have taken too much space here to explain why this is so. The interested reader can refer to Schilizzi 2014 (Can we design an equity metric when stakeholders hold conflicting views about equity?)

achieving higher levels of both. As a result, the WEM allows equity concerns to be traded-off on an equal footing with any competing decision or evaluation criterion.

In the process, a third aspect emerged revealing that many equity norms are underspecified, and to construct the WEM one needs to re-run the procedure outlined above to elicit definitional choices, in addition to eliciting preferences over defined norms. This is necessary for all those equity norms that require one or more parameters and/or one or more sub-rules to be completely specified. Although in our laboratory experiments we as experimenters dictatorially decided on their parameter values and sub-rules, in practice these should also reflect the constituency's views, because they form an essential part of what they consider as fair or equitable. Modern information technology can now allow for this at minimal cost.

Our solution to the multi-faceted equity problem does, however, have its limits. It leaves several questions unanswered, which are challenges for future research.

The WEM may appear very democratic, but it contains a hidden assumption that does not cater for, and would tend to exclude, ethnic or cultural minorities who hold value systems that differ radically from those held by an outnumbering majority. In Australia, Aboriginal people would be given a small weight due to their relatively small numbers. In many Asian countries, the preferences of ethnic minorities would also be swamped: for example, by the mass of Han (90%) people in China and by the Kinh (86%) in Vietnam. The solution to this problem is not straightforward, but could be analogous to saving the Amazon forest or its indigenous people in northern Brazil. Although Brazilians may not value them sufficiently to give them a 'voice' in local decision making (as per *Exit, Voice and Loyalty* by Hirschleifer 1970), people around the world may do so and would be willing to pay for forest conservation and indigenous culture preservation programs. Likewise, if the majority values a small minority's (equity) preferences, it could decide – as has been done in Canada for the Inuit and 'First Nations' – to give it a special weight; for example, an 'equal weight per cultural group'.

Another issue the DM may be faced with is whether stakeholders and 'impartial spectators' (Konow 2009) should be given equal weight; for example, for a community project, people within the community and people outside it. The solution is again not straightforward because it will depend on how 'uninvolved' people outside the community are. This is best understood in terms of externality: whether outsiders might benefit or suffer in any way from the community's internal choice. For instance, if the environmental projects are publicly funded, then in principle all taxpayers could feel to some degree involved and, as such, acquire the status of stakeholders. If the chosen project is likely to influence other similar decisions elsewhere, non-monetary externalities are present and may elicit reactions by outsiders, who might view the choice of this project as setting a good or bad precedent. Thus the status of 'stakeholder' is one of degree rather than kind. But giving stakeholders a weight in decision making that would be proportional to their degree of 'involvement' is easier said than done, and no clear procedure currently exists that can do this.

Although the WEM procedure is an empirical construct, it can be used in a normative way that reflects what may be termed '*empirical normativity*'. It is based on stakeholders', as well as, possibly, impartial spectators' equity preferences, but the DM can then use it to decide what is fair and what is not, and to rank different projects or policies according to this metric.

The WEM's empirical normativity is, however, partial, or may be biased by the decision-making entity itself. Indeed, between the extremes of a purely democratic solution (giving the WEM a 100% weight) and a purely dictatorial solution (giving the WEM a zero weight), intermediate solutions are possible. The DM's final decision can lie anywhere between these two extremes. The dictatorial component could be anything: the DM's own views about what should count as fair or equitable, or any other criterion, such as economic efficiency. In Australia, environmental projects often, but implicitly, include as a ranking criterion the benefits, in terms of portfolio size and political influence, expected by CMAs (Catchment Management Authorities) or by some individual staff. With reference to case study 1 in Chapter 13, carried out in Victoria in early 2015, it appeared that for the ranking of environmental projects in a given catchment, the decision-making group put a 79% weight on economic efficiency and a corresponding 21% weight on equity/fairness. This means that the importance of a WEM-based equity ranking of the list of projects, for the purpose of selection given limited funding, was valued five times less than 'value-for-money' considerations. As illustrated by case study 2 in Chapter 13, budget constraints tend to lead to a higher weight put on efficiency concerns.[12]

But this may not be the final word. If the constituency truly values the process of democratic decision making, the relative weight between its own (equity) preferences and those of the DM maker could itself be voted on and be subject to a WEM-like metric. In the end, the solution is political and depends on the relative negotiating power and degree of agreement between the decision-making entity and its constituency. The balance between the two is likely to be best understood in game-theoretic terms, given the nature of existing institutions and the relative costs and benefits to either party of changing them, if needed.

References

Becchetti L, Fiaschetti M, Marini G (2014) Card games and economic behaviour. *Games and Economic Behavior* **88**, 112–129. doi:10.1016/j.geb.2014.08.003

Cappelen AW, Hole AD, Sørensen EØ, Tungodden B (2007) The pluralism of fairness ideals: an experimental approach. *The American Economic Review* **97**(3), 818–827. doi:10.1257/aer.97.3.818

Cazorla M, Toman M (2001) *International Equity and Climate Change Policy*. Climate Issue Brief no. 27. Resources for the Future, Washington DC, USA.

Durante R, Putterman L, Weele J (2014) Preferences for redistribution and perception of fairness: an experimental study. *Journal of the European Economic Association* **12**(4), 1059–1086. doi:10.1111/jeea.12082

Faravelli M (2007) How context matters: a survey based experiment on distributive justice. *Journal of Public Economics* **91**, 1399–1422. doi:10.1016/j.jpubeco.2007.01.004

Frohlich N, Oppenheimer JA (1992) *Choosing Justice: an Experimental Approach to Ethical Theory*. University of California Press, Berkeley and Los Angeles, CA, USA.

Hirschleifer AO (1970) *Exit, Voice and Loyalty*. Harvard University Press, Cambridge MA, USA.

Konow J (2001) Fair and square: the four sides of distributive justice. *Journal of Economic Behavior & Organization* **46**, 137–164. doi:10.1016/S0167-2681(01)00194-9

12 One may therefore wonder what a government's increasing strictures on environmental funding might do to the importance of equity concerns relative to economic efficiency-related considerations and, more broadly, how implementable equity criteria might be in severely cash-strapped countries. Given the results of case study 2 in Chapter 13, an excessive weight on economic efficiency will also affect the level of environmental outcomes, because the most cost-effective policy could be to do very little conservation.

Konow J (2009) Is fairness in the eye of the beholder? An impartial spectator analysis of justice. *Social Choice and Welfare* **33**, 101–127. doi:10.1007/s00355-008-0348-2

Krawczyk M (2010) A glimpse through the veil of ignorance: equality of opportunity and support for redistribution. *Journal of Public Economics* **94**, 131–141. doi:10.1016/j.jpubeco.2009.10.003

Le Grand J (1990) Equity versus efficiency: the elusive trade-off. *Ethics* **100**(3), 554–568. doi:10.1086/293210

Rawls J (1971) *A Theory of Justice*. 2nd edn. Harvard University Press, Cambridge, MA, USA.

Ringius L, Torvanger A, Holtsmark B (1998) Can multi-criteria rules fairly distribute climate burdens? *Energy Policy* **26**(10), 777–793. doi:10.1016/S0301-4215(98)00032-9

Ringius L, Torvanger A, Underdal A (2002) Burden sharing and fairness principles in international climate policy. *International Environmental Agreements: Politics, International Environmental Agreements: Politics, Law and Economics* **2**, 1–22.

Rose A, Stevens BK (1998) Will a global warming agreement be fair to developing countries? *International Journal of Environment and Pollution* **9**, 157–177.

Rose A, Zhang ZX (2002) *Interregional Burden-Sharing of Greenhouse Gas Mitigation in the United States*. Mimeo, Department of Geography, The Pennsylvania State University, University Park PN, USA.

Rose A, Stevens BK, Edmonds J, Wise M (1998) International equity and differentiation in global warming policy. *Environmental and Resource Economics* **12**(1), 25–51. doi:10.1023/A:1008262407777

Schildberg-Hörisch H (2010) Is the veil of ignorance only a concept about risk? An experiment. *Journal of Public Economics* **94**, 1062–1066. doi:10.1016/j.jpubeco.2010.06.021

Schilizzi S (2014) Can we design an equity metric when stakeholders hold conflicting views about equity? Paper presented at the 58th Annual Conference of the Australian Agricultural and Resource Economics Society. 5–7 February, Port Macquarie, NSW, <http://bel-uwa.github.io/S15.AARES.pdf>.

Schilizzi S, Black J (2009) *Breaking Through the Equity Barrier of Environmental Policy*. PN-21249. Land and Water Australia, Canberra.

Schilizzi S, Iftekhar MS (2015) Context and self-serving bias in choosing equity norms. Paper presented at the 59th AARES Annual Conference, 10–13 February, Rotorua, New Zealand, <http://bel-uwa.github.io/SI.15.AARES.pdf>.

Ubeda P (2014) The consistency of fairness rules: an experimental study. *Journal of Economic Psychology* **41**, 88–100. doi:10.1016/j.joep.2012.12.007

16

Australian jurisprudence of justice in water management: present limitations, future issues and law reform suggestions

Jennifer McKay

Summary

This chapter describes several justice theories and counterposes these with a series of actual decisions in key water management cases. The decisions occurred under state laws and the *Water Act 2007* and hence the relevant laws differ. Hence, the first point that this chapter makes about justice is, that the differences between the relevant laws, creates differences and these may lead to injustice between residents of different states. This would be in the nature of a different outcome on the same facts. This arises from the constitutional settlement in 1901, which entrenched state pre-eminence in this domain and others such as criminal law. Hence, distributive justice is not achieved by this process of having different state laws. Procedural and informational justice are also a product of the different state systems too, and hence also differ from state to state. Therefore justice is not served well by having so many different rules about water management in Australia. For example, all urban water supply businesses provide relief to customers if they cannot pay their water bill, but use different criteria. Overall, the state level focus has impeded national thinking about water management.

The change to this was the radical adoption of the *Water Act 2007*, which provides limited oversight of state laws in the Murray–Darling region. This Act imposes a clear human rights statement that requires critical human water needs to be the first priority.

This chapter deals with rural water use for production and regional water allocation plans that attempt to create sustainable use that ensures ecologically sustainable development (ESD).

This chapter then considers the gaps between the theory and practice. It asks what has been the outcome of the laws and the processes and applies them to rural water management in Australia over the last 200 years, and then suggests some institutional and law reforms to make the future more just. It uses studies of key cases adjudicating disputes over powers to change consumptive water uses to achieve (ESD) in the new instrument, called 'regional water plans'. The Australian states have many names for these, and many hundreds of plans exist. The chapter decides that ESD is a conception of justice that incorporates present and future generations and finds that ESD has been operationalised very well through judges. The chapter however, suggests reform along the lines of Corporations Law: that is, that governments should pass identical state laws with uniform definitions.

This chapter is a personal reflection on our greatest natural resource, water, and how it is the basis for all wealth.

Wealth is evidently not the good we are seeking; for it is merely useful and for the sake of something else. (Aristotle[1])

My reflection on the legal system is that we must think like a nation with regard to water management. Water is our commons, and the greatest offence to common sense is to manage it using political boundaries. Water supply is a human right and it should be the first role of government to provide water in a sustainable way that preserves the environment for future generations.

They hang the man and flog the woman

That steal the goose from the common

But let the greater villain loose

That steal the common from the goose.

(English folk poem, Anonymous)

Introduction

Aristotle is clear that wealth is not the end but the means. A modern Aristotle may well say that wealth is the way to achieve ecologically sustainable development (ESD). All the pillars of ESD cover the gamut of environmental justice, such as to care for present and future generations and to apply the precautionary principle. The English folk poem beautifully points out that the foundation of the law was justice between people and protection of property. The villain of unbridled use of resources, which some call the tragedy of the commons, has been tackled by the adoption of ESD in many Australian state and national statutes relating to water management. So can ESD and its operationalisation achieve a new form of justice and reform the law from within? I would say yes it has and, depending on political will to not amend the laws, it will evolve. Every Parliament can change a law and the ESD object could be removed or tinkered with. The ESD pillars were first listed in Australia's National Strategy for Ecologically Sustainable Development (Ecologically Sustainable Development Steering Committee 1992), which defined ESD as: 'using, conserving and enhancing the community's resources so that ecological processes, on which life depends, are maintained, and the total quality of life, now and in the future, can be increased'. The goal is: 'Development that improves the total quality of life, both now and in the future, in a way that maintains the ecological processes on which life depends'.

The ESD core objectives are:

- to enhance individual and community wellbeing and welfare by following a path of economic development that safeguards the welfare of future generations
- to provide for equity within and between generations
- to protect biological diversity and maintain essential ecological processes and life-support systems.

1 *Aristotle Nicomachean Ethics*, translated by D. Ross (Oxford University Press, revised edition 1980, Book 1 section 5, p. 7)

The guiding principles are:

- Decision-making processes should effectively integrate both long- and short-term economic, environmental, social and equity considerations.
- Where there are threats of serious or irreversible environmental damage, lack of full scientific certainty should not be used as a reason for postponing measures to prevent environmental degradation.
- The global dimension of environmental impacts of actions and policies should be recognised and considered.
- The need to develop a strong, growing and diversified economy that can enhance the capacity for environmental protection should be recognised.
- The need to maintain and enhance international competitiveness in an environmentally sound manner should be recognised.
- Cost-effective and flexible policy instruments should be adopted, such as improved valuation, pricing and incentive mechanisms.
- Decisions and actions should provide for broad community involvement on issues that affect them.

These guiding principles and core objectives need to be considered as a package. No objective or principle should predominate over the others. A balanced approach is required that takes into account all these objectives and principles to pursue the goal of ESD. The ESD goal has many dimensions and has several aspects. These are challenging to any legal system but the powers of judges to interpret laws can make these objectives work.

There are several theories of justice and these derive from numerous scholars in disparate disciplines notably philosophy, history, law (criminal and equity), economics, sociology and psychology. There are arguments abounding that justice is concerned with fairness in the allocation of resources such as liberty, opportunity, income and wealth. The concept of fairness in liberty is absolute but, in respect of wealth and income, fairness does not always mean equality. The modern philosophers such as Rawls have said that any unequal distribution should be used to advantage the least well off (Rawls 1972). Modern tax laws are often cited as examples of redistribution of wealth, but, again, how these operate is not always seen as fair.

The 10 aspirations listed above have been enacted in some federal laws such as the *Water Act 2007* and several state laws (McKay 2006 and *Telstra Corporation Ltd v Hornsby Shire Council [2006] 146 LGERA 10* per Preston J). Once a decision is made then the doctrine of precedent ensures that similar cases are treated the same way. However, this only happens within a particular jurisdiction and this means that states such as New South Wales (NSW) and Victoria can have very different outcomes. Within the state, precedent ensures individual justice and hence promotes stability in the local society with the maxim *the rule of law and not of men*. However, this may not always promote the collective action required to achieve long-term management of water resources. As mentioned above, our river and aquifer systems do not follow the political state boundaries so we need a national system of laws administered by the states. The first section of this chapter, 'Ecologically sustainable development in Australia', examines ESD in relation to five justice theories: distributive, correctional, political, procedural and Sen's Idea of Justice. This section suggests that there is a sharp conflict between justice theories (which look at the current perspective for individuals) and ESD (which looks at longer term to the welfare of future generations and imposes a collective long-term response). So ESD may indeed be wider than justice theories for this reason. The justice theory of Sen copes with the future (see Table 16.1 below).

The first of the next two sections discuss the concept of non-justiciability, which refers to situations where judges will not adjudicate a political decision. The second section illustrates how procedural justice was implemented in several cases on water plans and discusses how strictly these plans were implemented by courts. In many cases, consumptive users endured massive losses of allocated water.

So, this chapter will look at the ESD concept and the administration of this by Australian law and judges with respect to surface and groundwater in a rural context. Justice may be wider than ESD, but ESD has moved the law in the direction of consideration of justice, with several attributes of fairness related to the long term, and strict requirements to administer the process in a fair way. However, the ESD concept has in many places reduced current wealth and incomes as it has reduced water available for consumptive uses. The fourth section turns to the *Water Act 2007* in detail and looks at potential justice issues in its implementation. One key issue is that the coverage applies only to one part of Australia: the Murray–Darling Basin. To be more just, it should apply to the entire nation. There is no legal reason to limit the Act, but in 2007 the drought was at its height and the Murray–Darling Basin was seen as the key region for production of irrigated agriculture.

The final section summarises this excursion through justice theories and water law in Australia. It finds that there has been a robust protection of the environment for future generations by the limitation of current water use for irrigation. This is because judges have applied the precautionary principle and accepted scientific evidence of a threat of serious or irreversible damage to the water supply. This is a profound change to legal reasoning.

Ecologically sustainable development in Australia in relation to five justice theories

Federation in 1901 did little to alter the state-level power over water and land. However, section 96 of the Constitution did give the federal government more fiscal influence over the state water and land laws and policy. The community demand for sustainability in water- and land-use decisions was consolidated through two waves of federal reforms in 1994 and 2004. These reforms required ESD to be applied to freshwater management and in other contexts such as mining. The duty to hear disputes has rested mainly with the state courts to date.

Figure 16.1 shows the system and it is important to remember that the legislative aims to achieve ESD is essentially non-justiciable. The executive of each state inserted this multi faceted obligation in several laws. The judges are then left to apply them to the facts and evidence before them.

This shows that ESD is embedded in state and federal laws through a political process and there is clear legal capacity in both levels of legislature to do that. No case has ever challenged ESD as being vague or uncertain (Preston 2005) although many judges have made that remark in passing, especially in the early part of the 1990s.

The ESD concept above has been included in national law such as the *Water Act 2007* and many state laws on water as a general objective. Unfortunately, the wording differs and this creates several nuances and a plethora of decision at state level. NSW has had most of the decisions and has developed nuanced understandings to examine the processes and evidence decision makers used to assess the ESD aspects of cases. However, these are

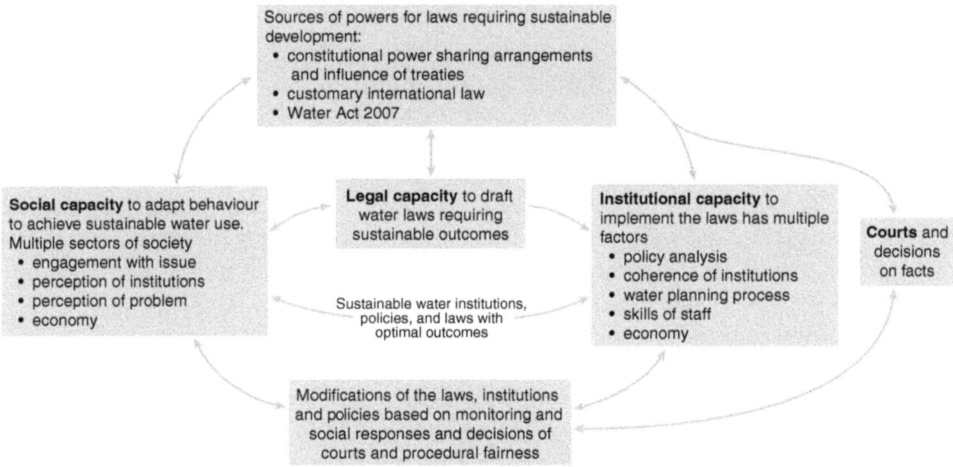

Figure 16.1. Framework in power and cases on sustainable water institutions, policies and laws (Source: McKay 2008; inspired by Daly and by adaptive management and the concept of strong sustainability, meaning that some things, such as green plants, water cycles and biodiversity, are critical and must be retained at all costs).

merely persuasive in the other States. So here is an aspect of the law that some people may consider impedes justice.

The final phase of reform from 2007 was built on several intergovernmental agreements (when the states referred some powers over water to the federal government) for the Murray–Darling Basin only and reflects a different balance in the use of political deal making. It is best embodied in the *Water Act 2007*. The architecture of the *Water Act 2007* relied on referred state power under section 51(37) and several sections of the Constitution. This creates a unique power of the federal government to accredit or adopt (or not accredit or adopt) state water plans relating to water allocation. This amounts to a considerable transfer of state power.

As yet, the *Water Act 2007* is untested, and testing it in the next few years will confirm a new federal legal architecture that will amend the entire notion of Australian federalism. The *Water Act 2007* created an independent body, the Murray–Darling Basin Authority (MDBA), and charged it with developing a Basin Plan for the consideration of the relevant Australian Government Minister (MDBA 2010). The sustainability duty of care was expressed as: 'the Authority is to determine the volume of water required to maintain and restore environmental assets, using best available science and the principles of ecologically sustainable development. Subsequently, the Authority investigated the optimisation of environmental, social and economic outcomes'. The politics of this is very fraught.

The *Water Act 2007* arose out of the dissatisfaction by the political leader of the day (Prime Minister, John Howard), enhanced by political commentators and independent think tanks, with the implementation of the earlier reforms, despite financial incentives being used to drive reforms (i.e. payment to the States for achieving ESD. It is fair to say that there have always been disputes between the parties about the meaning of the concepts of ESD as expressed in state laws and in the National Water Initiative and eventually in the *Water Act 2007*. As yet there are few cases on the Water Act 2007. In a significant amendment to the Water Act, the concept of critical human water needs (section 84) was added as a first priority. This human right statement and pre-empted the UN declaration on the human right to water by a few years.

The *Water Act 2007* was built, then, against a background where implementation issues were known to be a problem and the communication of science to the policy makers was fraught (Tomlinson and Davis 2010). Many of these problems arose from lack of knowledge, but also from lack of coordination between the states. The Guide to the Proposed Basin Plan specified that the reductions intake (interceptions and diversions) should be between 3000 and 7600 GL. It was recommended, however, that the reduction intake would be no more than 4000 GL because of socio-economic considerations. Even the 3000 GL proposed reduction resulted in considerable community angst and was also criticised for not taking into account social and economic effects (McKay 2010). The main dispute was over the assertions of projected job losses and the over weighting of the environment. The MDBA stated 800 job losses, but the NSW Farmers' Federation (NSWFF 2011) predicted 'a massive depopulation of the region'. Conargo Shire Council believed the Guide was extremely weighted towards the environment and requested equal weighting to environmental, economic and social (Conargo Shire Council 2010).

The MDBA took legal advice on the basis of the 3000 GL figure and finally suggested that this figure could be too high because of 'greater consideration of the social and economic needs of rural communities'. This was welcomed by irrigators but left environmentalists fuming. The advice would force a review, with less water returned to the river for environmental purposes. So, again, the question is how to interpret these terms, and as yet there has been no litigation on the Basin Plan. All of this is in the realms of administrative law and will be guided once there is some litigation. One acute issue is the financial risk for a litigant and has this impeded litigants in other areas of law (Australian Law Reform Commission 1995; Productivity Commission 2014). This leads to another suggestion to work on a way to litigate *public interest issues* with a fund to supports costs. This would require amendment to common law rules about champerty and maintenance. These prohibit a 3rd party from financing litigation.

Table 16.1 gives an outline of justice theories and cites cases on water management as enhancing or countering of the particular theory. See the introduction for details of the theories.

The analysis in Table 16.1 shows that the legal system at present inhibits the Rawls conception of distributive justice for four reasons. The law supports corrective justice and is built on procedural justice and the efforts made by judges to operationalise ESD in water plan cases attest to this. In summary, the legal system can review processes and require strict compliance and on occasions insert concepts such and natural justice in processes to augment the deficiencies of a Statute. The legal system cannot review priorities of government funding or priority lists because these are political decisions and non-justiciable.

Non-justiciability prohibiting the review of objects clauses such as ESD

Non-justiciability is a term that may conveniently be used to denote decisions where the court is of the view that 'the decision-making function lies within the province of the executive and that it is inappropriate that the courts should trespass into that preserve' (Mason 1994).

Table 16.1. Justice theories from literature and legal doctrines countering or enhancing those theories in water management law in Australia

Justice theory from literature	Legal concepts that counter a particular justice theory	Legal concepts that enhance a particular justice theory
Distributive justice. Distributive justice (Rawls 1972) concerns fairness, the distribution of benefits and burdens of social cooperation, needs to be equal unless an unequal distribution favours the least advantaged. Rawls is concerned about the fairness or rightness of the distribution of resources in society.	**ESD applied differentially due to the Constitution and the power residing in state laws and the lack of application of the *Water Act 2007* to the entire nation.** **National laws cannot change the status quo ante.** Under the *Australian Law Reform Commission Act 1996 (Cth),* the ALRC has a duty to ensure that the laws, proposals and recommendations it reviews, considers or makes: (a) do not trespass unduly on personal rights and liberties or make the rights and liberties of citizens unduly dependent on administrative, rather than judicial, decisions; and (b) are, as far as practicable, consistent with Australia's international obligations. **Narrow fact framed decision (as it must be) because judges are not legislators.** Adversary trial limits most proceedings to a contest between particular parties who rarely, if ever, have an esoteric interest in purely legal developments. They just want to win the case. The stimulus to new legal authority must often come from the judge's own reading and perceptions, encouraged by a sense of grave injustice to which the law or legal procedures seem directed. The adversary trial tends to limit the material available to the judge, particularly on the social and economic consequences of alternative solutions to the problem in hand. **Judges will not stray into political realm and will not re-write laws unless they are manifestly unjust The term non-justiciable applies.** It means the decision is non-justiciable because it is the role of Parliament to allocate resources and benefits and burden. Objects clauses and others are non-justiciable. Judges must interpret the law using statutory interpretation tools only. In *Minister for Arts, Heritage and Environment v Peko Wallsend*[b] the issues were complex economic, political and social in relation to a proposed mining project. The result was that the case may not be appropriate for judicial review.[c]	**Judges will not stray into political realm.** Non-justiciable as it is the role of Parliament to allocate resources and benefits and burdens (Finn 2002). So objects clauses and others are non-justiciable. However, the vague term ESD has led to judicial pronouncements and action, especially in NSW. The general formula for reductions of water under a water plan is to take the same percentage off every user. When a proposal came to reduce the allocations to reflect the amount used in the previous years, this was up held as fair because it was a reasonable process. *Arnold v Minister Administering the Water Management Act 2000*[a]. **Long-term ESD requires social cooperation currently and maybe foregoing current economic advantages to preserve future opportunities** (see ESD principles). Two ways to do this. The first is to use the law, says Mouritz (2010), to mandate behaviour and the second is the evolution of social norms through voluntary restraint as documented by Elinor Ostrom (1990) in relation to natural resource use, mainly in the developing world.

(Continued)

Table 16.1. (Continued)

Justice theory from literature	Legal concepts that counter a particular justice theory	Legal concepts that enhance a particular justice theory
Corrective justice. (Aristotle). Justice in voluntary transactions using contract law and legal protections from forced involuntary transactions (i.e. induced by fraud).	**Not applicable because the law protects individuals from fraud and has administrative law to require due process to change individual rights over resources.**	**This is the core of common law of torts and contract and administrative law and supported by several national and state-level statutes that have refined legal tests.** No cases on water market transactions being induced by fraud. For example, the Australian Consumer Law[d] would protect an individual from fraud by a water supplier. An example, the case of Tepko where negligent advice was given regarding costs of water supply to a business by government. Quote from the case: 'The only issue in this appeal is whether a statutory water authority owed a duty of care to a developer who wished to know, and was negligently told, how much approximately the authority might charge to bring town water to his land to enable him to subdivide it. It is crucial to what follows in these reasons for the limited follows in these reasons for the limited nature of the question for decision to be kept steadily in mind.' (Kirby and Callinan JJ).
Political justice. Use of the judicial process for the purpose of gaining (or upholding or enlarging) or limiting (or destroying) political power or influence.	**Not applicable.**	**Political decision making is non-justiciable.** In the *Tepko* case the judges did not dispute the political decision that the policy of NSW Government in allocating funds for urban water supply before rural water supply because it non-justiciable. However, there are examples see *Mabo*[f], which did rebuke an old legal concept of *terra nullius* and replaced it with a more modern view. This judge made decision was then followed up in the *Native Title Act 1993*.

Table 16.1. (Continued)

Justice theory from literature	Legal concepts that counter a particular justice theory	Legal concepts that enhance a particular justice theory
Procedural/legal. Sir Edward Coke[g] said that the wisdom of law is wiser than any man's wisdom and justice represents wisdom of the community. Also John Adams 1788[h].	**The doctrine of precedent requires a previous case be followed and this may be anachronistic. There is a limit to judicial innovation. Legal assumption that laws do not change core legal concepts.**	**The law and its focus on process in administering ESD has made a great contribution. This is through the concepts of administrative law.** Core competency of courts to look for procedural fairness and natural justice in the process.
Sen's theory of justice. Sen says that the Brundtland formulation is limited to the preservation and enhancement of human life (Sen 2009: 250). This is too limited.	**Sen says that the environment cannot be just a matter of what there is, but must also consist of the opportunities it gives to people and, second, the environment is not only for passive preservation but also for active pursuit.** Even though many human activities that accompany development may have destructive consequences, it is also within human power to enhance and improve the environment. Seeing people only in terms of their needs may give rather meagre view of humanity.	**This is a new theory and many of the concepts would be non-justiciable.**

[a]*Arnold v Minister Administering the Water Management Act 2000 (No 6) [2013] NSWLEC 73 (31 May 2013).*
[b]*Re Honourable Barry Cohen Minister of Arts Heritage and Environment; the Honourable Lionel Frost Bowen Attorney-General of the Commonwealth of Australia and the Commonwealth of Australia v Peko-Wallsend Limited; Peko Exploration Limited; Peko-Wall [1987] FCA 304 (7 September 1987).*
[c]*Minister for Primary Industries v Austral Fishing (1993) 112 ALR 211, 221 and 230 per Beaumont and Hill JJ.*
[d]Australian Consumer Law: see http://consumerlaw.gov.au/.
[e]*Tepko Pty Ltd v Water Board [2001] HCA 19; 206 CLR 1 (5 April 2001).*
[f]*Mabo v Queensland (No 2) ('Mabo case') [1992] HCA 23; (1992) 175 CLR 1 (3 June 1992).*
[g]Corkery J (2000) The rule of law. *The National Legal Eagle*, November 2000, pp. 3–7.
[h]John Adams 1788 quote 'A government of laws and not of men'. 'Novanglus papers,' no. 7. *The Works of John Adams* (Ed. Charles Francis Adams) vol. 4, p. 106 (1851). Adams published articles in 1774 in the Boston, Massachusetts, Gazette using the pseudonym 'Novanglus'. In this paper, he credited James Harrington with expressing the idea this way. Harrington described government as 'the empire of laws and not of men' in his 1656 work, The Commonwealth of Oceana, p. 35 (1771). The phrase gained wider currency when Adams used it in the *Massachusetts Constitution, Bill of Rights, article 30 (1780).*

Administrative law has been used at state level to test the boundaries of the water planning processes. Administrative law can only review the legality of decisions not their merits, as detailed by Brennan J:

> *The duty and jurisdiction of the Court to review administrative action do not go beyond the declaration and enforcing of the law which determines the limits and governs the exercise of the repository's power. If in doing so, the Court avoids administrative injustice or error, so be it; but the Court has no jurisdiction simply to cure administrative justice or error. The merits of the administrative action, to the extent that they can be distinguished from the legality, are for the repository of the relevant power and, subject to political power. (Attorney General v Quinn [1990]: 35–36)*

In the light of the above, decisions on state water plans will be discussed in the next section and the *Water Act 2007* in the following section.

Past water allocation plan – case examples where procedural justice has been paramount

These cases show that clear legislative direction will be followed but the process to change the status quo ante must be manifestly fair. These examples often heralded changes in procedure to enhance South Australian procedural fairness, particularly where the regional water plans made significant changes to the volume of water. The decisions show a case example where the law on ESD is in reality subordinate to procedural fairness.

Rowe v Lindner and Ors 2007 South Australian Supreme Court 189 – timely decision making

This case concerned sustainability of a proposal to establish a feedlot for 1500 cattle and the impact of the taking of 25 ML per annum of groundwater and surface water. The resource was not prescribed under the *Natural Resources Management Act 2004 (SA)* (NRM Act) and hence no licence was needed. The case used the *Development Act 1990 (SA)* and a plan under this that has an objective to protect all water resources from pollution or excessive usage that would threaten the long-term reliability of existing resources. The case also gave judicial guidance on the NRM Act and the precautionary principle. The outcome in the end was that the proposed development complied with many of the provisions of the Development Act but failed due to unsustainable use of water resources and was refused. The Development Act filled in where the NRM Act was deficient and could not apply because the area was not prescribed, and hence a water plan was not required. So the ESD general statement in the Development Act was interpreted to require the decision maker to be timely. The outcome of the case was that that planning approval authorities should seek advice from appropriate state agencies or experts to assist in the determination of impacted water courses, even though this is not a specific requirement of the NRM Act. The South Australian Environment Resources and Development Court (SAERD Court) also stated that when planning bodies are faced with a lack of data the precautionary principle should be applied. The evidence was insufficient to support a conclusion of unsustainable water use, but it was sufficient to support a conclusion of significant risk of serious harm due to water overuse coupled with current scientific uncertainty about the extent of environmental harm attracting the precautionary principle. The case also cited a NSW decision of a court at the same level. The justice consideration of future generations was implemented.

With regard to procedural justice, the SAERD Court held that the application for the water entitlement gives rise to entitlement that an administrative process will proceed within a reasonable period of time. So this reinforced the notion of temporal justice and this is enforceable by writ of Mandamus.

Murrumbidgee Groundwater Preservation Association v Minister for Natural Resources NSW [2004] NSWLEC 122 – justice as pro rata reductions

In NSW, the case of *Murrumbidgee Groundwater Preservation Association v Minister for Natural Resources* was about validity of a water plan. The Association challenged the plan on many grounds using administrative law. Administrative law deals with procedural fairness and the grounds alleged were:

extraneous purpose of the Minister in making the plan that was to avoid the community drafted plan, the formula for reserving waters for the environment contained a mathematical impossibility, uncertainty of timing of operation of the plan, and the imposition of uniform reductions in water allocation was irrational.

The area involved was within the power of the Minister to draft a groundwater plan himself. The plan deals with sustainable management of groundwater and identified limits on extraction; the overall aim of the plan was to reduce actual use over 10 years to the annual average recharge less a quantity preserved for the environment. Groundwater users were subject to pro rata reductions of entitlements over a 10-year period. All users were to be entitled by year 9 to only 52% of their original entitlements. There were adjustment mechanisms such as the creation of a market in access licenses and supplementary water access licenses.

All of the grounds for appeal were dismissed. The grounds were dealt with as follows:

- *Extraneous purpose.* The Appellant alleged that the Minister made the plan to avoid the notification, public exhibition and considerations as required under a plan made by a management committee. It was held that the power to establish a management committee to draft a plan is discretionary and a plan formulated by the Minister is valid.
- The literal construction of the clause did provide an absurd result, so the court applied a purposive construction that is to decide why the statute was passed and uphold that aim. The timing was considered to be capable of being certain and so valid.

The case here was based on the facts that it was irrational to treat the groundwater source as a single body of water, because aquifer recharge was site specific and that an activity in one area will result in changed conditions elsewhere. Historically, the groundwater system was managed in zones because in some areas use of entitlements would be unsustainable. The argument applied the precautionary principle to protect the resource in the absence of scientific data. The single system was argued to be irrational because it was not based on water availability. However, the court upheld the pro rata reduction on the grounds that the court has a confined role and it was for the Minister to balance the desired environmental outcome and the chosen method of achieving it with the beneficial and adverse social and economic consequences.

ICM Agriculture Pty Ltd v The Commonwealth [2009] HCA 51 (9 December 2009) – public interest as justice

This is a very powerful case, not only because it is in the High Court, but because the Attorneys-General for Victoria, Queensland, South Australia and Western Australia intervened and presented submissions generally supportive of those of the Commonwealth Government.

The case was about groundwater and surface water plans and a huge reduction of the past allocations of the bore licences by up to 70%. Successive governments of New South Wales have long monitored, regulated and restricted access to, and use of, both groundwater and surface water. Policies have been formulated and pursued so as to achieve equitable access among water users, to mitigate adverse effects on the environment, and to ensure that water, as a finite and fluctuating natural resource, is able to be replenished for future use. The extraction and use of water has been regulated by statute since 1896 and, in particular, from 1912 principally by the *Water Act 1912* (NSW) and in the modern period by the *Water Management Act 2000* (NSW), which repealed the 1912 Act. This Act includes an ESD requirement and under which the licences were reduced. The Schedule to the 2000 Act required the state to implement Water Sharing Plans over a 10-year period to reduce the water entitlements of licence holders and to make 'up-front *ex gratia* structural adjustment payments' to licence holders.

There are several issues in the case: the important one being whether the funding amounted to just compensation for the acquisition of property. It was not disputed that the Minister had the power in the public interest to make the amendments to reduce allocations. In fact on the public interest point the case states:

> *The term 'in the public interest' is one of broad import. When used in a statute, the term classically imports a discretionary value judgment to be made by reference to undefined factual matters confined only by the subject matter, scope and purpose of the statute in question [3]. As employed in s 45(1), the term must include, in its application to the circumstances obtaining when the Minister made the 2008 Order, the implementation of the Funding Agreement with respect to the LLGS. At that stage the State had not taken, within the times stipulated in the Funding Agreement, various steps required on its part to secure the federal funding in respect of the LLGS. Notwithstanding that delay, the Commonwealth intends to make payments in respect of the LLGS upon being satisfied of other matters stipulated in the Funding Agreement. These Commonwealth payments will be in the amount of $2.7 million.*

On the point relating to acquisition of property, the legal argument revolved around the fact that the licences were not property because the licences had always been subject to defeat by the Minister as water rights were vested in the Crown. This in fact occurred very early in Australian legal history to overcome the issues with the riparian rights doctrine. The *Irrigation Act 1886* (Vic) provided: '*The right to the use of all water* at any time in any river stream watercourse lake lagoon swamp or marsh *shall ... be vested in the Crown ...*' (emphasis added). The relevant Minister, Mr Alfred Deakin, explained that the provision was designed to overcome perceived difficulties with riparian rights developed in England by the common law. The *Water Rights Act 1896* (NSW) ('the 1896 Act') provided: 'The right to the use and flow and to the control of the water in all rivers and lakes ... shall ... vest in the Crown'. Section 6 of the 1912 Act retained this language. Similar language was adopted in water legislation in other parts of Australia.[2] From March 1984, the Department, and later the Ministerial Corporation constituted by section 7 of the *Water Administration Act 1986* (NSW), concerned to avoid resource exhaustion, adopted a policy of imposing on all licences (except those used to access domestic and stock requirements) a condition limiting the volume of water that could be extracted in a particular year to achieve ESD.

The judges stated:

> *Rather, and as these reasons have sought to demonstrate, it was a natural resource, and the State always had the power to limit the volume of water to be taken from that resource. The State exercised that power from time to time by legislation imposing a prohibition upon access to and use of that natural resource, which might be lifted or*

2 *Rights in Water and Water Conservation and Utilization Act 1910* (Qld); *Rights in Water and Irrigation Act 1914 (WA); Control of Waters Act 1919* (SA); *Control of Waters Ordinance 1938* (NT); *Lake Burley Griffin Ordinance 1965* (ACT).

> *qualified by compliance with a licensing system. The changes of which the plaintiffs complain implemented the policy of the State respecting the use of a limited natural resource, but that did not constitute an 'acquisition' by the State in the sense of s 51(xxxi) of the Constitution.*

The plaintiffs were ordered to pay the costs of the Australian Government.

Ashworth v Victoria [2003] VSC 194 – justice for the environment as water use was required to be licenced, status quo ante as law valid in respect to water use by farmers

In Ashworth's case, a farmer challenged the validity of the new laws in Victoria, which for the first time vested in the state the right to control water captured as overland flow in farm dams. The *Water Act 2007* (Vic) was at the time the most comprehensive in the nation and aims to set up a framework and management regime to ensure that upstream uses do not affect downstream users in Victoria (Pisaniello and McKay 2005).

The Government of Victoria in 2002 passed the *Water (Irrigation Farm Dams) Act 2002*. This act required the plaintiff to obtain a licence if he wished to continue with this practice. The statute was upheld as replacing common law rights. (Explanatory memorandum to *The Water (Irrigation Farm Dams) Act 2002 amending the Water Act.)*

The plaintiff had to pay the legal costs of the state of Victoria.

The *Water Act 2007* potential justice issues

The justice issue is that the Water Act imposes new tests to review water plans just for that region (see Fig. 16.2).

The *Water Act 2007* is unique and as yet untested. It has very broad objectives, including one to manage water in the national interest, which is a concept as yet undefined. It obliges those bound by it to implement relevant international agreements and, subject to the above, to promote the use and management of Basin water resources to optimise economic, social and environmental outcomes. Therefore it will generate a new jurisprudence in the areas where it applies (see Fig. 16.2). It aims for the new Murray–Darling Basin Authority (MDBA) to create a Basin Plan (for Murray–Darling Basin waters only). It makes the Basin Plan a legislative instrument, defining this (section 33) and impliedly allows review by courts exercising federal jurisdiction and specifically mentions the Federal Magistrates Court (section 139) and the Administrative Appeals Tribunal. This legislative instrument is a new type of form and it is unclear how that will operate and the legal boundaries of that will need to be debated.

The *Water Act 2007* requires state water plans to have certain content (section 22 and section 55) accredited (section 56) by the Commonwealth or adopted (section 57). It creates an overarching Basin Plan and requires the Murray–Darling Basin Authority, an agency of a Basin State, an operating authority, an infrastructure operator or the holder of a water access right not to do any act inconsistent with the Basin Plan (section 35). In sub-section 22 (8), The Basin Plan may also include any other matters prescribed by the regulations for the purposes of this subsection. This widens the potential ambit. Section 25 provides that the Commonwealth may specify the place at which the target is to be measured …

> *(b) may specify a target in terms of a particular level of salinity being met for a particular percentage of time.*

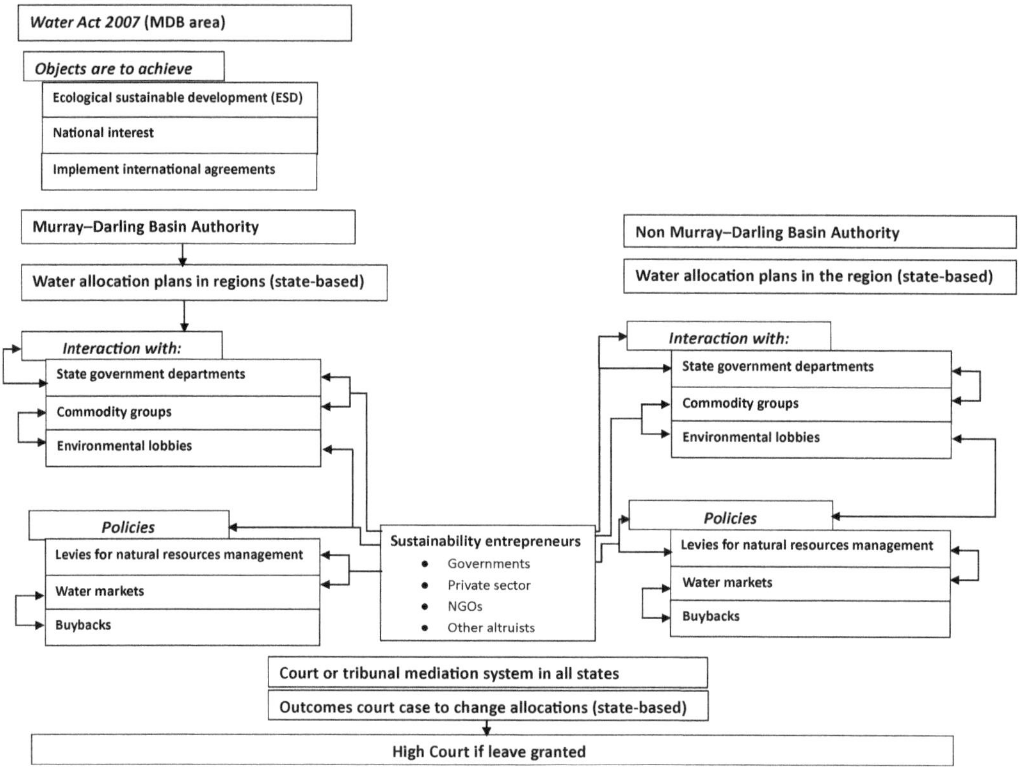

Figure 16.2. Justice regimes in Australian law with national interest in MDB and state laws in the rest of Australia (McKay 2011a).

It is notable that the Objects clause uses the term 'optimising economic, social and environmental outcomes', which are arguably narrower than ESD in that the long term is not mentioned. These achievements above are also subject to the national interest head and the treaties head. All of this will require judicial interpretation. Further analysis will need to be provided by judges in relation to the rank ordering of environmental outcomes as defined in the Act.

Environmental outcomes are defined as including:

(a) ecosystem function
(b) biodiversity
(c) water quality
(d) water resource health
(e) improve water security for all users of the basin.

Relevant international agreements are defined in clause 4, and include the Ramsar Convention, the Biodiversity Convention, the Desertification Convention, the Bonn Convention, CAMBA, JAMBA, ROKAMBA and the Climate Change Convention (Explanatory Memorandum to Water Act)[3] After this, the Kyoto Protocol was added to the mix. These are broad and raise the spectre of importation of lots of international law. The

3 *Water Act 2007* see explanatory memorandum (http://www.austlii.edu.au/au/legis/cth/bill_em/wb200793/memo_2.html viewed 22 January 2016.

act does define ESD in the terms of Box 1 in Section 4(2) but this is limited in its application to the Basin Plan. The relevant section provides a list of factors and this could be seen as limiting ESD as ESD is subject to consumptive uses of the basin. The list also includes many reviewable factors such as using best available socio-economic analysis.

The Basin Plan must include and account for (s 21) ESD:

(a) take into account the principles of ecologically sustainable development
(b) act on the basis of the best available scientific knowledge and socio-economic analysis
(c) have regard to the following:
  (i) the National Water Initiative
  (ii) the consumptive and other economic uses of Basin water resources
  (iii) the diversity and variability of the Basin water resources and the need to adapt management approaches to that diversity and variability
  (iv) the management objectives of the Basin states for particular water resources
  (v) social, cultural, Indigenous and other public benefit issues
  (vi) broader regional natural resource management planning processes
  (vii) the effect, or potential effect, of the Basin Plan on the use and management of water resources that are not Basin water resources
  (viii) the effect, or the potential effect, of the use and management of water resources that are not Basin water resources on the use and management of the Basin water resources
  (ix) the state water-sharing arrangements.

Once again, this long list is indeterminate and will require judicial rationalisation of it the operations of the Act and the Basin Plan. In any event, there are many options here for administrative law review of the consideration by the MDBA of state laws. This means the processes will be examined for fairness and other attributes. Courts are accustomed to fleshing out vague terms in statutes and constitutions, and the doctrine of precedent will give certainty to all engaged (Epstein 1995).

Section 22 provides a list of elements of the Basin Plan. Sub-section 22(10) provides that a provision of the Basin Plan has no effect the extent to which the provision directly regulates land use, the management of other natural resources or pollution. This opens up a justiciable issue around direct or indirect, despite the next subsection attempting to define direct regulation of land use.

The Commonwealth through the Murray–Darling Basin Authority accredits or adopts state water plans. These are defined as water resource plan for a water resource plan area means a plan that:

(a) Provides for the management of the water resource plan area; and

(b) is either:

  (i) accredited under section 63; or

  (ii) adopted under section 69;

but only to the extent to which the water resource plan:

(c) relates to Basin water resources; and

(d) makes provision in relation to the matters that the Basin Plan requires a water resource plan to include.

This last clause, and the fact that the Basin plans are legislative instruments, means that the Commonwealth is able to set a protocol for the states to adhere to and that any disputes would be non-justiciable. This is a huge change from the paradigms above, when the Commonwealth could only fiscally influence state government water-planning processes.

In addition to the above, the Commonwealth is able to regulate water charges by all water supply businesses in the states (section 91).

The duty and jurisdiction of the Court to review administrative action do not go beyond the declaration and enforcing of the law, which determines the limits and governs the exercise of the repository's power. If, in doing so, the Court avoids administrative injustice or error, so be it, but the Court has no jurisdiction simply to cure administrative justice or error. The merits of the administrative action – to the extent that they can be distinguished from the legality – are for the repository of the relevant power and subject to political power. The *Water Act 2007* applies a legal regime to state water plans and does allow them to be reviewed. At the federal level, the Administrative Appeals Tribunal can clearly review the legality of decisions, and the High Court would ultimately hear matters referred from state courts.

Human rights addition to the *Water Act 2007*

In 2012, the concept of critical human water needs was added to the Water Act. This is a clear statement of water priorities in response to clear community angst[4] and deals with distributive justice. The wording is yet to be tested and there are several under defined terms such as prohibitively high.

> Water Act 2007 – *Section 86A*
>
> *Critical human water needs to be taken into account in developing Basin Plan*
>
> *(1) Without limiting section 21, the Basin Plan must be prepared having regard to the fact that the Commonwealth and the Basin States have agreed:*
>
> *distributive (a) that critical human water needs are the highest priority water use for communities who are dependent on Basin water resources; and*
>
> *distributive (b) in particular that, to give effect to this priority in the River Murray System, conveyance water will receive first priority from the water available in the system.*
>
> *(2) Critical human water needs are the needs for a minimum amount of water, that can only reasonably be provided from Basin water resources, required to meet:*
>
> *distributive (a) core human consumption requirements in urban and rural areas; and*
>
> *(b) those non-human consumption requirements that a failure to meet would cause prohibitively high social, economic or national security costs.*

Conclusions and law reform suggestions

Justice has many interpretations and has been the subject of much literature in several disciplines around the world. It is a difficult concept to grasp, and it can and will change

4 Penny Wong statements; see for example http://www.abc.net.au/lateline/content/2008/s2353504.htm.

over time and place. In relation to the natural environment, concepts of justice at law were concerned only with property rights. The poem at the beginning illustrates that. Concepts of ESD were brought into the world stage recently and now have a place in Australian law. Australian judges have enforced the precautionary principle and made several important decisions to support ESD as an expanded notion of justice. This has great potential, but it is still limited by legal impediments such as the operations of the Acts and also the finances of litigants. We cannot hear all the cases and hence cannot be advised as to how to alter the laws.

Since 1994, Australia has used ESD to cover some aspects of justice and has been innovative in changing the status quo ante to incorporate concepts related to future generations, and preserving ecological balance. This makes us a world leader in the implementation of ESD. Once the Murray–Darling Basin Plan is implemented, there will be further litigation and this should clarify the meaning of some of the terms. An important addition to the existing *Water Act 2007* would be to impose a duty to cooperate (McKay 2011b) on all state and private sector bodies. This would be a short-term solution that may eventually lead to a system of identical state laws covering the entire nation.

In addition, a key reform would be to amend the Act so that it applies to whole nation, thus reducing the inconsistencies and injustice of having a critical human water need and a national interest test apply to only the Murray–Darling Basin. These innovations would enhance the distributive and procedural justice aspects of water management in Australia.

References

Australian Law Reform Commission (1995) *ALRC 75 Costs Shifting – who Pays for Litigation?* Commonwealth of Australia, Canberra, <https://www.alrc.gov.au/sites/default/files/pdfs/publications/ALRC75.pdf>.

Conargo Shire Council (2010) *Council Submission to the Guide to the Proposed Basin Plan.* Conargo Shire Council, <www.aph.gov.au/...Business/.../House_of_Representatives_committees>.

Ecologically Sustainable Development Steering Committee (1992) *National Strategy for Ecologically Sustainable Development.* Endorsed by the Council of Australian Governments, Canberra.

Epstein R (1995) Simple Rules for a complex world. Harvard University Press, Cambridge MA, USA.

Finn C (2002) The justiciability of administrative decisions: a redundant concept? *Federal Law Review* **30**(2), 239–263.

Mason A (1994) 'The importance of judicial review of administrative action as a safeguard of individual rights'. An Address to the Australian Bar Association Fifth Biennial Conference, 4 July 1994. The Hon Sir Anthony Mason AC KBE, <http://www.austlii.edu.au/au/journals/NSWBarAssocNews/1994/4.html>.

McKay JM (2006) Issues for CEOs of water utilities with the implementation of Australian water laws. *Journal of Contemporary Water Research & Education* **135**, 120–136.

McKay JM (2008) The legal frameworks of Australian water: progression from common law rights to sustainable shares. In *Water Policy in Australia: The Impact of Change and Uncertainty.* (Ed. L Crase) pp. 44–61. RFF Press, Washington DC, USA.

McKay JM (2010) Some Australian examples of the integration of environmental, economic and social considerations into decision making – the jurisprudence of facts and context. In *Global Justice and Sustainable Development.* (Ed. D French) pp. 327–431. Brill Martinus Nijhoff Publishers, Brill, Netherlands.

McKay JM (2011a) Australian water allocation plans and the sustainability objective-conflicts and conflict-resolution measures. *Hydrological Sciences Journal* **56**(4), 615–629. doi:10.1080/02626667.2011.580456

McKay JM (2011b) Evidentiary issues with the implementation of the sustainability duty of care in the Basin Plan. In *Basin Futures: Water Reform in the Murray-Darling Basin.* (Eds D Connell and RQ Grafton) pp. 233–242. ANU E Press, Canberra.

MDBA (2010) *Murray-Darling Basin Authority Annual Report 2010.* MDBA, Canberra, <www.mdba.gov.au/sites/default/files/pubs/report_2011.pdf>.

Mouritz T (2010) Comparing the social contracts of Hobbes and Locke. *The Western Australian Jurist* **1**, 123.

NSWFF (2011) 'No balance in Basin Plan, say peak agricultural groups'. NSW Farmers' Federation, Sydney, <www.nff.org.au/read/2319/no-balance-in-basin-plan-say.htm viewed 25 April 2016>.

Ostrom E (1990) *Governing the Commons.* Cambridge University Press, Cambridge, UK.

Pisaniello J, McKay JM (2005) Australian community responses to upgraded farm dam laws and cost effective spillway modelling. *Water Resources Development* **21**(2), 325–340. doi:10.1080/07900620500036455

Preston BJ (2005) The role of the judiciary in promoting sustainable development: the experience of Asia and the Pacific. *Asia Pacific Journal of Environmental Law* **9**, 109–111.

Productivity Commission (2014) 'Access to justice arrangements – inquiry report'. Report no 72, Productivity Commission, Canberra, <http://www.pc.gov.au/inquiries/completed/access-justice/report>.

Rawls J (1972) *A Theory of Justice.* Clarendon Press, Oxford, UK.

Sen A (2009) *The Idea of Justice.* Allen Lane, London, UK and Harvard University Press, Cambridge MA, USA.

Tomlinson M, Davis R (2010) Integrating aquatic science and policy for improved water management in Australia. *Marine and Freshwater Research* **61**, 808–813. doi: 10.1071/mf09224

17

The Social Justice Framework: untangling the maze of justice complexities

Anna Lukasiewicz

Summary

This chapter provides details of the Social Justice Framework, which I have been developing for several years as an aid to environmental management, especially decision-making processes about natural resources. I discuss the Framework's origins, components, audience and intended usage. Although the Framework has already been described in several publications, this chapter is the most complete guide, detailing all the Framework components and using multiple case studies to explain and illustrate them. The three justice dimensions of distribution, process and interaction are dealt with in turn and individual justice principles within these dimensions are explained and illustrated with examples. The chapter ends with a discussion of some important caveats about what the Framework can and cannot achieve in decision making and points out the aspects that still require further development.

Introduction

Although there are many aspects to environmental management (as well as natural resource management and conservation), all inherently contain elements of justice: who benefits from the resource and why? Yet the justice and fairness dimensions of resource management are often ignored or simply deemed too hard. As public acceptance of environmental policies is highly influenced by stakeholder perceptions of justice and fairness, there is a need for 'legibility' in how justice considerations are articulated and conceptualised by decision makers, observers and stakeholders.

This chapter is based on my research to develop a Social Justice Framework (SJF): a comprehensive list of justice principles that (theoretically) leads to fairness. I began this journey by studying water allocation conflicts in the Murray–Darling Basin (MDB) and my initial focus was on the Australian Government as the ultimate decision maker in natural resource management (NRM) conflicts. What should 'the government' do to ensure justice? My initial research led me to believe that there are certain principles that, if absent or violated, can act as catalysts for perceptions of injustice in different stakeholders and thus create conflicts.

As an academic (and therefore an outsider to the environmental conflicts I've studied), I believe that these principles are an essential consideration for a fair-minded decision maker (i.e. one who intends to implement fairness) and the initial development of the SJF was geared towards outlining the bits of 'justice' that need to be thought about by the decision maker in any NRM process. I have since moved from my original focus, as further case studies proved that the decision maker is not the only actor that can affect perceptions of justice because injustice can equally come from other stakeholders in the process, and in consensus-based decision making there is no single decision maker. Thus while I continue to believe that the SJF is very useful to the decision maker, it can be equally useful for other stakeholders to establish the elements of a fair process or to guide expectations of what a fair process might look like.

Preliminary work on the SJF has resulted in a structured set of justice principles categorised into three dimensions of justice: resource distribution, decision-making process and personal interactions. The initial set of principles (described in Lukasiewicz *et al.* 2013a) forms the core of the SJF. These justice principles are not abstract concepts: each has a definition based on the literature and a comprehensive description of how it is operationalised within the resource conflict. Further case studies have expanded the initial set of principles and focused my attention to different areas, such as environmental water and marine conservation. The SJF thus continues to be a work in progress and this chapter sets out my current findings and issues.

How the Social Justice Framework was developed

The principles of the SJF (Fig. 17.1) are grounded in the literature on social psychology and public engagement and have been tested through empirical application to different case studies. While the basic structure of the SJF has remained the same, each case study has refined it through greater illustration of justice principles and the nuances between them. The four case studies that the SJF is based on so far include: the water reform process undertaken by the Australian Government in the MDB in the time period of 2004–2010 (Lukasiewicz *et al.* 2013a,b,c); water re-allocation decisions undertaken by the New South Wales (NSW) and Queensland state governments in specific floodplains around the same timeframe (Lukasiewicz and Baldwin 2014); the current management of environmental water in NSW (Lukasiewicz and Dare 2016) and ongoing negotiations by the Commission for the Convention of Antarctic Marine Living Resources (CCAMLR) to create marine protected areas (MPAs) in the Southern Ocean around Antarctica.

The first and largest case study (my PhD thesis) dealt with justice issues directly and sought to establish the justice principles in the Australian water reform process. This research resulted in the subsequent creation of the SJF. The research included a semi-quantitative content analysis of key policy and legal documents, as well as semi-structured interviews with key respondents. The resource in question was surface water in the MDB river system and its relevance to social and economic life on two floodplains – one in South Australia and the other in NSW. The focus was to establish the justice intentions of the Australian Government in the water reform process, how those intentions were implemented at the state level and how the intention and implementation was perceived by three stakeholder groups: the environment (through its recognised government representatives), floodplain landholders and Aboriginal communities.

Following this large-scale inquiry into water reform, a more focused case study looked into two floodplains in NSW and Queensland to test the newly developed SJF at a more local level. This project retrospectively applied the SJF to interview transcripts with different types of landholders on the two floodplains during state government-led negotiations over valley-based water management plans. This was done through a secondary analysis of stakeholder interview data. The main aim of the project was to compare the state governments' processes in engaging stakeholders.

The SJF was next retrospectively applied to a project evaluating Environmental Water Advisory Groups (EWAGs) in NSW. As stakeholder advisory groups, EWAGs are an indispensable part of environmental water decision making that serves multiple purposes: they act as a sounding board for the implementation of environmental watering, provide community endorsement for watering actions and function as a forum for information exchange. The two aims of the project were to evaluate how EWAGs fulfil stakeholder and government expectations of local participation and how effective they are as vehicles for democratic participation and localism in environmental water management. Methods included semi-structured interviews with EWAG members, document analysis and participant observation.

The last case study discussed differs from the others in several respects. It is an analysis of international negotiations by CCAMLR to establish a series of large-scale Marine Protected Areas (MPAs) in the Southern Ocean around Antarctica. These negotiations have been ongoing for over a decade and include several MPA proposals. The SJF was applied to a thematic content analysis of the publicly available records of CCAMLR meetings. This is the first time that the SJF has been applied outside of the MDB to a resource other than freshwater in a process of international negotiations among nation-states.

The Framework

Empirical development of a justice framework ensures that the included justice principles are comprehensively grounded in reality rather than abstract concepts. There are three dimensions to justice recognised in social psychology: distributive justice, the oldest and most studied dimension (Kymlicka 2002); procedural justice, which relates to the decision-making process (Thibaut and Walker 1975); and interactional justice (also sometimes

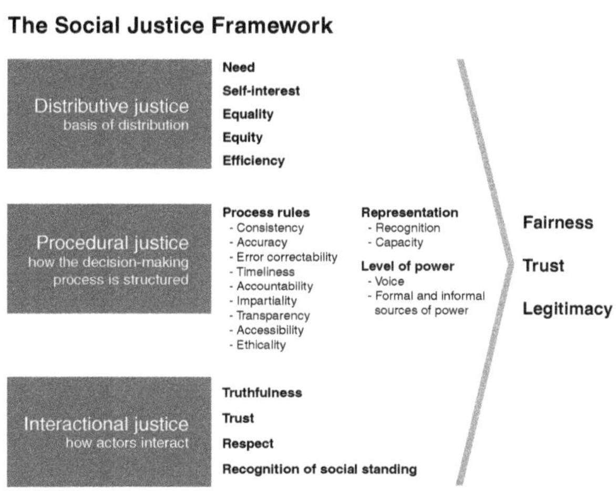

Figure 17.1. The Social Justice Framework.

called interactive justice), which deals with the interpersonal interactions between actors as opposed to the 'structures' of the decision-making process (Cropanzano *et al.* 2002). Some researchers in this book have further split interactional justice into informational justice, although in the SJF informational aspects are covered through a mixture of interactional and procedural principles.

Distributive justice

There are two ways of thinking about distributive justice: as the outcome of a decision; and as the intention of distribution. In the SJF, I use the latter because the actual decision outcome is usually an interplay of principles across the three justice dimensions.

How then can natural resources be distributed and what do those principles look like in practice? When I started looking through the literature on distributional principles, six principles stood out: equity, equality and need (Deutsch 1975), and self-interest, efficiency and fairness (Wilke 1991). The first set, described by Deutsch in 1975 is universally known and always quoted by justice researchers. However, Deutsch presented the three principles as the theory of cooperative social relations, which is more concerned with ultimate social goals, rather than a specific distributional decision. So equity would be the dominant principle in relations that strive for economic productivity, equality will dominate where social relations are most important and need will dominate where personal development and welfare is the common goal (Deutsch 1975). It is interesting to note that, when it comes to making decisions about the environment and natural resources, equality really isn't an important principle. Because there are no real empirical examples to base discussions on,[1] I will not discuss it further. The second set of principles comes from Wilke (1991) who wrote specifically about Australian NRM and constructed the Greed–Efficiency–Fairness (GEF) hypothesis: that people primarily focus on maximising their own outcomes but are restrained in this to some extent by concerns over efficiency and fairness. Fairness does not appear as a principle in the SJF because Wilke's definition of fairness is the same as Deutsch's equality. Below I look at the remaining principles in more detail, illustrating how these principles are operationalised using empirical data from my four case studies.

Equity

This is the most difficult principle to pin down. Equity is synonymous with fairness (and therefore, justice) in the minds of many people. However, equity as a justice principle can be used to support radically incompatible stakeholder demands. What does it mean? Equity 'conceptualises discrepancies as a perceived mismatch between inputs and outputs' (Törnblom and Vermunt 2007: 312). It is generally taken to mean that resources are distributed in proportion to the input made by each stakeholder. That input can be the: effective utilisation of a resource (as Deutsch 1975 suggests); magnitude of inputs (Diekmann *et al.* 1997), effort or deservedness (Syme *et al.* 1999); or strength of claims (Diekmann *et al.* 1997). This simple explanation is complicated because in reality a stakeholder can make multiple inputs (and outputs).

Equity can actually lead to favouring already dominant groups (as discussed in Lukasiewicz and Baldwin 2014). During the Millennium Drought, the NSW state

1 In his announcement of the National Plan for Water Security in 2007, the then Prime Minister John Howard flagged equality as an important distributional principle when he stressed that water savings obtained by better infrastructure would be evenly split between landholders and the environment. However, the expected water savings from infrastructure upgrades were largely insignificant compared with the 'water buyback' program.

government sought to protect the rights of irrigators in water distribution (an established and relatively wealthy group of farmers) rather than those of croppers and graziers by recognising their history of water use through the licensing system. This gave the irrigators as a group an advantage in the newly created water market (Lukasiewicz 2014; Lukasiewicz and Baldwin 2014) and was seen as fair by government managers because it acknowledged the contributions that the irrigation industry has made to communities. When it comes to managing environmental water, the NSW Office of Environment and Heritage speaks of equity as ensuring that people are not negatively harmed or benefitting unduly (such a definition is not common in the literature). This is another interesting example of equity as keeping the status quo. Equity can also be defined as deservedness, which can be based on rights or on basic needs (in which case it would come under the need principle in the SJF).

If equity is interpreted as deservedness, it could potentially lead to greater recognition of less powerful stakeholders, such as Indigenous people who have historically been dispossessed of land and natural resources. In this context, a focus on equity can lead to a major re-distribution of resources, rather than the preservation of the status quo. In Australian water reform, this was most evident in the concern of government officials towards ensuring that Indigenous people are included in water management. But, in terms of water distribution, discussions of equity were actually about ensuring that the needs of Indigenous people are met, so this will be discussed under the need principle.

Need

Distribution of natural resources according to need sounds very fair and straightforward. In practice, need as a distribution principle means that the prioritisation of stakeholders changes. For example, the environment was recognised as a stakeholder in need of water early on in the Millennium Drought. However, as the Drought progressed and water shortages loomed, the *Water Act 2007* was amended to prioritise 'critical human needs' over the environment. The progression of the Drought thus dictated a change in the prioritisation of needs and political will to act on behalf of the environment faltered.

Need usually means 'basic needs' – so how much does a stakeholder need to survive, rather than to thrive, prosper and multiply? Need is the most common distributive principle applied to the environment. Incidentally, equity is never applied to the environment in any of the case studies researched. Those familiar with the torturous process of the Murray–Darling Basin Plan (Gale *et al.* 2014) are familiar with the problems of resource distribution according to need – how much water needs to be allocated to the environment? What are the water needs of the environment? To achieve what effect? Is the goal mere survival or rehabilitation to some desired state? Answers to these questions are closely connected to a more fundamental question: 'what is the environment?' and the answer to that rests on one's values and beliefs (Lukasiewicz *et al.* 2013c).

The environment is a complicated and diverse stakeholder, and there are many people who claim to speak for the environment who have agendas of their own. Also, catering to the environment's needs is not just about a volume of water. How and when this water is delivered is also important because water needs of fish, birds and vegetation are not the same and catering to one may cause deleterious effects to the other. For example, pooling water through a regulator in order to wet red gum forests may promote carp breeding, which is bad for native fish, and releasing an environmental water flow from a dam to provide for a bird breeding event may cause cold-water pollution, which is also harmful to native fish (Mallen-Cooper *et al.* 2008). A focus on need in relation to the environment leads decision makers to rely on science, and the precise science is often lacking. This can, and does, lead to conservation measures being countered on the basis that they are environmentally unnecessary, as is the case in the CCAMLR negotiations where the size and scope of MPAs has been consistently reduced as proponents struggle to precisely define and defend why measures are needed. Quite simply, a focus on need sets the bar too high when it comes to catering for environmental needs. The precautionary principle would seem to be applicable but in reality is often ignored or actively undermined.

What of other stakeholders? My interviews with various farming enterprises found that distribution according to need is not desirable, because the goal of a profit-making enterprise is to thrive, not merely survive. Farming businesses thrived or failed during the drought by participating in the water market, which is an operationalisation of efficiency, rather than need (Lukasiewicz *et al.* 2013a). The needs of Indigenous people were seriously considered by government managers but there was, and to some extent still is, confusion over what these needs are and whether they are any different to the needs of non-Aboriginal stakeholders. Special provisions have been made in NSW and Australian legislation to cater for the cultural water needs of Aboriginal people and, later, their economic needs. However, few of these opportunities have been taken up (Jackson 2011) due to persistent economic disadvantage (Chapter 9). This raises an interesting point about restorative justice. Asking Aboriginal people what their needs are now does not consider the temporal consequences of the past. So in my interviews with Aboriginal elders in the water reform case study, their biggest complaint was lack of access to the land and rivers because landholders fenced these off (Lukasiewicz 2014). The historic dispossession of land came up here, even though the elders were not actively calling for land to be given back, but simply for access to the public areas of the river. Thus, although Aboriginal people are finally (belatedly) recognised as stakeholders, they are still disadvantaged by the past and the introduction of the water market, which, in the case of the Australian water market, tends to disadvantage new entrants.

Self-interest

Self-interest is defined as the 'individual's pursuit of one's own goals, regardless of, or at the expense of, other people or the group' (Rachlin 1997: 88). Although it has obvious negative connotations, it is mostly considered in social sciences to be a dominant, normal and legitimate motive, even if there is no direct empirical evidence for it (Montada 1998; Ingram *et al.* 2008). Self-interest has become a central assumption in many social sciences and there is a belief that acting in one's own interest is a social norm in western culture and people feel pressured to conform to it, believing that to do otherwise would lead to their exploitation (Miller 1999). Pursuing self-interest then becomes the rational, socially acceptable and even desirable thing to do (Miller and Ratner 1996; Müller *et al.* 2008).

Self-interest is widely perceived to be both negative and ever-present. This perception that resource conflicts simply boil down to self-interest is why justice is often put in the 'too hard' basket, as one Australian Government respondent in the water reform case study stated: 'some people will win and some people will lose and everybody who loses will believe that it's been unfair'. Establishing just procedures for decision making requires time, effort and political will, and the belief that it's all just about self-interest diminishes political will for procedural justice (Syme and Hatfield-Dodds 2007) can lead to dismissals of other stakeholders' viewpoints and arguments as self-interest masquerading as other concerns.

Should self-interest even be a principle of justice? After all, self-interest is perceived as the antithesis to justice. One answer to this is to say that the SJF is not a normative framework prescribing what should be, but rather an empirical framework describing what could be. In this case, self-interest in resource distribution simply cannot be ignored. More importantly, self-interest is neither evil nor as straightforward as people think. For example, in natural resource management, the owners and managers of natural resources are governments bound to act in the interests of their constituencies. However, what that interest is will be shaped by those very constituencies. Thus, it is accepted that many of the environmental problems in the MDB are a historical legacy of individual state governments' pursuit of their self-interest, which was the promotion of irrigation and water resource exploitation, because those pursuits were seen as beneficial to the state's growth and development (Connell 2007). On the other hand, currently the Australian Government is a proponent of MPAs in the CCAMLR negotiations that would restrict fishing around some parts of the Southern Ocean, despite its status as a fishing nation. In this case, Australia chooses to define its self-interest as acting for conservation for the sake of sustainability (L. Goldsworthy *pers. comm.*, 8 October 2015).

Efficiency

Why would an efficient distribution of natural resources be a justice principle? In Wilke's conceptualisation, efficiency is described as the 'intelligent management of the resource' (Wilke 1991: 169). It can also be defined in terms of not deteriorating the resource (Eek and Biel 2003), which is a desired policy principle in a world of finite resources, especially those such as water, on which all other life depends.

Two types of efficiency are usually discussed in terms of natural resource: economic efficiency (i.e. choosing the most valuable activities for the resource); and allocative efficiency (doing more with less). The latter is well illustrated in the EWAG case study, where water managers revealed that watering sites are chosen based on the ease and feasibility of being able to get sufficient water to the site, as well as their ecological significance and needs (Lukasiewicz and Dare 2014). In the water reform case study, allocative efficiency manifested quite differently: in the form of irrigation infrastructure upgrades (Lukasiewicz 2014). Proponents of this manifestation of efficiency assume that eliminating water losses in activities such as irrigation results in more water being available for other things. However, this assumption has been challenged, with sceptics pointing out that changes in irrigation efficiency do not create water, they merely change existing flow patterns from one recipient to another (Ward *et al.* 2007) and ignore that irrigation water 'losses' may in fact be benefitting the environment (Wolff and Stein 2003).

Market-oriented efficiency disadvantages new entrants, is not designed to consider restorative justice and tends to de-value intrinsic environmental assets and ecosystem services (Wegner and Pascual 2011). The water reform process did attempt to ensure the participation of Aboriginal stakeholders, but the market continues to be a barrier to Aboriginal communities becoming water owners. Theoretically, Aboriginal communities are able to buy water as anyone else is. However, as a group that has been historically disposed of land and resources in the MDB, most Aboriginal communities lack financial capital to compete in the market or to establish efficient and productive irrigation enterprises (Lukasiewicz 2014).

Thus the questions to ask about distributive justice an environmental conflict are:

- Which principles are most relevant to the management goal?
- What are the different principles at play?
- How are they manifested and are they incompatible?
- Who is espousing which principles?

Procedural justice

Procedural justice is the most prolific justice dimension for research. The SJF's conceptualisation of procedural justice is based on Paavola's (2007) categorisation of different aspects of procedural justice: representation, level of power and process rules. In theory, these different aspects of procedural justice are relevant irrespective of the decision-making methods used to obtain a decision. A decision-making process over natural resources could include submissions to a management or decision-making body, public meetings, advisory groups, targeted consultation or all of the above. However, as previously stated, some principles may be less applicable than others, depending on the decision-making method.

Representation

This category establishes who the stakeholder groups are (i.e. who is owed justice) and who is representing those who are recognised. Groups of people may be directly affected by a decision, they may be indirectly affected or they may hold an interest in the decision (e.g. Australian taxpayers may want some say in how NRM funding is distributed).

However, it's not as simple as running a stakeholder analysis. A stakeholder group or a representative may be recognised by the decision makers or through provisions in the legislation but not be accepted by other stakeholders. For example, in the EWAG case study, the legitimacy of the environmental representatives was questioned by some landholders on the basis that they did not live in the local area.

An intractable problem is the extent to which representatives on committees and advisory groups actually represent their constituencies. In the three MDB-based case studies, all of which had decision-making processes that involved committees and advisory groups, there were two types of representatives: members of a formal stakeholder body who were chosen by their constituents as a representative; and people who were informally representing a particular group because they were seen by the decision makers as being representative of that group. The issues around representation of environmental and Aboriginal interests have been covered in this book and elsewhere.

The questions that need to be asked about representation are:

- Who is recognised as being owed justice?
- Who represents recognised stakeholder groups, (why?) and are they accepted by other stakeholders?
- Which stakeholders are not represented and why?
- Are there questions about legitimacy of some stakeholders or their representatives?

Level of power

The level of power aspect refers to 'voice' or the ability of a stakeholder to affect the decision outcome. While this concept originates with Thibaut and Walker (1975) and Arnstein (1969), it has since evolved to an understanding that not all stakeholders need the same level of influence (Loux 2011) and that stakeholders don't necessarily desire total control (McCreddin *et al.* 1997). For example, stakeholders who are directly affected by a NRM decision may be afforded more power by the decision maker in a decision-making process than those indirectly affected, while an 'interested' stakeholder may be given 'observer' status to witness the process but not formally influence it.

The power to influence decisions can come from: how the process has been structured; formal sources of power (giving all stakeholders an equal say or prioritising some stakeholders over others); and the capacity of stakeholders to take action on their own behalf, their connections, assets and abilities (informal power). Capability comes into this, because those with greater capacity for involvement or specialised knowledge of the issue in question may be intentionally or unintentionally given more indirect power in the decision-making process (see also Chapter 6).

In the three MDB-based case studies, there is evidence that all stakeholder groups tended to perceive their opponents as having relatively more power and downplaying their own capacities and influence. Thus, irrigators often lamented that governments are dictated to by urban-based environmentalists despite the environmentalists decrying the lack of power they perceived they had relative to established interests such as irrigation. The methods of decision making inadvertently bestow different power relations among participants. For example, in the CCAMLR negotiations, any decision of substance, such as the establishment of an MPA, must be taken by consensus (i.e. all members must agree to the proposal). This puts the proponents of MPAs at a relative disadvantage because opponents have been able to extract significant concessions by holding out. However, if a proposal is passed, then it will be equally difficult for opponents to discard an MPA in the future.

The questions that need to be asked about level of power are:

- What formal and informal avenues and capabilities do stakeholders have to exert influence?
- What perceptions do stakeholders have about their and others' level of power?
- Are decision-making processes promoting the equalisation of power?

Process rules

These principles are about the actual process of decision making. The list outlined in Table 17.1 originates with Leventhal's (1980) six rules of procedural justice: consistency, impartiality, error correctability, accuracy, ethicality and representation. These have been incorporated into the SJF; however, representation features as a separate aspect of procedural justice (Paavola 2007) and several more process rules have been added through empirical development. The second case study added the last three process rules: accountability, timeliness and accessibility (see Lukasiewicz and Baldwin 2014). Table 17.1 provides definitions and illustrations for each process rule.

The application of these principles has been described in the past (Lukasiewicz *et al.* 2013a; Lukasiewicz and Baldwin 2014); however, I will briefly talk about each principle here.

Ethicality is a principle that I have struggled with. Leventhal's idea that for decisions to be fair, they must be compatible with fundamental moral and ethical values (what Skitka 2002 termed 'moral mandates') is certainly important. Although it has been difficult to identify what these fundamental values are, some are clearly evident. For example, the belief in the 'sanctity of private property,' described by Gunningham (2009: 155) as the 'widely held view that a landowner is free to do whatever he or she wants with their land' is very powerful in rural areas and successive state and Australian governments have acquiesced to this belief opting for financial incentives over regulation to promote environmentally friendly NRM behaviour on private property. The issue with ethicality is that different groups of stakeholders may hold mutually incompatible values and any decision will violate either one set of values or another (see Chapter 15 for a possible solution to this dilemma). Thus it is not always possible for a decision to be compatible

Table 17.1. Process rules and their application in an NRM context (Sources: Leventhal 1980; Lukasiewicz and Baldwin 2014)

Principle	Meaning	Application in NRM
Consistency	Decisions consistent between jurisdictions and across time	National, regional approaches to dealing with environmental management (e.g. Basin-wide management of the MDB)
Impartiality	Suppression of decision maker's bias	Preference for 'independent' chairperson for management and advisory groups; Ministers remain in control of resource decisions and are directed to consider all sides
Accuracy	Informed opinion with minimum processing error	Decisions are to be made on the basis of best available science
Error Correctability	Ability to redress mistakes	Focus on monitoring programs and adaptive management
Ethicality	Decisions being compatible with fundamental moral and ethical values	Sanctity of private property: conservation management seeks to minimise regulation and encourages financial incentive schemes on private property
Transparency	All available information made available to all stakeholders. This includes feedback on why and how decisions were made	Information provision to all stakeholders; dissemination of scientific information
Accountability	Representatives must be accountable to their constituents	Preference by government managers for formal processes with well-defined constituencies or, at a local level, for nominating active locals to sit on stakeholder groups
Timeliness	Adequate time being given for the process without undue delays	Set timeframes given to management committees to produce plans or for governments to act on their commitments
Accessibility	The engagement of stakeholders must be done in a way that enables the stakeholders to participate	Multiple methods of public engagement employed; special processes for Aboriginal stakeholders employed

with fundamental moral values of all stakeholders; however, the lack of this compatibility must certainly be considered in the process.

Accountability has already been flagged as an issue in the discussion on representation. There must be clear avenues of two-way communication between a representative and their constituency group, because NRM decisions force stakeholders to acquire technical expertise and broad socio-environmental understanding that may change their initial positions and beliefs. When a representative of a constituency is chosen – either by peers or appointed by the government – they will enter into a negotiation process where new learning will take place and compromises will be reached (ideally). The newfound knowledge, learning and understanding gained by the representative as a result of being directly engaged must be effectively transmitted to the constituency to ensure that the resultant decision will be understood and therefore more likely to be accepted. If the outcomes of this learning are not communicated down to the constituency group, the representative may lose legitimacy.

Accuracy and error correctability are important in NRM decision making and often manifest as a desire for science-based decisions and regular monitoring or adaptive management. These two principles are usually triggered by the distributive principles of need and efficiency. The issue is that, while science is usually viewed as objective, definitive and value-free, it is in fact contested, especially when scientific understanding diverges from local knowledge. The added problem for science-based decision making is that the necessary scientific data may not be present or may be highly uncertain.

Transparency is about making all information available to all stakeholders. The availability of easily digestible, accessible and timely information is important and goes a long way to producing procedural satisfaction, as was evidenced in the EWAG case study. However, transparency is not just about making decision-relevant information available to stakeholders. It is also about frank and honest admissions of uncertainties, knowledge gaps, risks and feedback about decisions taken. Transparency was added as a process rule to the SJF following analysis of the Australian water reform case study because it was an important focus for the reforms due to government acknowledgement that it was a serious deficiency in the past.

Accessibility to decision-making processes has long been recognised in international development, where studies often point out instances where marginalised groups such as women are constricted by social and cultural customs from attending public forums and having access to decision makers. It can also an issue in developed countries (see e.g. Lukasiewicz and Baldwin 2014). For example, I observed that in the three MDB case studies, adequate literacy was a common problem preventing some stakeholder groups in participating, because a lot of NRM decisions involve reading and comprehending very technical reports.

Impartiality of the decision maker is a justice principle supported to some extent in law, which legislates against decision makers having a conflict of interest or a pecuniary interest in a decision. Violation (or the perception of violation) of this principle inevitably causes perceptions of unfairness. However, this principle is seen as less relevant in consensus-based decision making arrangements where there is no single decision maker.

Consistency of decisions between jurisdictions and over time is a very legalistic process principle that was most important in the water reform case study, where the Australian Government sought to harmonise water management through consistent principles and definitions (such as water licenses, Sustainable Diversion Limits and over-allocation), rather than in details, in order to develop a common language of discussing water management problems between the jurisdictions.

Timeliness comes up often as a justice issue. Decisions either take too long or not enough time is given to stakeholders. When the Murray–Darling Basin Authority released its 2008 Guide to the Basin Plan, they started a tour of regional communities to explain it. Once of the most vocal complaints in the first few meetings was the lack of time given between the Guide's release and the first meeting. The guide was released on a Friday afternoon, with the first public meeting held on the following Monday, giving interested people less than 3 days to read and digest the lengthy document in order to fully participate in the ensuing discussion. On the other hand, the development of a water management plan for the Lowbidgee floodplain took 9 years during the Millennium Drought, which forced many graziers to cease their operations and leave farming altogether.

It is important to note that not all of these process rules apply equally in all situations and there can be inherent contradictions between them. For example, there can be a tension between timeliness (the need to quickly obtain a resolution) and accuracy (the need to obtain scientific data to inform a decision). The questions that need to be asked about process rules are:

- How are process rules manifesting in different decision-making processes?
- Which principles appear to be absent and why? How does this affect the decision-making process?
- Which principles are emphasised by which stakeholders?
- How do the principles influence each other?

Interactional justice

Interactional (also known as interactive or interpersonal) justice deals with how people as individuals interact and the basis for the following principles come from the considerable volume of work in organisational justice (e.g. Cropanzano *et al.* 2002; Masterson *et al.* 2005). Interactional justice considers decision-making processes to be important, not for the outcomes they produce but for the intangible benefits (social capital) that such processes bestow on those who participate: confirmation of status, a symbolic sense of belonging, positive self-worth, or fulfilment of moral beliefs (De Cremer *et al.* 2008). The interactional justice elements are the least developed and analysed in the four case studies so far and thus it is difficult to discuss them separately.

Respect is perhaps the most important interactional principle, and it was important in all four case studies. In NRM, respect seems to be manifested as the consideration of others' views and values, which automatically confers interpersonal respect (Tyler and Blader 2003). Recognition of social standing is closely related to respect. Social standing refers to the status an individual holds within society. Recognising that status is not just about respecting an individual but also the social group that this individual identifies with. As an example, several times in my research I've observed NRM-related meetings where a stakeholder would voice negative views about Indigenous knowledge, arguing that Indigenous communities had nothing positive to offer to the NRM issue under discussion. The Indigenous elders that I have met in my research take their obligations of looking after country very seriously and such views devalue not just their worth as individuals, but also their status as traditional custodians of the land.

The perceived absence of respect and status recognition can exacerbate feelings of injustice. This was most clearly demonstrated in the second case study, where Queensland graziers felt personally disrespected by the government's chosen chairperson, which heightened their negative perceptions of the decision-making process. On the other hand, there were no such perceptions in the NSW floodplain. On the contrary, despite several procedural justice issues (such as timeliness), the NSW croppers and graziers reported that they were satisfied with their interactions with the state government representatives. It is possible that the favourable interactional justice perceptions negated procedural problems dissatisfaction; however, there is not enough evidence to verify that hypothesis. In the EWAG case study, the landholders felt that their local knowledge of the land (an integral part of their identity as farmers) was recognised, needed and appreciated. These feelings could have helped to develop the largely positive interactions between stakeholders and government representatives.

Trust is a complex principle to define. In relation to NRM agencies, trust relates to people's perceptions about the agencies' capabilities, values and legitimacy (Beierle and Konisky 2000). Trust also refers to intentions of being fair and ethical in the immediate situation and in the future (Lind and Tyler 1988). The lack of trust is possibly correlated with perceptions of self-interest; this was an indirect link in the EWAG case study, where some stakeholders did not trust the motivations of others. Trust is also something that is dependent on past experiences. For example, Aboriginal stakeholders may be wary of becoming involved in government-run NRM processes due to the distrust they feel towards 'government' based on past experiences. In that case, the positive steps that government decision makers take to engender trust may not be enough to fully overcome these historical experiences, although a focus on establishing fair and accessible procedures is a good starting point. In the SJF, trust is a justice principle and the expected outcome of applying the framework, recognising the catch-22 situation where trust is both an essential input and an outcome of the process.

Truthfulness, being honest and free of deception (Masterson *et al.* 2005), is similarly a difficult concept because it relates to the intentions and motivations of participants (not just decision makers). It is very difficult to objectively measure and observe, and yet it is a fundamental starting point that participants enter a decision-making process free of deception. Earlier iterations of the SJF contained other interactional justice principles, such as propriety. These have been omitted from this version because of an absence of empirical data to support their inclusion. However, because this dimension is the least developed, future case studies may reveal more principles and provide greater depth to the existing ones.

Questions to ask about the interactions during a decision-making process:

- How is interactional justice operationalised?
- What steps need to be taken to assuage any historical distrust in stakeholders?
- Are identified issues with lack of respect and recognition coming from specific stakeholder groups?

A couple of caveats

Now that the reader is familiar with the justice principles of the SJF, there are a couple of caveats about the framework itself. It is not as clear cut as a simple checklist. Each justice situation is different and inherently difficult (see Chapter 1) and because of that, existing justice principles manifest differently and their relative importance differs depending on the circumstances. For example, in the EWAGs case study, the perceived impartiality of the group chairperson was very important, especially in those groups that had a high level of tension between stakeholders; however, it was not even mentioned in CCAMLR negotiations where there is no overall decision maker because decisions are made by consensus between members. Given this, can the relative importance of a principle be predicted? The four case studies clearly show that individual justice principles become important only when they are perceived to be absent. For example, the water reform process saw government put a lot of attention on transparency, because this was recognised as a weakness in the past. In the comparison of state-based water plans, lack of respect was perceived as a great injustice in Queensland, where graziers felt disrespected, but a non-issue in NSW.

So far, analysis of existing or past conflicts using the SJF has been mainly retrospective (see Chapters 5, 8 and 11 for other applications of the SJF). In retrospective analysis, there is some inevitable subjectiveness on the part of the analyst as to how an action or event is categorised and differentiated. As with any framework, categorising real-life events is as much an art as it is a science. The development of illustrative examples of each principles (through diverse case studies and authors) will aid future researchers.

It is important to remember that these principles are not just applicable to the decision maker. The SJF was formed to guide the decision maker through the maze of justice complexities and it could form a basis of some form of evaluation of how well a decision maker considered justice. However, it also applies to the stakeholders. Procedural and interactional principles such as transparency and respect are required not just of the decision maker towards stakeholders but of all actors involved.

Although it is convenient to separate the different dimensions of justice, in reality they are connected. For example, the range of distributive justice principles in operation will depend on the range of stakeholders involved. Perceptions that a process was unfair seem to require a pattern of principle violations. For example, in the floodplain case study, the Queensland process had a chair that graziers felt was not impartial – the committee rules effectively were skewed against them (lack of accessibility) – graziers felt disrespected throughout the process and consequently felt that the process was deeply unfair. In contrast, the NSW example identified very few issues, which were relatively minor; most importantly, the delay in getting a management plan finalised was seen as unfortunate, but not intentional, and therefore it did not invoke feelings of injustice.

The principles themselves are also connected. For instance, a strong focus on the distributive principles of efficiency and (environmental) need, leads to concerns for accuracy and error correctability. In all the case studies, a focus on the needs of the environment and the wise use of resources led to debates over the state of the science and the scientific basis of conservation measures, such as the Sustainable Diversion Limits in Australian water reform and the scientific basis of MPA proposals in the CCAMLR negotiations. It also leads to demands for monitoring programs. Stakeholders in the EWAG case study and in CCAMLR negotiations strongly supported the establishment of rigorous and continual monitoring programs to validate the scientific basis of management decisions. Finally, any analysis quickly reveals that multiple (and sometimes incompatible) principles are pursued at the same time at different levels. For example, in the water reforms, the Australian Government tried to focus on implementing efficiency, but had to temper this with concessions to equity due to pressure from state governments (Lukasiewicz 2014).

Things to remember:

- The mix of principles at play varies with each situation.
- Principles usually become an issue when they are perceived to be absent.
- Principles are inter-related: the pursuit of some principles automatically triggers others.
- Multiple, and often incompatible, principles are pursued by multiple actors at the same time.

This chapter has covered all the potential justice principles that could lead stakeholders to perceiving an environmental decision as unfair. This presentation has been rather basic and certainly much more could be said about each principle, as well as how they influence each other; however, the aim was provide an overview of the Social Justice Framework in the hope that others will make use of it in researching, resolving or avoiding environmental conflicts.

References

Arnstein S (1969) A ladder of citizen participation. *Journal of the American Institute of Planners* **35**(4), 216–224. doi:10.1080/01944366908977225

Beierle TC, Konisky DM (2000) Values, conflict and trust in participatory environmental planning. *Journal of Policy Analysis and Management* **19**(4), 587–602. doi:10.1002/1520-6688(200023)19:4<587::AID-PAM4>3.0.CO;2-Q

Connell D (2007) *Water politics in the Murray-Darling Basin*. The Federation Press, Sydney.

Cropanzano R, Prehar CA, Chen PY (2002) Using social exchange theory to distinguish procedural from interactional justice. *Group & Organization Management* **27**, 324–351. doi:10.1177/1059601102027003002

De Cremer D, Cornelis I, Van Hiel A (2008) To whom does voice in groups matter? Effects of voice on affect and procedural fairness judgments as a function of social dominance orientation. *The Journal of Social Psychology* **148**(1), 61–76. doi:10.3200/SOCP.148.1.61-76

Deutsch M (1975) Equity, equality and need: what determines which values will be used as a basis of distributive justice? *The Journal of Social Issues* **31**, 137–149. doi:10.1111/j.1540-4560.1975.tb01000.x

Diekmann KA, Samuels SM, Ross L, Bazerman MH (1997) Self-interest and fairness in problems of resource allocation: Allocators versus recipients. *Journal of Personality and Social Psychology* **72**(5), 1061–1074. doi:10.1037/0022-3514.72.5.1061

Eek D, Biel A (2003) The interplay between greed, efficiency, and fairness in public-goods dilemmas. *Social Justice Research* **16**, 195–215.

Gale M, Edwards M, Wilson L, Greig A (2014) The boomerang effect: a case study of the Murray-Darling Basin Plan. *Australian Journal of Public Administration* **73**(2), 153–163. doi:10.1111/1467-8500.12051

Gunningham N (2009) The new collaborative environmental governance: the localisation of regulation. *Journal of Law and Society* **36**(1), 145–166. doi:10.1111/j.1467-6478.2009.00461.x

Ingram H, Whiteley JM, Perry RW (2008) The importance of equity and the limits of efficiency in water resources. In *Water, Place, and Equity*. (Eds JM Whiteley, H Ingram and RW Perry) pp. 1–32. MIT Press Ltd, Cambridge MA, USA.

Jackson S (2011) Aboriginal access to water in Australia: opportunities and constraints. In *Water Resources, Planning and Management: Challenges and Solutions*. (Eds Q Grafton and K Hussey) pp. 601–628. Cambridge University Press, Cambridge, UK.

Kymlicka W (2002) *Contemporary Political Philosophy: an Introduction*. 2nd ed. Oxford University Press, New York, USA.

Leventhal GS (1980) What should be done with equity theory? New approaches to the study of fairness in social relationships. In *Social Exchange: Advances in Theory And Research*. (Eds KJ Gergen, MS Greenberg and RH Willis) pp. 27–55. Plenum Press, New York, USA.

Lind EA, Tyler TR (1988) *The Social Psychology of Procedural Justice*. Plenum Press, New York, USA.

Loux J (2011) Collaboration and stakeholder engagement. In *Water Resources Planning and Management*. (Eds Q Grafton and K Hussey) pp. 251–273. Cambridge University Press, Cambridge, UK.

Lukasiewicz A (2014) Fair water distribution. In *Opportunities for the Critical Decade: Enhancing Well-being within Planetary Boundaries*. (Eds AY Lo, LJ Pearson and MC Evans). The University of Canberra and Australia New Zealand Society for Ecological Economics, Canberra.

Lukasiewicz A, Baldwin C (2014) Voice, power, and history: ensuring social justice for all stakeholders in water decision-making. *Local Environment*. doi:10.1080/13549839.2014.942261.

Lukasiewicz A, Dare L (2014) 'Evaluation of NSW environmental water advisory groups'. Report to NSW Office of Environment and Heritage, University of Canberra, Canberra.

Lukasiewicz A, Dare L (2016) When private water rights become a public asset: stakeholder perspectives on the fairness of environmental water management. *Journal of Hydrology* **536**, 183–191. doi:10.1016/j.jhydrol.2016.02.045

Lukasiewicz A, Bowmer K, Syme GJ, Davidson P (2013a) Assessing government intentions for Australian water reform using a Social Justice Framework. *Society & Natural Resources* **26**(11), 1314–1329. doi:10.1080/08941920.2013.791903

Lukasiewicz A, Davidson P, Syme G, Bowmer K (2013b) How the social construction of the environment affects people's reactions to water policy. *Australasian Journal of Environmental Management* **20**(3), 179–192. doi:10.1080/14486563.2013.816641

Lukasiewicz A, Syme G, Bowmer K, Davidson P (2013c) Is the environment getting its fair share? An analysis of the Australian water reform process using a Social Justice Framework. *Social Justice Research* **26**(3), 231–252. doi:10.1007/s11211-013-0186-y

Mallen-Cooper M, Koehn J, King A, Stuart I, Zampatti B (2008) *Risk Assessment of the Proposed Chowilla Regulator and Managed Floodplain Inundations on Fish*. Fishway Consulting Services and Arthur Rylah Institute for Environmental Research, produced for the Department of Water, Land and Biodiversity Conservation, Melbourne.

Masterson SS, Byrne ZS, Mao H (2005) Interpersonal and informational justice. In *What Motivates Fairness on Organisations?* (Eds SW Gilliland, DD Steiner, DP Skarlicki and K van den Bos) pp. 79–105. Information Age Publishing, Greenwich CT, USA.

McCreddin JA, Syme GJ, Nancarrow BE, George DR (1997) 'Developing fair and equitable land and water allocation in near urban locations: Principles, processes and decision making'. Division of Water Resources Consultancy Report 96–60. CSIRO, Perth, WA.

Miller DT (1999) The norm of self-interest. *The American Psychologist* **54**(12), 1053–1060. doi:10.1037/0003-066X.54.12.1053

Miller DT, Ratner RK (1996) The power of the myth of self-interest. In *Current Societal Issues About Justice* (Eds L Montada and MJ Lerner) pp. 25–48. Plenum Press, New York, USA.

Montada L (1998) Justice: just a rational choice? *Social Justice Research* **11**(2), 81–101. doi:10.1023/A:1023299119352

Müller MM, Kals E, Maes J (2008) Fairness, self-interest, and cooperation in a real-life conflict. *Journal of Applied Social Psychology* **38**(3), 684–704. doi:10.1111/j.1559-1816.2007.00322.x

Paavola J (2007) Institutions and environmental governance: a reconceptualization. *Ecological Economics* **63**(1), 93–103. doi:10.1016/j.ecolecon.2006.09.026

Rachlin H (1997) Self and self-control. *Annals of the New York Academy of Sciences* **818**(1), 85–97. doi:10.1111/j.1749-6632.1997.tb48248.x

Skitka LJ (2002) Do the means always justify the ends or do the ends sometimes justify the means? A value protection model of justice reasoning. *Personality and Social Psychology Bulletin* **28**(5), 588–597. doi:10.1177/0146167202288003

Syme GJ, Hatfield-Dodds S (2007) The role of communication and attitudes research in the evolution of effective research management arrangements. In *Managing Water for Australia: The Social and Institutional Challenges*. (Eds K Hussey and S Dovers) pp. 11–22. CSIRO Publishing, Melbourne.

Syme GJ, Nancarrow BE, McCreddin JA (1999) Defining the components of fairness in the allocation of water to environmental and human uses. *Journal of Environmental Management* **57**, 51–70. doi:10.1006/jema.1999.0282

Thibaut JW, Walker L (1975) *Procedural Justice: a Psychological Analysis.* Lawrence Erlbaum Associates, Hillsdale NJ, USA.

Törnblom KY, Vermunt R (2007) Towards an integration of distributive justice, procedural justice, and social resource theories. *Social Justice Research* **20**(3), 312–335. doi:10.1007/s11211-007-0054-8

Tyler TR, Blader SL (2003) The group engagement model: procedural justice, social identity, and cooperative behavior. *Personality and Social Psychology Review* **7**(4), 349–361. doi:10.1207/S15327957PSPR0704_07

Ward FA, Michelsen AM, DeMouche L (2007) Barriers to water conservation in the Rio Grande Basin. *Journal of the American Water Resources Association* **43**(1), 237–253. doi:10.1111/j.1752-1688.2007.00019.x

Wegner G, Pascual U (2011) Cost-benefit analysis in the context of ecosystem services for human well-being: a multidisciplinary critique. *Global Environmental Change* **21**, 492–504. doi:10.1016/j.gloenvcha.2010.12.008

Wilke H (1991) Greed, efficiency and fairness in resource management situations. *European Review of Social Psychology* **2**, 165–187. doi:10.1080/14792779143000051

Wolff P, Stein TM (2003) Improving on-farm water management – a never-ending challenge. *Journal of Agriculture and Rural Development in the Tropics and Subtropics* **104**(1), 31–40.

18

Exploring environmental justice and public policy

Stephen Dovers

Summary

It is said that politics is the art of the possible, and so the discipline of public policy is the art and craft of implementing the possible. Especially in resource and environmental management, where allocation of 'environmental goods and bads' is the core task, public policy has focused on procedural justice (although rarely using those words) in terms of seeking greater and more constructive modes of public participation and community engagement. This has produced a very large body of theory and practice. Although modelling of distributional outcomes are often a large part of input into decisions, the fairness of eventual outcomes is less a focus. This chapter looks at how concepts of environmental justice can inform public policy in the environmental domain, concluding that a more nuanced and robust lexicon and knowledge of justice would be a positive contribution. Conversely, public policy can guide justice research towards points in the policy and institutional system where justice might be pursued to greater or lesser effect.

This book presents a range of Australian perspectives, applied across different resource and environmental management contexts, from the relatively new and still fluid and contested field of environmental justice. Environmental justice seeks fairness – variously in process or outcome – across groups in a human population in the policy and management that decides on and assigns access (or not) to natural resources and environmental assets. At times, it explores justice for the environment itself, or living parts of the environment. Pragmatically, as noted in Chapter 2, environmental justice more often seeks to identify and mitigate *injustices* or actual or perceived unfairness. We can postulate that this is simply because such instances will be brought to the attention of governments and the public rather than just or fair outcomes (no headlines there!), and because 'perfectly' just outcomes are at once unknowable and certainly always contestable depending on one's perspective and interests.

Fundamentally, environmental and natural resource management policy is, like all public policy, about the allocation of resources, access and privileges – whether of water, timber, fish stocks, health care, open space, recreational opportunities, transport routes, emergency services, minerals, or protection from the negative products of human activities such as pollution or noise. Public policy, as both a discipline and practice, seeks to provide structure, transparency and defensible processes underpinning such allocations of 'goods and bads' (generally, see Howlett *et al.* 2009; specific to environment, Dovers and Hussey 2013). Public policy deals with fairness and justice in ways qualitative and fungible, or quantitative and rigid, and sometimes leaves judgement to the courts when procedures to cater for fairness are deemed unfulfilled. Public policy also leaves significant justice and other value decisions to elected governments and their constituent parts (first ministers, ministers, Cabinet), in a parliamentary liberal democracy. The warning of Davis and colleagues, emphasis in original) is apposite:

> *Politics is* the *essential ingredient for producing workable policies, which are more publicly accountable and politically justifiable ... While some are uncomfortable with the notion that politics can enhance rational decision-making, preferring to see politics as expediency, it is integral to the process of securing defensible outcomes. We are unable to combine values, interests and resources in ways which are not political. (Davis* et al. *1993: 257)*

Values, interests and resources: the last of these, *resources*, can be in most cases measured; *interests* can be at least identified and possibly managed; but *values* are more political or moral, difficult if not impossible to measure, individually or socially formed and negotiated, and always contestable and resistant to the analyses and trade-offs that are the stock-in-trade of attempts at 'rational' public policy processes. Justice is a value matter, and thus fundamentally a matter of politics, in whatever social arena those politics are played out.

This introduction, and the role of politics, prompts a crucial distinction. Environmental justice – for the moment recognising just the two predominant forms discussed – concerns both the *process* of reaching decisions around allocation (procedural justice) and the fairness of the *outcomes* (distributive justice). Public policy, and the environmental and resource management regimes thus shaped, is concerned overwhelmingly with *process*: the institutions, organisational settings, rules, norms and practicalities of policies as they are formulated and implemented within the political system. Public policy may shape, but does not determine, political decisions. Whatever judgements may arise from normal procedures of review of policy implementation – Senate inquiries, independent evaluations or investigations by an Auditor-General – these will only judge fairness of process or outcomes against what the policy stated would happen. If the outcome is demonstrably unfair, or strongly argued to be, but is consistent with the original intent, then that may be remarked upon but not judged as wrong. An unfair allocation policy, or one with skewed distributional impacts (and there are many such), is not, from a public policy perspective, a failed policy: it may be as intended by the government or key political figure who set the parameters, supported by sufficiently numerous or powerful actors in the wider policy and political system. To that extent, public policy is a less normative enterprise than environmental justice, whatever individuals in the process may personally believe or desire.

Scholars and practitioners of public policy tend not to use the term 'justice' but rather less strict enunciations of fairness. If justice is considered, it tends to be in the hands of what public policy would view as the proper arbiters – the court that overturns a decision or process, the electorate who ejects a government or politician they think made unfair decisions, or the media pressure that reacts to or inflames public rage and forces the reversal of a decision.

In this chapter, I first note how environmental and natural resource policy seeks to allow as wide as possible or permissible input into formulating policy, and how it seeks to achieve at least tolerable outcomes in allocation, which are not open to the perception or proof of unfairness. The question is then asked: how can the ideas from the field of environmental justice inform questions of public participation and of tolerable, or at least transparent, fairness in the outcomes of decisions that seek to consider values, interests and resources in allocation of goods and bads? That is, what might environmental justice contribute to public policy in this domain? Along the way, we might ponder the converse, too: what can environmental justice learn from public policy?

Environmental policy thus far

In resource and environmental policy and management (and elsewhere in public policy), the basic challenges of considering and achieving fairness – or at least tolerable unfairness – have been complicated by two factors over recent decades. The first is an increasingly 'full world', where population growth, rising material consumption and a globalised economic system have resulted in greater use of limited, or only slowing growing, resources and often finite or diminishing environmental assets. There is more competition – in many resource contexts there is less to go around – and so we see greater contest and friction over allocation and decision making. If a 'good' or resource has sufficient spare capacity (e.g. an abundant fish stock) or the environment's assimilative capacity of a 'bad' is large enough (e.g. to safely absorb a pollutant) then tolerably just outcomes, totally unremarkable decisions or justified inaction, are not too hard to achieve. There is less of that luxury nowadays. Scarcity matters. More decisions now, and many more in the future, will confront the unpleasant fact that one groups' benefit from environmental assets or natural resources will inevitably be at the expense of others, and the latter will deem this unjust.

The second shift is in the kinds of policy styles and instruments used in the resource and environment field, and in the broader sustainability domain. The toolkit has expanded considerably from when simple end-of-pipe regulation, mild public education campaigns and reservation of a few natural areas were the principal options. The implications for fairness and justice of this can be profound. Market mechanisms such as tradable allocations of water, fish or carbon are a transformative policy approach entailing the revision of property rights, and a redefinition of the underlying policy logic: away from social distribution, towards ecological limits as the outside parameter and economic efficiency as the allocative principle (see Chapter 7 in Connor and Dovers 2004). Policy formation processes that engage diverse stakeholder groups in public processes are a relatively recent experiment, and one in which Australia has played a once leading, but now declining, role – for example, Regional Forest Agreements, the Resource Assessment Commission process, catchment management and regional organisations, and the National Strategy for Ecologically Sustainable Development (see various chapters in Dovers and Wild River 2003). Direct participation in environmental and resource management has become standard – for example, community-based boards of regional organisations, and the world-leading example of Australia's Landcare program – even if both have been critiqued on their impact, resourcing and genuineness in terms of actual empowerment (Lane *et al.* 2009).

Participation in public policy and environmental management has been a particularly significant and well-documented shift, whether or not one believes the advances in environmental or social outcomes, or fairness in process, are sufficient. Environmental and resource management has, in many countries, deviated away from top-down, regulation-based approaches to ones where there are diverse collaborations between policy and management agencies, communities, resource users, other management agencies, NGOs and industry. This is in line with broader politics pushes regarding the role of citizens and the sharing of power, and a move from direct government action being the main game to a 'governance' including more parties (e.g. Rhodes 1997). This has been particularly noticeable in environmental and resource management, and this field has been the site of many experiments in participatory policy and management (e.g. Healey 1997; Dobson 2003; Paehlke 2003; Dryzek and Niemeyer 2010; Holley *et al.* 2012). As well as these broader trends in participatory democracy, there is a cleanly pragmatic imperative: the participation of communities, firms and NGOs promise better purchase on problems and delivery of solutions than relying on the actions of the state alone.

There have been significant advances in allowing 'stakeholders' to be part of policy formation, resource allocation and actual environmental management – in part at least considering questions of justice and fairness – some of which may now be almost taken for granted. Modern environmental planning and development laws across most jurisdictions, such as the breakthrough (but now somewhat diluted) *Environmental Planning and Assessment Act 1979* (NSW), provided for community notification of and some degree of input into planning proposals, and specialised courts to adjudicate conflicts (in that case, the New South Wales Land and Environment Court). Unthinkable a few decades ago, such access to decision making is now commonplace, whether considered sufficient or not. The 'corporatist turn' in public policy, whereby key representative groups are recruited into government-run or -coordinated policy processes, emerged in the 1980s, and, while somewhat diminished, it replaced a much more opaque tradition of who had input. The globally remarkable, but still-struggling, Australian experiment of community engagement in the Indigenous domain, such as Indigenous Protected Areas and some degree of co-management, was similarly unimaginable even in the wake of the 1992 *Mabo* decision. Tests for public interest and environmental protection were placed in the processes underpinning major economic and social reform, such as the National Competition Policy. Whether these justice-related provisions were sufficiently or rigorously applied, such provisions were non-existent not so long ago, and not even thought of. Imperfect as all these examples may be, they represent an inclusion of the broader community that previously did not exist.

These cases are about the rights and protections of people. The very idea that species, ecosystems or the more general notion of 'the environment' having rights, and thus being afforded consideration in policy, is not new, but it is constantly evolving. Australian water policy now has defined 'environmental water rights' and even a statutory office of the Commonwealth Environmental Water Holder – the environment cannot be a buyer or seller in the neoliberal and ecologically constrained tradable water market, so the state acts in its stead. Community groups can buy water allocations for 'the environment', too, or at least for their perception of the environment's needs. Rights for 'the environment' to some degree cater also for the rights of future generations, who, in accord with sustainability principles, should be considered in decision making. Chapter 10 explored non-human species and justice. Shifting policy styles and instruments redefine who gets to have a say, and who can get what, and thus can fundamentally alter the social, legal and institutional landscape that defines the possibilities for (un)fairness and (in)justice.

At a broader scale – post the 1992 UN Conference on Environment and Development and the resulting Rio Declaration – issues of equity, fairness and justice have been strongly present in the international, overarching construct and policy agenda of sustainable development. The rights of the world's poor and the need for particular consideration of the rights and conditions of women, youth and Indigenous people have been central to that agenda since then, most recently in the outcomes of the Rio +20 and the ambitious and comprehensive Sustainable Development Goals.[1] In the ever more closely related area of disasters, gender equity and differential vulnerability to the impacts of hazards are deeply woven throughout the 2015–2030 international program of action and collaboration expressed in the Sendai Declaration.[2] To be unduly exposed to deadly risk when that need not be the case is neither just nor sustainable. Across these global agreements and their thousands of national and sub-national expressions in government policy, including in Australian policy and even law, a core principle of sustainable development is recognition that citizens and communities should have a say in decisions that affect them.

In many such international agreements, there is a presumed, and often real, division between developed (i.e. rich) and developing (i.e. poor) countries. While it is undoubtedly true that environmental injustices are heaped upon the poor much more than the rich, cases recorded in the literature (e.g. Washington *et al.* 2006) concern the poorer or otherwise marginalised in rich countries. And countries change, with 'emerging economies' that were in the early stages of poverty-ridden development during the early years of modern environmental awareness now middle-income and rising. International policy and law – over intellectual property in biological resources, toxic wastes, and so on – have attempted to prevent environmental harms, whether effectively or not.

A proposition: it is not that governments, singly or collectively, do not recognise the need for fairness and justice – they constantly promote and inscribe the principle in policy and process, even if they may be suspected of empty rhetoric. Generously, perhaps they just are not very good at implementing the principle, or are blocked or frustrated when they try. This is like the sincerely pledged, but challenging, New Year's resolution. Certainly they could do better, and maybe ideas from environmental justice can suggest how.

Why would literally hundreds of nation states pledge justice publicly and clearly in international agreements if they do not believe it? (Of course, when the stakes are high enough, justice will not be part of an international agreement: witness climate change, the vague recognition of 'common but differentiated responsibilities', and ongoing, intractable discussions of who should act and who will be affected.) What blocks or frustrates? Chapter 2 pointed out (and a great many others accuse directly) the commonly identified culprit is a justice-blind neoliberal, global economic system. Is the global economic and trade system unjust by intent, or does it seek its own forms of economic justice via means that in many commentators' eyes just don't work (trade liberalisation, a market economy, reducing the role of the state and protection of key industries)? Efficiency, optimisation and productivity are dominant drivers of public policy and political decision making, and do tend to override equity, let alone equality. Governments may value equity and justice, but the overriding goals and tools of managing a society and economy might be ill-suited to achieving justice. They can't do both, and people will fall on one side or the other of the argument of which should be viewed as more important. Of course, powerful vested interests may simply block.

Inviting justice into the policy tent

As largely a process-oriented discipline and practice, environmental policy and management has focused largely on *participation* in decision making and in management

1 See: http://www.un.org/sustainabledevelopment/sustainable-development-goals.
2 See: https://www.unisdr.org/we/inform/publications/43300.

actions, including monitoring. That is the focus of the bulk of the rest of this chapter, but a note on outcomes first. Returning to the distinction made above, this indicates the first possible lesson from environmental justice to policy: that a robust, legal and even highly participatory policy and process may produce fundamentally unfair outcomes. The establishment of markets in water or fish may disadvantage small-scale farmers or artisan fishers; the promulgation of pricey renewable energy sources in homes may leave a residual rump of poor households stuck on an inefficient grid and paying more and more; carbon pricing may impact relatively more on rural or outer suburban households without matching transfer payments to compensate; or an industry might be restructured but the structural adjustment may not reach those indirectly impacted such as suppliers and secondary component or input suppliers.

An inherent messiness of policy reform is that new policy instruments that disadvantage some groups as a side-effect of a larger goal then require transfer payments or other compensatory measures, reducing the effect of the policy and creating administrative transactions. Such concessions might be granted with due goodwill or as a matter of political expediency: it matters not because the inefficiencies remain the same. The public policy system of advisers, lobbyists and bureaucrats can predict such outcomes, and indeed endless modelling is done to that effect: they can point this out to the eventual decision maker (Minister, Cabinet) and compare options, but that is as far as it may be able to go. Political decisions are political decisions.

What environmental justice can do is to provide a perhaps more nuanced and comprehensive language around impacts, and the reasons why a process, and thus by implication the outcomes of that process, may be problematic. This may alert politics to an impact otherwise unforeseen (and they may or may not care or consider it significant), or politics might be self-interestedly glad to be forewarned of an oncoming potential friction that must be managed. Chapter 7 demonstrated that local government – a particularly pragmatic and constrained component of the state – was better at comprehending 'justice' than many might expect, but identified one particular dimension (informational justice) that posed a challenge. The ability to identify more dimensions of (in)justice, as in this book – relational, interpersonal, spatial, procedural, and so on – may (repeat *may*) enable public policy processes to explore impacts more deliberately and accurately, and thus lead to more fair outcomes. That would be an improvement on a well-meaning but simple recognition and pursuit of a vague notion of fairness that hides multiple dimensions.

Returning to process, Fig. 18.1 is a conveniently simple characterisation of issues and challenges in catering for public participation in environmental policy and management:

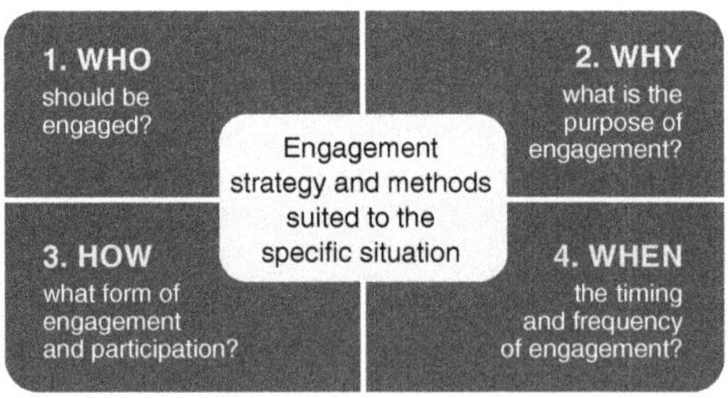

Figure 18.1. Framework to inform design of an engagement strategy (Source: Dovers *et al.* 2015).

in this case from the context of protected area management (a large field transitioning from state management to more engaged and multi-actor forms).

We can consider briefly where the environmental policy and management is up to on the four questions in this framework:

1. *Who* should be engaged? Identifying the matter/s at hand and assessing the realistic, but fair, range of 'stakeholders' who have a right to be involved, who is prescribed as having such a right in policy or law, or who might be engaged because they would be helpful to the process.
2. *Why:* what is the purpose of the engagement? Clarifying what is to be debated and decided, and what is not, and ensuring that all participants understand the 'game' they are playing and the limits and rules of that.
3. *How:* what form/s of engagement and participation? Following (1) and (2) above, choosing specific participatory designs and tools to use to engage people (e.g. public forums, stakeholder advisory committees, public inquiries, surveys, deliberative forums such as citizens juries, statutory rights of co-management: the list is long).
4. *When:* what should be the timing and frequency of engagement? Identifying and using the relevant points in the policy process – from problem identification through design, implementation and evaluation – when interest groups and communities are most likely to wish to have, or should be desired to have, input.

At the risk of over simplification, and of enraging theorists and practitioners in the field, a summary judgement on how well we are doing in a mature field of participatory environmental policy and management, that has absorbed many insights from other disciplines and arenas of practice such as social psychology and community development (a point well demonstrated in Chapter 6).

On '*who?*', we are well advanced in theory, resources, knowledge and at times in good and successful applications. That said, the use of our knowledge, and genuineness of applying it, is at times quite poor. On '*why?*', the situation is less advanced in practice (as opposed to theory and knowledge) and confusion and misinterpretation of the scope and purpose of participatory policy and management mechanisms abounds. On '*how*?', there are endless lists and guidebooks that demonstrate a full, diverse and well-tested toolbox of options to use, whether we choose the right one or use it well notwithstanding. On '*when?*', despite a common rule of 'consult or engage early', that is often not done, thus alienating communities, and even more often interested parties are not engaged well in implementation and evaluation – the tap of engagement is switched on and off seemingly at convenience.

What can ideas of environmental justice add to this situation that might help? Broadly, and to repeat a point made above, the more nuanced and detailed identification of the multiple elements of justice can provide a better lexicon and identify aspects for scrutiny in formulating, implementing or evaluating policy. Chapter 6 identified that potential, and the further lessons that identifying and reducing injustice is the reasonable and possible aim, as perfectly just processes and outcomes are likely out of reach. There is an unfortunate persistence in environmental management of advocating the alluring possibility of 'win–win' processes and outcomes, increasingly unlikely as these are in an ever more full world. Disappointment and alienation are predictable when that is not achieved, and an (in)justice frame promotes recognition of the reality of the spectrum of acceptable, imperfect but likely win–lose–tolerate-but-retain-respect outcomes. Even with necessarily imperfect outcomes (from a justice standpoint), the finer scale understanding offered by subsets of procedural justice offers environmental policy and management greater insights into how to cater for and manage expectations, and an enhanced ability to develop common understandings between agencies and stakeholders.

In this vein, it is not simply the several, main dimensions of justice that matter. Chapter 8, applying the Social Justice Framework, investigated one dimension: procedural justice, via nine principles, allowing a greater level of explanatory purchase on a public policy process. The case study – the Tasmanian Forest Agreement (TFA) – reveals other important relationships between justice and public policy, and especially politics. In the TFA, participation was far from ideal and complete, but Chapter 8 exposed the reasons in a manner too rarely done. The deep and persistent conflicts in Tasmania over forests meant that exclusion and careful management of who knew what – even when it very much affected them – was a necessity, albeit one that would be judged as procedurally unjust. But the authors point out that interactional justice was enhanced: evidencing the reality of trading off different dimensions of justice, a finer point that could be done more often. If the proof of the policy pudding is in the eating, then the TFA was a remarkable feat, and only failed in the end (for now at least) by a political kill.

Procedural justice is easily enough recognised by environmental policy and management practitioners, and indeed the four-element engagement framework of who-why-how-when above is largely about this dimension of justice, and the definition of a policy *process* (often with back-drop regulatory or legal procedures): the main game for public policy. But the relationship of procedural with interactional or relational justice could often be improved. Chapter 6 stressed the crucial importance of relational dynamics. In environmental policy and management contexts that typically run over extended time periods of decision and allocation processes, relational dynamics over time, not just at the start, are something too often overlooked. However, that highlights a difficult issue for public policy: set and forget is the ideal of public policy, so that the scarce assets of resources and time can be turned to the next task. The ideal of justice – an ongoing, respectful, well resourced and inclusive engagement – may well be resisted by those who have to resource and maintain it over election cycles and the time-limiting strictures of forward budget and program estimates. In such cases of oversight and reluctance, as with outright and explicit resistant of justice, those wishing to maintain access to decision making will need to agitate and actively seek engagement.

Chapter 7 similarly presented a more detailed view of interactional justice, with the more commonly recognised interactional being divided into the often locally critical *interpersonal*, and the second subset of *informational* that emerges as particularly important in their coastal adaptation case study. The capacities for generating information, and for running highly engaged processes of discussion and negotiation, are also evident in this case. More justice, particularly procedural, informational and interactional, as a rule means more time, personnel and resources: something that especially local governments have little of. Chapter 7 also identified the overarching role that local and state planning laws and procedures play in shaping the ways in which justice considerations play out – or cannot be played out – in policy and management processes. The possibilities that agencies of the state have to adapt to different constituencies and contexts are often, or in fact almost always, constrained by law, regulation and government, even if officials might wish it otherwise. Calls for flexibility, long-term commitment and inclusion by theorists, commentators and activists tend to overlook these very real constraints (Wyborn and Dovers 2014). The realities of public policy may be instructive, if sometimes dispiriting, for proponents of environmental justice.

Especially in federal systems, there is generally a hierarchy of laws, policies and processes, and the expression of and implementation of justice principles and related practices may not translate easily across such administrative and geographical scales. State-planning or resource allocation policy and law will set out general principles and requirements, and we would generally agree that these should be evenly and fairly adhered to across a society, with some discretion to vary these to accommodate local variation and needs. The finer details of justice may vary significantly by time, place and community. But too much flexibility risks the vagaries of local power politics, and the most powerful antidotes to injustice – recourse to statute or common law in the courts – may be out of reach of unfairly treated individuals or minorities.

Spatial and administrative scale is important in environmental management, in a somewhat different sense to the spatial justice discussed in Chapter 7. Resource complexes and environmental functions operate over multiple scales, begging the questions of: (i) at what scale to best construct policy and undertake management (locally, catchment, regional or national?); and (ii) how to ensure that the inevitably multiple scales of policy and management operate synergistically, or at the very least not in conflict. The first affects the prospects of justice fundamentally. Once a decision on scale is made, some issues and some people are defined and thus included, generally in an explicit fashion. But others will be excluded, and often not as transparently: it is in silence or ignoring that injustice often occurs. To construct stakeholder input to water management principally at the catchment scale may exclude various localised issues and interests while including other issues and stakeholders. Call-in powers, where a state government overrides a local government's planning decision, may ensure probity and adherence to higher level policy and legal principles, but completely disempower local democracy and interests. Should the rich, occasional visitor to a beautiful natural tourist destination have an equal say in its management as next-door residents, when they are equal as voters in that jurisdiction? There are deep justice questions there, and everyday headaches for public policy and politics. Procedural and interactional justice are very much defined by the scale of policy making and management, and by who is thus allowed inside the tent.

Discussion and concluding comments

From this scant survey of the intersections, we can conclude that environmental justice might offer a more nuanced and comprehensive lexicon and potential understanding of the subsidiary dimensions of justice. A consistency in that lexicon – the variants of justice – would help, because the justice literature can be confusing to an outsider in its contestation of terms and meanings. That may be powerful or ineffective depending on the political context and stakes. It can also serve to remind that vague notions of 'fairness' or of good policy process in fact do involve issues of justice, and more pointedly of potential *injustice*. Unfair is a common, everyday complaint: injustice has much sharper and deeper implications.

Public policy is a process of compromise and trade-offs, where an outcome that is somewhat criticised, but not violently objected to, by all participants is taken as a sign of success – of a tolerable, least-bad outcome. If politics is the art of the possible, and public policy the art and craft of implementing the possible, then perhaps the key contribution of environmental justice is to promote a more sophisticated understanding of what might be possible and the unintended consequences of pursuing a narrower view of the possible. In return, public policy and environmental management can teach justice about the realities of policy, and the (limited, but real) possibilities that exist in policy systems. That sounds like a mutually beneficial relationship.

That brings forward a conclusion and proposed domain of activity for research, and for possible positive policy change. During moments of policy formation and change, clear articulations of environmental justice are rarely present. It would be constructive for there to be a greater communication of justice understanding and concepts informing the ways in which the constraining frameworks of law and policy might, when the time is ripe, be reformed in ways that would allow public policy practitioners and environmental managers to admit justice that little bit more in what they (are allowed) to do. Justice advocates might be surprised how pro-justice many officials would *like* to be. Public policy cannot supplant political (value) decisions, but it can inform those decisions better, in a justice sense.

To close, contests and conflicts over environments and resources will continue. It is interesting to note that the focus of contests – and thus inevitably the focus of participatory policy and environmental justice discussions – has shifted over time. Scholarship and politics are closely related: they attend what matters at the time, and generally the former follows the latter. Environmental justice research has been overwhelmingly focused on natural resources and on rural, non-urban situations, with water a strong focus and now unconventional gas extraction (e.g. Chapters 11 and 12). We focus on what society is worried about. Impacts of decisions on Indigenous people, and their access to and involvement in decision making, have been a constant undercurrent, featuring both dispiriting failures and inspiring stories (e.g. Chapter 9).

What of the future: where will contests and possible injustices be located? Some sectors covered in preceding chapters will doubtless continue to generate heat: water, mining, Indigenous empowerment and benefit from resources, and climate adaptation on the coast. Let us finish with three predictions of arenas that will be deserving of a justice lens in coming years. First, in an increasingly urban world, questions of access to environmental amenity and tractable livelihoods in cities will become more pronounced as urban structure and form change and marginalise more people. Second, as carbon constraints shift energy systems, watch for those not treated unjustly by deliberate decisions but simply left behind without the resources to access new technologies and opportunities. Injustice is often an unintended, unexpected consequence of an otherwise desired social change. Third, vulnerability to climate extremes (e.g. floods, heatwaves and storms) tends to be greatest among the already vulnerable and lacking in response capacity, and increased extremes will exacerbate that vulnerability: surely a justice issue.

It is clear enough that environmental justice researchers will have no shortage of opportunities to engage with society and public policy.

References

Connor R, Dovers S (2004) *Institutional Change for Sustainable Development.* Edward Elgar, Cheltenham, UK.

Davis G, Wanna J, Warhurst J, Weller P (1993) *Public Policy in Australia.* Allen & Unwin, Sydney.

Dobson A (2003) *Citizenship and the Environment.* Oxford University Press, Oxford, UK.

Dovers S, Hussey K (2013) *Environment and Sustainability: a Policy Handbook.* 2nd edition. Federation Press, Sydney.

Dovers S, Wild River S (Eds) (2003) *Managing Australia's Environment.* Federation Press, Sydney.

Dovers S, Feary S, Martin A, Morgan D, Tollefson M (2015) Engagement and participation in protected area management: who, why, how and when? In *Protected Area Governance and Management.* (Eds GL Worboys, M Lockwood, A Kothari, S Feary and I Pulsford) pp. 413–440. ANU Press, Canberra.

Dryzek JS, Niemeyer S (2010) *Foundations and Frontiers of Deliberative Governance.* Oxford University Press, Oxford, UK.

Healey P (1997) *Collaborative Planning: Shaping Places in Fragmented Societies.* Macmillan, Macmillan, London.

Holley C, Gunninham N, Shearing C (2012) *The New Environmental Governance.* Routledge/ Earthscan, London, UK.

Howlett M, Ramesh M, Perl A (2009) *Studying Public Policy: Policy Cycles and Policy Subsystems,* 3rd ed. Oxford University Press, Don Mills, Ontario, Canada.

Lane M, Robinson C, Taylor B (Eds) (2009) *Contested Country: Local and Regional Natural Resources Management in Australia.* CSIRO Publishing, Melbourne.

Paehlke RC (2003) *Democracy's Dilemma: Environment, Social Equity and the Global Economy.* MIT Press, Cambridge MA, USA.

Rhodes RAW (1997) *Understanding Governance.* Open University Press, Maidenhead, UK.

Washington SH, Rosier PC, Goodall H (2006) (Eds) *Echoes from the Poisoned Well: Global Memories of Environmental Injustice.* Lexington Books, Lanham MD, USA.

Wyborn C, Dovers S (2014) Prescribing adaptiveness in agencies of the state. *Global Environmental Change* **24**, 5–7. doi:10.1016/j.gloenvcha.2013.11.020

19

Current status and future prospects for justice research in environmental management

Anna Lukasiewicz, Stephen Dovers, Libby Robin, Jennifer McKay, Steven G.M. Schilizzi and Sonia Graham

A space for justice research in environmental management

A central aim of this book has been to provide a space where researchers from diverse disciplines can come together to discuss a common topic, which they approach from vastly different starting points. Given the interdisciplinarity of this field, there is very little space in academia where a social psychologist can talk to a geographer about the river red gum's right to water and where the classical philosopher can read about what they said and the economist can contribute to the discussion. The workshop preceding this book was a first step in creating such a space in Australia.

This book represents a collection of contemporary research in the field of justice and the environment. As such, it is a first attempt at mapping out the diversity of a vaguely defined interdisciplinary field and the semantics of justice approaches within it. While we have deliberately avoided imposing an organisational structure or overarching definitions in this book, the act of collecting and bringing together different justice concepts is a first step towards greater consolidation and synthesis. The different justice typologies presented in Chapter 1 should go a long way to clear up confusion over the names and meanings of different justice strands that a researcher can pursue. It should also help better 'manage' the differences in how justice is viewed by stakeholders, as highlighted in Chapter 15, which can lead to the unexpected or unwanted consequences described in Chapter 13.

Doing justice research

An important contribution of this book is starting a discussion on how to do justice research. Synthesis and analysis of the various methods used by our contributors reveal a preference for mixing qualitative and quantitative methods for complementarity and validation. The variety of methods used in environmental justice research reflects the diversity of disciplinary backgrounds of contributing scholars and once again highlights the interdisciplinarity of the field, as stated in Chapter 4:

> *to evaluate distributive justice effectively, researchers will benefit from an understanding of both physical and social science ... To understand and influence procedural justice, researchers will benefit from legal, institutional and social science expertise.*

An important contribution of the workshop that preceded this book has been a discussion of motivations for doing justice research (summarised in Chapter 1) and the provisional classification of the types of justice researchers that contribute to the field based on the centrality of the justice focus in their research (with 'core' researchers explicitly focusing on justice, 'affiliated' ones working in a similar space, 'incidental' ones who do not set out to study justice but find it crops up in their results and 'interested' researchers who find justice relevant and necessary to their research but do not pursue it themselves). All of these researcher categories are represented in this book and the fact that most of the contributions are from people who describe themselves as 'affiliated' or 'incidental' explains the fragmentation and disconnect in the field.

Focus on process as an entry point into distribution

This book places justice as a central concern of environmental conflicts and tensions. A common source of such tensions is the question over who benefits and who pays the costs of using natural resources. Indeed all the chapters in Part 2, and most in Part 3, deal with this question, whether the resource is water, mining or forestry. However, while the contexts differ greatly, all of these case studies are concerned first and foremost with aspects of procedural and interactional (including relational and informational) justice. Indeed it seems that the entry-point to dealing with issues of injustice is by looking at processes. Thus Chapter 5 looks at procedural justice principles and public participation more broadly in assessing the shortcomings of public–private partnerships in building water infrastructure provision; Chapter 6 focuses on relational justice to understand how injustice can be minimised in the context of mineral exploration; Chapter 7 exposes mismatches in expectations between the theory and practice of informational, temporal and procedural fairness in adaptation planning; while Chapter 8 focuses directly on how the failure of integrating procedural justice principles ultimately derailed the Tasmanian forestry negotiations. Chapters 9 and 10 both look at recognition of previously (and currently) excluded stakeholders: Indigenous Australians and animals.

The focus on procedural and interactional justice partly comes from the understanding that perfect distributive justice (where everyone is satisfied with their allocation) may be impossible or impractical (because there may not be win–win solutions), but also from the realisation that the pursuit of distributive justice (however imperfect) relies on having just processes (as is pointed out in Chapter 15). This does involve giving greater attention and protection to marginalised and politically weaker stakeholders (as pointed out in Chapters 9 and 14). However, as Chapter 3 points out, it is not enough to simply mandate good processes and public participation, because the history of participative planning in NRM has tended to promote advocacy through its recourse to stakeholder advisory groups. In practice, the distributive and procedural aspects of justice are intertwined; however, focusing on process is a common entry point to distributive justice.

While the importance (and complexity!) of distributive justice must not be over-looked, neither is the focus on procedural (and interactional) justice unwarranted. We must continue to examine the ownership and management arrangements subsisting over natural resources and examine if the models selected to achieve procedural (and other) justice mentioned in the introduction (see Fig. 1.1) are adequate. We in modern Australia are in a position to further develop world leading governance arrangements. Are the new governance models trusted? It is essential for the community to have trust in the organs of government and the rule of law, and this needs to be a priority to build on models that exist at all levels of government.

Putting justice at the centre of management

Contributors to this book place justice at the centre of environmental management. We believe that justice has a right of residence alongside other legitimate aspects (such as sustainability), rather than being an interesting afterthought or an addition to current management approaches. Environmental conflicts can, and should be, approached from a holistic justice perspective because justice is inherent in our conceptualisations of sustainability, including ecologically sustainable development (Chapter 16) and sustainable development (Chapter 2), as noted in Chapter 3 'it seems evident that the search for justice is an essential and integral part of achieving sustainability'.

As various contributing authors have stated, a key way for justice research to have impact in the real world (which is the explicit intention of justice researchers) is the continued development of a justice 'discourse' (Chapter 3), 'legibility' (Chapters 11 and 17) and a 'lexicon' (Chapter 18). Justice concepts need to be discussed by decision makers and stakeholders, both of whom currently shy away from expressing their management goals and policies in terms of justice due to the pervading influence of economics, or at least a certain type of economic interpretations (Chapter 3) and a lack of understanding of justice (Chapter 18). As many contributors here have pointed out, at the very least, consideration of justice in NRM can lessen unintended and incidental injustices and thus improve decision making. Chapters 11, 14 and 17 bring justice into management through their use of empirical case studies and frameworks: approaches familiar to managers and policy makers. As stated in Chapter 18, 'justice advocates might be surprised how pro-justice many officials would *like* to be'.

Predictions for the future

We would like to end by commenting on possible future loci for justice research. As Australia's reliance on natural resources continues, we foresee that conflicts and tensions in water, mining and forestry will continue as current and new stakeholders press their claims for recognition and fairer (re)distribution of benefits. Similarly, the role of the environment as both an object and subject of justice will continue to be debated and the implications of treating the environment as a resource or a stakeholder will continue to lead to different management approaches (as illustrated in Chapter 10). The exploration of the inherent complexities of justice (such as the meanings of equity discussed in Chapter 15) will also continue, eventually leading to better manage different and potentially conflicting views of justice across communities that, economically and culturally, are increasingly heterogeneous.

Let us reiterate the three areas for justice research outlined in Chapter 18. First, in an increasingly urban world, questions of access to environmental amenity and comfortable livelihoods in cities continue as urbanisation marginalises more people. This includes access to clean air, green spaces and public parks within cities. Second, as carbon constraints shift energy systems, some financially and socially vulnerable communities may be unintentionally left behind, without the resources to access new technologies and opportunities, as opposed to being treated unjustly by deliberate decisions. This includes the intended and unintended consequences of new schemes such as carbon farming and access to alternative sources of energy. Third, those dealing with the distribution of environmental benefits must equally consider the distribution of environmental harms, especially because the social vulnerability to natural hazards tends to be greatest among those who are already marginalised and vulnerable in other aspects of life (i.e. precisely the groups that justice researchers tend to focus on).

Finally, what have we learned about researching justice through our original workshop, and through the formal and informal collaborative processes in developing this book? First, we note that justice is a concept that demands integrative research. No single disciplinary framework can fully capture its legal, moral and practical implications. Among our (six!) editors and (many) contributors, we have training in many disciplines: ecology, geography, governance, economics, history, law, public policy and sociology, to name just a few. What we share is our commitment to getting beyond the limits of the methods of our original disciplines to new insights that draw on collaboration *across* disciplinary frameworks. Justice research is read by people in many places. It is most often generated in strongly interdisciplinary environments, through either workplaces that are integrative in style or individuals who read and write across several disciplines. The contributors here all fit this description, and the book itself is a product of such conversations. Justice is practical stuff. We are seeking findings that are relevant to practical situations in the real world. NRM is also a complex, multidisciplinary craft, with a focus on actual outcomes. Ensuring justice in NRM practice is demanding on many levels. Above all, it demands consultation with the people on the ground in each situation, especially those who suffer *in*justice now or are at risk of suffering injustices in the future. Specific case studies and individual stories are as important to justice research as frameworks that summarise its overarching structures. This book has worked to put the case studies and frameworks together in new conversations about justice in NRM.

Index

ability to pay 172
Aboriginal and Torres Strait Islander peoples 234
  equity metrics and 212
  market efficiency and 239
  social injustice and 16
  stakeholders 242
  see also Indigenous Australians
Aboriginal Land Rights Act 1983 (NSW) 127
Aboriginal Water Trust 128
Academy of the Social Sciences in Australia v, vi
access benefits, sharing of 27
accessibility 110, 241, 242, 243
accountability 62, 65, 66, 86, 241, 242
  Adelaide water survey and 69, 70
  Tasmanian forest peace process and 110
accreditation system, animal property rights 134
accuracy 241, 242, 243, 244, 246
  Tasmanian forest peace process and 111, 114–15
adaptation planning 264
  sea-level rise and 91–103
adaptive management 106, 242, 243
Adelaide Desalination Plant (ADS) 61 78
Administrative Appeals Tribunal 227, 230
administrative law 222, 259
  groundwater plan and 224–5
  water management and 223, 230
adversary trials, water law and 221
advisory groups, stakeholders 33
advocacy 28, 31, 33
affiliated researchers 264
AGL Sales (Qld) P/L v Dawson Sales P/L & Ors [2009] 157
agriculture, water rights and 32
Alcoa, Juruti Velho Lake community and 80, 81, 83–4, 85, 86–7
allocation process 239, 252
alternate energy 266
altruism 68, 133
aluminium production 81
Amazon region 79, 80, 81–8, 212
analytical framework, innovation politics 186–9
animal property rights 51, 133–42
Antarctica 234, 235
anti-coal seam gas industry protests 160–2
anti-cruelty codes of practice 137, 138
applications, justice research 27–38
APT Petroleum Pipelines Pty Ltd v Western Downs Regional Council [2014] 157, 162–4
aquifers 150, 225
  contamination of 148, 156
ARC Linkage project 93, 94
arguments evaluation 51
Arnold v Minister Administering the Water Management Act 2000 221
Arnstein, ladder of participation 28
Ashworth v Victoria [2003] 227
Association of the Communities, Juruti Velho 84
Australia 1, 144, 185
  environmental movement 18
  innovation politics 186, 190–1, 193–7
  integrated water cycle management 190
  justice research 4–5
  liquefied natural gas supplies 190–1
  Millennium Drought 61
  water management 31, 121–32, 215–32
Australian Government 109, 112, 233
Australian Heritage Commission (Australian Heritage Council) 18
Australian Law Reform Commission (ALRC) 64, 221
Australian National University v, vi, 190
Avalon 140

Baker v Minister for Employment, Skills and Mining & Another [2012] 158–60
Barrington-Gloucester-Stroud Preservation Alliance Inc v Planning and Infrastructure [2012] 159
basic needs argument, animal property rights 135–8, 139
Basin Plan, Murray–Darling 126, 219, 220, 227–8, 229, 230
Bates, Marston 15
bauxite mining 80, 81
behavioural economics 52
Beijing 191, 193, 194, 195, 197
benefit–cost analysis (BCA) 171, 173
benefit sharing 34, 86
bias 48, 110, 113, 242
  see also self-serving bias (SSB)
biodiversity protection 134, 186, 228
bore licences 225
Boulding, Kenneth 18
Brazil 79, 80, 81–8, 212
Bretton Woods Conference 1944 22
Brisbane floods 2012 (Rocklea floods) 146, 147–8, 151
Brown, Ray 162, 163
Brundtland Report (*Our Common Future*) 19, 23
budget constraints 53
  conservation projects 173–4
  Pakistan rebuilding project 175–6
  social equity goals 178, 179
Builders Labourers Federation 18
Bulgaria, innovation politics 185, 186, 191, 192, 193–7
Burke, Tony 109
business owners, water management and 44
Button, Justice 160, 161, 162

call-in powers 259
canal estates (Gold Coast), sea level rise and 148, 150–1
Cap, the 127
capabilities 81, 83, 86, 87, 88, 241
carbon emissions reduction 171, 256, 266
care, ethics of 14
case studies:
  conservation projects funding 172–4
  eastern water dragons (Sydney) 140
  environmental water management (NSW) 234
  floodplains water re-allocation (NSW and Qld) 234
  innovation uptake 190–2
  Marine Protected Areas (MPAs) Southern Ocean 234
  Murray–Darling Basin (MDB) water allocation 234
  Pakistan house reconstruction support 175–7
  Social Justice Framework 234
  water rights 145–52
catastrophe economies 22
Catchment Management Authorities (CMAs) 172, 173, 213, 253
catchment restoration innovation 193
Central Condamine Alluvium case study 146, 147
chairperson, impartiality of 245
change inhibition 197
Charmahal 192, 194, 195
China 185
  ecosystem services 186, 191–2, 193–7
  equity metrics 212
circuit breakers 112
citizen science 45
clean energy 186
clients, politicians as 35
climate change 1, 2, 144, 150, 193, 228, 260
  coal seam gas industry and 155
  denial of 96–7, 99–100
  justice and 92, 151
closed negotiations, Tasmanian forest peace process and 109, 110, 111, 112
closed questions 47
coalition politics 187–8, 195, 198
coal seam gas (CSG) extraction industry 49, 148, 149, 260
  Condamine Alluvium (Surat) 150, 151
  justice and 155–67
coastal adaptation 94, 258
co-evolution 187, 198
Coke, Edward 223
cold-water pollution 237
collaborative justice research 253–4, 266
colonial history
  Indigenous law and 49
  injustice and 129, 188
  water rights 121, 123, 126–8
commercial-in-confidence information 69
commercial values, public utilities and 63, 64
commodification 122

common law 259
  water rights and 126, 227
communication channels, Juruti Velho Lake mine 85, 86, 88
community groups 44, 85, 87, 160, 197
  coal seam gas industry and 159, 160–2
  committees 196, 253
  flood management 191
  law debate 49
  mining companies and 79–89
  water management and 216, 217, 254
community justice 149
Community Reference Panel 147
community relations office, Alcoa 83
compensation 99, 256
computer-aided qualitative content analysis 49–50, 66–8
conceptual analysis 46, 50–1, 67
Condamine–Balonne Alluvium case, water allocations and coal seam gas industry 147, 149, 150, 151
confidence creation 34–5
conflict mitigation 49, 138, 192
conflict of interest 243
Connell, Rachel 160–1
Connell v Santos NSW Pty Limited [2014] 157
consensus-based decision-making 46, 241, 243, 245
conservation justice 178
conservation movement, history of 16–17
conservation projects funding, case study 172–4, 178
conservation theory, animal property rights 138–9, 141
consistency 242, 243
Constitution 62, 218, 219, 221
Constitutional Amendment 1967 16
Constitutional Settlement 1901 215
consultation 69, 194
content analysis 40, 41, 42–3, 54
context-dependence 7, 52, 208, 205, 206, 209–10
conventional fuels 155
Convention of Antarctic Marine Living Resources (CCAMLR) 234, 235, 238, 239, 241, 245, 246
core negotiating group, Tasmanian forestry peace process 107
core researchers 264
corporations 81, 149, 216
corrective justice, water management and 220, 222
costs 99
  Adelaide Desalination Plant 67, 68, 69, 72, 74
  distribution 171
Council of Australian Governments (CoAG) 61
  Indigenous water rights and 126, 130
  water markets and 127, 128
council rates, coal seam gas industry and 163
counter-dependence, diversity and 206–9
critical human needs 220, 237
cross-scale ramifications, innovation 186, 197, 198
cultural minorities, equity metrics and 212
cultural values, inhibiting effect of 196
custodial rights and responsibilities, Indigenous 122, 123, 125, 128

Dalton, Jean 158–9
Day, Max 18
Deakin, Alfred, water reforms and 126–7
debate clarity 51
decentralisation 45, 194
decision making
  equity norms and 206, 207, 211
  evidence-based 32
  exclusion from 100, 101
  imbalances 50
  impartiality and 113, 212, 243
  innovation and 188
  interactive justice and 244
  justice-based analysis 34, 107
  lack of 72, 196
  local government 91, 92
  natural resource management and 233–4
  outcomes 236
  procedural justice and 29, 240, 241
  self-interest and 238
  social justice and 30, 235, 236, 246
  sustainability and 254
  water management and 145, 150, 217
defamation suits, coal seam gas industry and 157
deliberative democracy 46
'democratic-like' justice 169–70
democratic participation
  consensus and 211
  dictatorial solutions and 212–13
  outcomes distribution 189
  restriction of 63
  Tasmanian forest peace process and 111, 112
Department of Infrastructure and Regional Development 62, 65, 66

Department of Natural Resources and Mines 158, 159
desalination plants 44, 62, 64, 190
deservingness 31, 33, 237
Design–Build–Operate–Maintain (DBOM) contract 65
desire theory of wellbeing, animal property rights and 136
detachment, planning and 28
developing countries, developed countries and 255
Development Act 1990 (SA) 224
developmental justice 23
dictator, decision-maker as 210, 212–13
direct negotiations, mining companies and 80
disaster risk management 22
discourse analysis 43
discussion forums 195
dispute settlement 21
distribution adjustments 79, 197
distributional equity 197, 205–14
distributional justice 124, 171
  conservation projects and 172
  innovation and 192
  mining industry and 80
distribution intention 236
distributive fairness 47
  sea-level rise and 94–5, 97, 98, 99, 101
distributive justice 2, 3, 30, 42, 63, 66, 92, 246, 252
  coal seam gas industry and 155, 157, 162–4
  evaluation of 53–4
  innovation and 185, 186, 188, 195
  Iran and 196
  legal system and 215
  natural resources efficiency and 169–72
  process and 264–6
  Social Justice Framework and 235, 236
  Tasmanian forest peace process and 105, 110, 113
  water management and 145, 149–50, 151, 152, 220, 221, 230
diversity, counter-dependence and 206–9
doctrinal analysis 40, 42, 48–9, 54
document analysis 146, 190
dominant groups, advantage of 236
doughnut diagram (Raworth) 21, 23
drought 67, 144, 145, 190, 193
duty of care, sustainability and 219
eastern water dragons (Sydney) 140
East Gippsland Shire Council (EGSC) 91–101
eco-compensation (China) 191, 193, 195
ecological justice (definition) 2, 231
ecological limits 253
ecologically sustainable development 61, 228, 231, 265
  definition of 229
  goals (water resources) 215–23
  procedural fairness and 224
  requirement (NSW) 226
economic benefits, coal seam gas industry 163, 164
economic decline, Tasmanian forestry industry 107
economic efficiency 53, 169–81, 207, 211, 213, 239
economics
  experiments 172
  modelling 21
  social justice and 34, 35, 179
  water management and 122, 123, 217, 229
ecosystem risk, coal seam gas industry and 156
ecosystem services, case study 191–2
efficiency 65, 66, 236, 246
  Adelaide Desalination Plant 67, 68, 71, 72–3
  Australian Government 246
  criteria 169, 177, 178, 255
  Social Justice Framework 239
efficiency–equity trade-offs 170–4
egalitarian norm (EG) 176, 177
electorates 252
emerging economies 255
empirical normativity 212
employment, Tasmanian forest industry and 107, 114
empowerment 45
endangered species 138, 140
endowment effect 32
engagement strategy 256
entitlement 31, 33
environment
  irrigation and 239
  Murray–Darling Basin plan and 220
  natural resources and the 3–4
  stakeholder status 237–8
environmental allocation 29
environmental benefits, innovation 197
environmental disasters 22

Environmental Flow Rules 47
environmental harm
  coal seam gas industry and 155, 157
  liquefied natural gas production 191
  prevention of 255
environmental justice 1–10, 14, 23
  coal seam gas industry and 155–67
  definition 1
  democratic consensus and 211
  ecologically sustainable development and 215–16
  equality and 236
  governments and 171
  ideas 13–25
  injustice and 195
  innovation and 192
  irrigation and 218
  practical assessment of 205–14
  private water ownership and 63
  public policy and 251–61
  Tasmanian forestry peace process and 105
environmental management 1
  global history 14–15
  justice and 265
environmental movement 18, 28, 29, 111, 193, 197
environmental non-government organisations (ENGOs) 106, 110, 115
environmental outcomes
  fairness and 210
  natural resources industry and 169–81
  Water Act 2007 228
environmental projects 212, 213
environmental representatives, legitimacy 240
environmental rights 30
environmental river flows 171
environmental science 15, 17
Environmental Water Advisory Groups (EWAGs) (NSW) case study 234, 235, 239, 240, 243, 244, 245, 246
environmental water management 122, 237, 254
Environment Planning and Assessment Act 1979 (NSW) 254
Environment Protection Authority (EPA) 73
equality 209, 236
  fairness and 217
  private–public partnerships and 65
Equality and Efficiency: The Big Tradeoff (Okun) 170
equal weight per cultural group 212
equity 65, 246, 265
  definitions 209–10
  policy decisions and 177
  social justice and 29, 236–7
  water rights and 33
equity–efficiency trade-offs 170–1, 177, 178, 206, 208, 210
  Pakistan 175–7
equity metric 172–3, 179, 205, 206, 211
equity norms 176, 177, 178, 206, 209, 210, 211, 212
error correctability 241, 242, 243, 246
ethical approval 106, 190
ethicality 241, 242
  interviews 44
  Tasmanian forest peace process and 111, 113–14, 115
  trust and 245
ethics 14, 22, 64
ethnographic models 79, 81
European Union 20, 191, 195
evaluation, water advisory groups 235
evidence-based decision making 32
experimentation 40, 42, 52–3
expertise 17, 54
exploration pilot studies 160
external actors 50, 86–7
extreme events studies 151

fairness 43, 65, 236
  coal seam gas industry and 156
  definitions 2, 209–11
  environmental justice and 233, 251–2
  equality and 217
  equity and 236
  governments and 255
  interpretations of 6, 98–101, 256, 259
  local government decision making 91, 92
  mining industry and 79
  private–public partnerships and 65
  sea-level rise adaptation and 94–101
  Tasmanian forestry peace process and 106–7
  water management law and 218, 221
family size 209
farmers
  coal seam gas industry and 155–6
  colonial era 127
  Millennium Drought and 238

federal government
  environmental justice and 259
  intervention 197
  Tasmanian forest peace process and 115
  see also Australian government
Federal Magistrates Court 227
feedback 243
feedlot water supply 224
Fenner School for Environment and Society v, vi
field surveys 171
financial incentives, environmentally friendly activity and 241
'first do no harm' 14
First Nations, water access 125
fishing quotas 171, 239
floodplains 149, 244, 246
  water re-allocation (NSW and Qld) case study 234
flood prevention 93, 146, 150, 191, 193, 194, 197
  funding 195
  Pakistan 175–7
floodwater provision 147
focus groups 40, 41, 44–5, 146
food justice 2
forestry issues 19, 105–18
forestry peace process, Tasmania 105–18
formal interviews 93, 94
formal processes 85–6, 240, 242
forum theatre process, Iran 192
freedom-of-information requests 157
freshwater management 31, 218
Fullerton Cove Residents Action Group Inc v Dart Energy Ltd [2013] 157, 160
Future Earth 13, 21
future scenarios 33, 210

game theory 170
gas extraction 163, 191
gender differences 172
geographic information systems (GIS) 41, 48
Gladstone 158, 190, 191
global governance, environmental issues and 13–25, 217
Gold Coast 148
Golder v Maranoa Regional Council & Ors [2014] 162
Google 20
governance 23, 254
  Adelaide water survey and 66, 69–73
  fragmentation 196
  good governance concept 61–78
  innovation and 186
  natural resources management 185, 186
  newspaper reports and 67–8
  systems of 265
government ownership
  public support for 73
  water utilities 63
governments
  centralisation 193
  China 192, 196
  coal seam gas industry and 150
  flood planning and 150
  justice decisions 32, 252
  lack of accountability 72
  levels of 194
  partiality perception 149
  water management and 44
graziers
  coal seam gas industry and 158
  decision-making and 244, 246
  floodwaters and 147
  Lower Balonne region 151
Great Barrier Reef 191
Great Extermination, The (Marshall) 18
Greed–Efficiency–Fairness (GEF) hypothesis 236
green bans 18
greenhouse gas emission reductions 50, 206, 209
groundwater 147, 218
  reliance on 146
  visualisation tool and 48
groundwater allocations 33, 34, 224
  coal seam gas industry and 148, 150
  validity of 224–5
group dynamics 33, 43, 45, 195
growth doctrine 15
guardianship, animal property rights 134, 140, 141
Guide to the Basin Plan 2008 220, 243
guiding codes 106
Gunns Limited pulp mill proposition 107, 109

hazards impacts 255
hierarchical society 193
High Court of Australia 52, 230
  water allocation reductions decision 225–7

historical analysis 42
historical injustice 123, 129, 130
holistic justice 265
Holling, C.S. (Buzz) 18
honest brokers 152
horizontal equity 209, 211
house reconstruction support, Pakistan 175–7
Howard Government 62, 219, 236
human intervention, animal property rights and 136, 138–9, 140–1
Human Research Ethics Committee (HREC) 106
human right to water supply 68, 74, 215, 230
humans, basic needs 19, 137
Hurricane Katrina 22
Huxley, Julian 16
hypothesis testing 52

ICM Agriculture Pty Ltd v The Commonwealth [2009] 225–7
ideal theory, animal property rights 139
Idea of Justice, The (Sen) 80
impartiality 206, 212, 241, 242, 243
  justice and 28, 29, 30
incidental researchers 264
inclusiveness 21, 45
  Adelaide water survey and 69, 70
  innovation and 188
incremental achievement 28, 116
independent chairperson 242
Independent Verification Group (IVG) (Tasmania) report 108, 109, 111, 114, 116
in-depth interviews 148
Indigenous Australians
  colonial law and 49
  environmental decisions and 260
  equity norms and 212
  Indigenous customary law 128
  injustice and 188
  research and 46
  social standing and 244, 245
  stakeholders 43, 134, 254, 264
  water access rights 121–32, 237, 238
  see also Aboriginal and Torres Strait Islander peoples
Indigenous Land Corporation 128
Indigenous Protected Areas 254
indirect rights 139
individual equity 209, 210
individual interviews 45
inequity, water resources provision 148–9
informal power, stakeholders 241
informational fairness, sea-level rise and 95–6, 98, 99, 101
informational justice 3, 92, 256, 258, 264
  legal system and 215
  Social Justice Framework and 236
information sharing 94, 96
infrastructure projects 62, 65, 66, 190
injustice 13–14, 259
  colonial era water rights 129
  concepts of 31
  Indigenous water rights and 121–32
  minimisation of 81, 251
  public policy and 256
  social change and 260
  social exclusion and 244, 259
innovation politics, justice ideas and 185–203
inputs–outputs, equity and 236
institutions
  structural injustice and 130
  water allocations and 144, 149, 219
instrumentalism, avoidance of 35
insurance companies 96, 149, 196
integrated water cycle management (IWCM) case study 190, 193, 195, 197
integrative research 266
interactional fairness 47, 97
interactional justice 2, 3, 42, 92, 194, 258, 264
  innovation and 185, 186, 188, 194
  Iran 197
  mining industry and 80
  Social Justice Framework and 235–6, 244–5
  Tasmanian forest peace process and 112
  water rights and 145, 151
interactive justice 30, 63–4, 244
  Tasmanian forestry peace process and 106–7
interactive mapping 146
interdisciplinary research 54
interest group politics 35, 135, 240, 264
interests of the future 209–10
intergenerational equity 19, 30, 100, 172, 195, 209
internal feedback 87
international agreements 13, 255
  water rights and 227, 228
international competitiveness, water management 217

International Council of Scientific Unions (ICSU) 13, 16
International Court of Justice 21
international law 219, 221
  water allocations and 150, 151
international negotiations 143–4, 235
international project partners 195, 196
International Union for the Conservation of Nature (IUCN) 13, 16
interpersonal justice 92, 94, 101, 151, 244, 258
inter-species conflict 6
interviews 40, 41, 42, 43–4, 49, 54, 235
intrinsic assets 239
Inuit 212
inverse proportionality (IP) 176, 177, 209
involuntary resettlement 79
involvement, spectrum of 4–5
Iran, water conflict transformation 185, 186, 192, 193–7
irrigation 146, 241
  dams 147
  Lower Balonne region 151
  Murray–Darling Basin and 220
  procedural justice and 150
  upgrades 239
  urban groups and 241
  water rights and 31, 122, 218, 237
Irrigation Act 1886 (Vic) 226
Isfahan 192, 194, 195

judicial system 220, 221
  coal seam gas industry and 164
Judt, Tony 14
Juruti Velho Lake community 81, 82–4
justice
  culture 63, 64
  definitions 209–10
  distribution 186
  environmental management and 185–6, 233, 265
  governments and 255
  innovation politics and 186, 187
  levels of 31–2
  self-interest and 238
  trust and 245
justice equity or fairness (JEF) criteria v, vi, 172, 205, 206–7
justice jurisprudence, water management and 215–32
justice regime changes 188–9
justice research 2, 4–5, 260, 263–6
  applications of 27–38
  collaboration and 266
  environmental 39–57
  evaluation of 33–4
  future of 265–6
  methodology 39–57
  public policy and 251
justice theories 2, 30, 265
  animal property rights 133, 135–8
  applications 29–30
  eastern water dragons and 140, 141
  ecologically sustainable development and 217–20
  economics and 170
  innovation politics and 185–203
  water rights and 129–30
justiciable issues, Murray Darling Basin Authority 229

*kacha*s 175, 177
Kelty, Bill 108–9
keystone species 139
Kyoto Protocol 206, 228

laboratory experiments 52, 207, 209
labour market interventions 172
ladder of participation (Arnstein) 28
Lakes Entrance 93, 94, 96
land access information 158, 159
Land and Environment Court (NSW) 159, 254
landholders 123, 234, 238
  animal property rights and 134
  coal seam gas industry and 150, 158–60
  Indigenous people 121, 127
land-use planning systems, animal property rights and 135
land values protection 94–5, 96
legal concepts
  coal seam gas industry and 156
  use of 48–9
  water management and 219, 221–3
legal geography 157–8
legibility 265
Legislative Council Select Committee (Tas.) 108, 109, 113
less-carbon-intensive fuels 191

Leximancer 67, 68
licensed private diversions (water) 126, 127
Limits to Growth (Club of Rome) 18
liquefied natural gas (LNG) industry 186, 193, 194, 195, 197, 226
  carbon reduction and 198
  case study 190–1
literacy levels 82, 83, 243
literature, justice theories from 148, 221–3
litigant financial risk 220
local government 197
  coal seam gas industry and 157, 163, 164
  disempowerment and 259
  environmental justice and 256, 258
  forums 94
  sea-level rise and 91–103
  water management and 235
locality, water rights 144
lock-in 187
Lock the Gate movement 151, 155–6
Lockyer Valley irrigators case study 146–7, 148, 149, 150
long-nosed bandicoots 140
Lowbidgee floodplain 243
Lower Balonne River stakeholders case study 146, 147, 148–9, 150
low-income households 68

Mabo 1992 decision 49, 222, 254
macro-context analysis 189, 193
Man and the Biosphere (MAB) 18
Manns Beach 93, 94
Maori rights 129
map scale 158–9
marginalised groups 43, 49, 53, 144, 266
Marine Protected Areas (MPAs) 235, 238, 239, 241, 246
  Southern Ocean case study 234
market economies 253, 255, 256
market-oriented efficiency 239
Marshall, A.J. (Jock) 18
maxmin 208, 209
McLoughlins Beach 93, 94
Meadows, Denis 18
media pressure 99, 252
Metgasco Limited v Minister for Resources and Energy [2015] 160–2
methodology
  Australian justice research 41–53
  mining company–community relations 81–2
micro-contest analysis 189, 194
Millennium Development Goals 144
Millennium Drought (2001–2007) 61, 62, 146, 236–7, 243
mineral resources 164, 264
  company–community relations and 79–89
Minister for Arts, Heritage and Environment v Peko Wallsend 221
modelling, politics and 256
monitoring 242, 243, 246
Montreal Protocol on Substances that Deplete the Ozone Layer 20
moral issues 4, 6, 30, 133, 241
Mosley, Geoff 18
mothballing, Adelaide Desalination Plant 67, 68
motivations, justice research 264
multi-faceted equity 212
multi-level governance 185
  innovation and 197, 198
  relevance theories and 187–8
multi-method evaluation 49
multinational companies, infrastructure projects and 62
multiple criteria analysis (MCA) 41, 45, 146, 147, 149
multiple owner problem, animal property rights 138
multi-scalar systems 187
Mumford, Lewis 15
Murray–Darling Basin Authority (MDBA) 147, 219–20, 227–8
  Indigenous water rights 125, 128
  management 242
  needs principle and 237
  reform process 43
  states and 229, 239
  timeliness and 243
  water rights and 122, 123, 127, 215, 218, 233, 234
Murray Lower Darling Rivers Indigenous Nations (MLDRIN) 125
Murrumbidgee Groundwater Preservation Association v Minister for Natural Resources NSW [2004] 224–5

National Action Plan for Salinity and Water Quality (NAPSWQ) 172
National Competition Policy 254
national interest, water management and 227, 228, 231
nationalisation, water rights 130
National Plan for Water Security 2007 236
National Strategy for Ecologically Sustainable Development 1992 216, 253
National Water Commission 61–2, 122
National Water Initiative (NWI) 61, 122, 127, 128, 219, 229
native animal species 138
native forest harvesting 107
Native Title Act 1993 127, 128, 130, 222
natural capital 6
natural goods, animal property rights 135–6
natural justice, water management and 220
natural resources
  economic efficiency and 169–81
  environment and 3–4
natural resources management (NRM) 1, 125, 236
  fairness and 210
  innovation uptake and 186, 187–9, 192, 198
  justice and 185
  outcomes 266
  participative planning 264
  public policy and 251–61
  self-interest and 239
Natural Resources Management Act 2004 (NRM Act) 224
Needham, Joseph 16
need principle 236, 237–8
neoliberal economic model 31, 32, 61, 255
  historical water rights 127–8
network analysis 40, 50
neutrality, Tasmanian forest peace process and 110, 113–14, 115
newsletters 84
New South Wales (NSW) 62, 64, 147, 159–60, 197, 235, 236–7, 244
  coal seam gas industry and 155, 162
  environmental water management 234, 238
  water management law and 127, 216, 217, 218, 222, 225
newspaper articles, content analysis 66–8
non-critical human interest, animal property rights and 136, 137
non-elected representatives 113
non-governmental organisations (NGOs) 21, 194
non-human rights 230, 254
non-ideal theory, animal property rights 139
non-justiciability
  ecologically sustainable development and 220, 221, 222, 223
  Murray–Darling River Basin and 230
  water management law and 218
non-material values, climate change and 99
non-persons, persons and 137–8
non-positional context 207, 208
non-selfish behaviour 170
non-technical innovation 191, 193
normative justice 33, 133, 144
  coal seam gas industry and 156
northern Australia, water rights 125
Northern Basin Aboriginal Nations (NBAN) 125, 126
Northern Territory 128, 197
nostalgia, social justice and 32
'not in my backyard' (NIMBY) activists 28
nuns group, Juruti 86–7

observer-only status 112
observer stakeholders 240
O'Connor & O'Connor v Arrow (Daandine) Pty Ltd [2009] 159
one-on-one settings 43
one-way communication 84
online survey, Adelaide Desalination Plant and 68–73
open coding 43
open-ended questions 46, 47
opinion destabilisation 196
options mapping 50
organisational justice 244
Organisation for Economic Co-operation and Development (OECD) 63, 65, 68
other-regarding preferences 172
outcomes
  alternatives 50
  distribution 94
  imperfections 256, 257
  process 252
out of court mediation sessions 134
overconsumption 15
overland flow waters 227

ownership 66
  Adelaide Desalination Plant 71, 72, 73
Oxfam 21, 23

Pakistan, house reconstruction support 175–7
Pannell, David 173
Pareto Compensation (PC) 176, 206, 207
parliamentary democracy 193
partially controlled experiments 52–3
participant observation 40, 42, 47
participatory action research (PAR) 40, 41, 42, 45–6, 48, 50
participatory innovation, Iran 197
participatory policy 31, 33, 50, 193, 194, 195, 253–4, 255, 257
Partnerships Victoria 62, 65, 66
peace, conservation and 17
Pepper, Justice 159
perceptions, justice and equity 34, 43, 205, 211
perpetual water rights 127
personal interactions 234
personal welfare 236
petroleum exploration licences 160, 162
philosophy, justice research and 50–1
photovoice interviews 48, 146, 147
Pilbara, fresh water rights 125
pipeline compressors 163
planning alternatives, Adelaide Desalination Plant 72–3
planning authorities, water allocations and 224
planning, challenges to 28–9
Police v Rankin [2013] 157, 160
Police v Roberts [2013] 157, 160
policy decisions, equity and 177
policy networks 187, 188, 253
political context 259, 260
political economy 40, 42, 53
political process
  justice and 34, 116, 222
  modelling and 256
  public policy and 35–6, 252
  Tasmanian forestry peace process and 105–18
pollution 79, 156, 172, 193
poor, needs of the 23
population density 193
population growth, environmental policy and 253
Port Albert 93, 94, 96
positional context 207
potential Pareto improvement (PPI) 171
power levels 158, 240
  decision making and 107
  focus groups and 44
  Juruti Velho Lakes region 87
  mining industry and 80
  Social Justice Framework and 240–1
  Tasmanian forest peace process and 112–13
power trade-offs 115
practical assessment, environmental justice 205–14
precautionary principle 19–20, 224, 231, 238
  water rights 216
precedent, doctrine of 217, 223
preference intensity 211
prior rights 30
private dams 147, 149
private property 134, 241, 242
private sector participation (PSP), Adelaide Desalination Plant and 61–78
problematic species problem, animal property rights 138
procedural fairness 47
  denial of 158–9
  sea-level rise and 95, 96–7, 98, 99, 101
procedural justice 2, 3, 30, 42, 45, 63, 92, 124, 251, 252, 258, 264, 265
  China 197
  coal seam gas industry and 155, 157, 158–60, 164
  innovation and 185, 186, 188, 195
  Iran 197
  legal system and 215
  public participation and 161, 162
  Social Justice Framework and 235, 238, 240–4
  Tasmanian forestry peace process and 105–18
  water allocation and 145, 149–51, 152, 218, 220, 223, 224–7
process rules 107, 240
  distributive justice and 264–6
  public policy 258
  Social Justice Framework and 241–4
procurement methods 62
Productivity Commission 65, 69, 72, 73
profit objectives, water supply 63
property prices 94–5, 96
property rights 30, 123, 231, 253
  animals and 135, 137

protected area management 16, 257
protestors, anti-coal seam gas industry 159–60
public accountability 62
public benefit 32, 114
  Murray–Darling Basin Plan 229
public interest 62, 65, 143
  Adelaide Desalination Plant and 61–78
  litigation of 220
  Tasmanian forest peace process and 116
  water allocations reductions and 225–6
public meetings 195
  Adelaide water survey and 69
  mining companies and 83
public participation 28, 29, 66
  Adelaide water survey and 69, 70
  politicians and 35
public policy, environmental justice and 251–61
public–private partnerships (PPPs) 63–5, 66, 264

QCLNG Pipeline Pty Ltd 158–9
qualitative research methods 41, 42, 46, 49, 50, 53, 263
  case studies 146–8
  newspaper articles and 66–8
  Tasmanian forestry peace process and 106
  water allocations studies 152
quality of life, water management and 216
Qualtrics 68–9
quantifiable injustice 32
quantitative research methods 41, 42, 46, 49, 53, 263
Queensland 146, 147, 148, 162, 190–1, 244, 246
  coal seam gas industry and 150, 155, 158–60
  floodplains 234
questionnaires 40, 42, 46–7, 146

rainwater capture 190
random sampling 67
Ratcliffe, Francis 18
'rates roads and rubbish' 99
rational argument 40, 42, 50–1
Raworth, Kate 21
re-allocation, water rights 123
recycled water, opposition to 62
Regional Councils Victoria Annual Summit 94
Regional Forest Agreements 253
regional water plans 61, 215
regulatory agency, animal property rights and 141
relational justice 3, 258, 264
  mining companies and 79–80, 81, 85–8
relative deprivation 31
relative significance, persons and animals 137–8
relevance theories 187–8
remote communities 83
remunicipalisation 73
renewable energy sources 32, 256
reparative justice 124, 188
representation 107, 240, 241, 242–3
  mining companies and 84, 87
  Tasmanian forest peace process and 110
research methods 40, 66–73, 152
  innovation uptake and 189–90
Resource Operation Plan (Qld) 147
resources 106
  access 32
  allocation 251–2, 253
  definition 4, 6
  distribution justice 234
  exploitation 193
  management of 22
  redistribution of 171
respect, lack of 245, 246
responsible innovation 188
restoration measures 127
restorative justice 3, 124
  efficiency and 239
  innovation and 188
  lack of 130
  Social Justice Framework and 238
Revenue Statement 2012–13 (Western Downs) 162–3
Rights in Water and Irrigation Act 1914 (WA) 127
Rio Declaration on Environment and Development 1992 15, 19–20, 254
Rio+20 Conference 21, 255
riparian rights doctrine 126, 130, 226
risk avoidance 22, 208, 209
risk management 144, 195
Rocklea (Brisbane) flooding, case study 146–51
Roma–Brisbane Gas Pipeline 163
Rosella Well 160, 161, 162
Rowe v Lindner and Ors [2007] 224
rural water management 215, 217

Sackville, Ronald 156
Saipem Australia Pty Ltd v GLNG Operations Pty Ltd [2014] 157

salinity targets 227
Sauer, Carl 15
saved benefits 210
scarcities 53, 253
scientific data 196
  environmental decisions and 229, 237, 246
  distributive justice and 264
  groundwater allocation plan and 225
  justice research and 32, 41
  natural resources management and 242, 243
  Tasmanian forest peace process and 114, 115
  water management and 217
sea-changers 93
sea-level rise 149
  local government and 91–103
  Sunshine Coast case study 146, 148, 151
Second World War, human rights and 15
selection criteria, conservation projects funding 172
self-interest principle 236
  Social Justice Framework and 238–9
  social preferences and 170
self-serving bias (SSB) 206, 208
  see also bias
semi-structured interviews 43, 44, 106, 146, 234
Sen, Amartya 80–1
Sendai Framework for Disaster Risk Reduction 2015–2030 22, 255
sensitivity analyses 50
SEQWater 149
Shoebridge v The Office of the Minister for Police and Emergency Services [2014] 157
Shrivastava, Paul 21
Silent Spring (Carson) 18
simulations 40, 42, 49–50
single-party governments 193
Smith, Adam 170
Snow, C.P. 15
social activism 1960s 28
social capacity, sustainable water use 219
social capital theory 170
social discounting 32, 33
social equity criteria 53, 169–70, 174, 178
social impact assessment 32, 99
social injustice 16, 186
social justice 1–3, 14, 35–6, 64
  coal seam gas industry and 155–67
  mining industry and 81
  water disputes and 143
Social Justice Framework (SJF) 30, 64, 105, 106, 107, 143, 233–49, 258
  water allocation case studies and 144, 145–52
social norms, evolution of 221
social preferences, self-interest and 170
social psychology 52, 80
social research 29, 39, 50
social responsibility of science 16
social standing, recognition of 244
social welfare 171
socio-economic impacts
  coal seam gas industry 156, 157
  Murray–Darling Basin Plan 220
South Africa, water allocation law reform 130
South Australia 61–78, 224, 234
Southern Ocean 234, 235, 239
southern Queensland 146, 148, 152, 162, 190
Space Age, environment and 17–19
spatial justice 2, 3, 5–6, 92, 158, 259
  coal seam gas industry and 155
  legal system and 159, 160
  sea-level rise and 98, 100, 101
species-blind animal property rights 138
spirit of justice 185–203
sponsors 185, 198–9
staged adaptation policies, sea-level rise 99, 100
stakeholders 14, 207
  animal property rights and 135, 139–1
  coal seam gas industry and 162
  computer simulations and 49–50
  distributive justice and 246
  diversity of 169–70
  documents 43
  engagement of 257
  equity metric and 206, 236
  equity preferences and 179, 211, 212
  ethicality and 241
  fairness criteria and 210
  governments and 194
  groups 33, 110, 185, 195, 209, 235, 240, 253, 254
  Indigenous Australians 264
  information availability to 242, 243
  interpersonal relations 46, 151
  interviews with 146, 235
  justice research and 39, 40, 41, 53, 54, 233, 234
  needs and 238
  participant observation and 47
  resource dependence 6

statutes and 51
Tasmanian forestry peace process and 106–7, 108, 111, 112
water management and 44, 45, 145, 147, 152, 259
state governments 194
colonial water access history 126–7
ecologically sustainable development and 218
information availability 96
resource allocation and 259
water management laws 215
Statement of Principles, Tasmanian forest peace process and 108, 111, 114
statutory interpretation 40, 42, 51–2
steel industry, Iran 194, 195, 197
Stockholm conference 1972, Declaration 17–19
Stolen Generation, apology to 122
Stone, Christopher 16
Stoner v Jones [2015] 157
stormwater harvesting 67, 69
Strategic Regional Land Use Policy (NSW) 161
structural adjustment 226, 256
structural injustice 129–30
structure, animal property rights 133–5
structured quantitative surveys 46
sub-rules 209, 210, 212
Sunshine Coast 148, 149
Surat Coal Seam Gas (CSG) development, case study 146, 148–51, 158
surface water 218, 224
surveys 146, 149
sustainability 4, 20, 23, 29, 138–9, 265
innovation and 186
Lockyer Valley 151
self-interest and 239
water rights and 33, 190
sustainable development 15, 19, 21, 81, 255
Sustainable Diversion Limit, Murray–Darling Basin 122, 123, 243, 246
Sustainable Juruti Council 83–4
sustainable yield 19
Sydney 18, 140
synoptic planning 28

tacit agreements 31
Tasmanian Forest Agreement (TFA) 105–18, 258, 264
Tasmanian forest peace process (TFPP) 105–16
Tasmanian government 108, 109, 111, 112
technical culture 21, 193
technical solutions, flood and drought risk minimisation 191
technological innovation uptake 186, 189
Telstra Corporation Ltd v Hornsby Shire Council [2006] 217
temporal justice 2, 3, 92
sea-level rise and 97–8, 100
Tasmanian forest peace process and 111
Tepko Pty Ltd v Water Board [2001] 222, 223
territorial behaviour 135
thematic analysis 67, 93
Theory of Moral Sentiments, The (Smith) 170
three-dimensional (3D) visualisations 41, 48, 146, 147
threshold level of significance 136–7
timber industry 19, 110, 115
see also forestry; Tasmanian Forest Agreement (TFA)
timeliness 241, 242, 243–4
public policy and 258
stakeholder engagement 257
Tasmanian forest peace process and 114–15
toolbox, participatory justice 257
tourism 93
trade-offs 169, 173, 211
public policy and 259
Tasmanian forest peace process and 115–16
transactive planning 28, 33
transition theories 187
transparency 65, 66, 242, 246
Adelaide water survey and 69, 70, 73
Tasmanian forest peace process and 110
trust 65, 81, 83, 87, 88, 97
Social Justice Framework and 245
Tasmanian forest peace process and 111
truthfulness principle 245
typologies, justice research 3, 5

unconventional gas 155, 157
see also coal seam gas industry
Underground Water Impact Report 148
unions 110, 111
United Nations (UN) 13, 15, 19, 20, 21
United Nations Conference on Environment and Development (UNCED) 19, 23, 254
United Nations Educational, Scientific and Cultural Organization (UNESCO) 16, 17, 18

United Nations Environment Program (UNEP) 14, 17–18, 21
United Nations High Commissioner for Refugees (UNHCR) 21
United Nations Scientific Conference on the Conservation and Utilisation of Resources (UNSCCUR) 17
United Nations World Conference 2015 22
United States of America (USA) 29, 63
Universal Declaration of Human Rights, The 15–16
unsustainability, water use 224
upstream communities 194
urban environment 260, 265
urban water sector 144
  privatisation and 62–4
  public interest and 69
Uruguay Round of Trade Talks 1986–94 20
utilitarian theories 30

value-for-money principle 172, 174, 176, 178, 213
values
  environmental justice and 6–7
  justice research and 54
vegetation planning systems 134
veil of ignorance 207, 208, 209
verbal communication, rural communities and 83, 86
vertical equity (VE) 176, 177, 209
vested interests, negotiation between 113
Victoria 127, 173, 213
  local government and sea-level rise 91–103
  water resource law 126
Vietnamese minorities, equity metrics and 212
virtue theory 30
visual/spatial methods 40, 41, 42, 48
voice 81, 83, 85, 86, 87, 88
  stakeholders 240
  Tasmanian forest peace process and 111

Waldron, Jeremy 129, 130
waste disposal 186
wastewater recycling 67, 69, 190
wastewater storage, coal seam gas industry 162
water access 123
  canal estates 150–1
  Indigenous-specific rights 121–32
Water Act 1912 (NSW) 127, 226
Water Act 2007 (Vic) 62, 215, 217, 218–20, 221, 223, 227–31, 237
  review of 2014 125, 130
Water Administration Act 1986 (NSW) 226
water allocations 151–2, 225, 254
  committees 48
  Indigenous peoples 121, 123
  irrigators and 237
  Murray–Darling Basin case study 233–4
  neoliberal 144
  politics of 121–32
  procedural justice and 224–7
  reduction 149–50
water bureaucracy 190
water buyback 236
water conflicts 143–4, 193
Water Framework Directive (EU) 191
water governance, colonial history and 126–8
water grid 195, 197
water institutions 61, 144
Water (Irrigation Farm Dams) Act 2002 (Vic) 227
water licences 126, 127, 148, 225, 237
water management 147, 243, 260
  consistency and 243
  innovation politics and 189
  justice and 2, 126, 143–54, 215–32
  Lower Balonne 151
  public ownership and 73
  questionnaires and 47
  semi-structured interview questions 44
  stakeholders and 259
Water Management Act 2000 (NSW) 127, 128, 226
water market 64, 122, 128, 190, 237, 238
water monopolies 62–3, 194
water nationalisation 126
water needs, human rights and 215
water pollution 224
water pooling 237
water prices 62, 67, 171, 197, 215, 230
water quality 193, 228
water reform
  allocative efficiency 239
  case study 243
  floodplains 234
Water Resource Plan 126, 229–30
water rights 31, 32, 33–4, 144
water savings 62, 236

water-sharing agreements 226, 229
water supply 228
  equality and 198
  outsourcing 63
  security 228
water sustainability, legal framework 219
wealth distribution 208, 209, 217
Wealth of Nations, The (Smith) 170
weighted equity metric (WEM) 211–13
Wellington Shire Council (WSC) 91–101
Western Australia 29, 127
Western Downs Regional Council 148, 162, 164
western living patterns 14–15
Westrex Services Pty Ltd & Anor v Maranoa Regional Council & Anor [2014] 162
wet gum forests, water supply 237
win–lose negotiations 195, 197
win–win outcomes 257
Wivenhoe Dam 148, 150
women, lack of accessibility 243
workable policies, politics and 252
workshops 4, 42, 45, 93, 146, 266
World Charter for Nature 19
World Heritage Commission 18, 108, 109
World Trade Organization (WTO) 20–1
written submissions 83, 84, 86, 87, 195, 196

www.ingramcontent.com/pod-product-compliance
Ingram Content Group UK Ltd.
Pitfield, Milton Keynes, MK11 3LW, UK
UKHW062008041026
15591UKWH00017B/1234

9 781486 306374